International Review of
Cytology
A Survey of
Cell Biology

VOLUME 134

International Review of

Cytology

A Survey of

Cell Biology

Edited by

Kwang W. Jeon
Department of Zoology
The University of Tennessee
Knoxville, Tennessee

Martin Friedlander
Jules Stein Eye Institute
UCLA School of Medicine
Los Angeles, California

VOLUME 134

Academic Press, Inc.
Harcourt Brace Jovanovich, Publishers
San Diego New York Boston London Sydney Tokyo Toronto

This book is printed on acid-free paper. ∞

Academic Press, Inc.
1250 Sixth Avenue, San Diego, California 92101-4311

United Kingdom Edition published by
Academic Press Limited
24–28 Oval Road, London NW1 7DX

Library of Congress Catalog Number: 52-5203

International Standard Book Number: 0-12-364534-4

PRINTED IN THE UNITED STATES OF AMERICA
92 93 94 95 96 97 EB 9 8 7 6 5 4 3 2 1

CONTENTS

Role of Calcium/Calmodulin-Mediated Processes in Protozoa

Leonard William Scheibel

The Molecular Biology of Intermediate Filament Proteins

Kathryn Albers and Elaine Fuchs

CONTRIBUTORS

Numbers in parentheses indicate the pages on which the authors' contributions begin.

Kathryn Albers (243), *Howard Hughes Medical Institute, and Department of Molecular Genetics and Cell Biology, The University of Chicago, Chicago, Illinois 60637*

A. I. Backer (1), *Gesellschaft für Biotechnologische Forschung, D-3300 Braunschweig, Germany*

Elaine Fuchs (243), *Howard Hughes Medical Institute, and Department of Molecular Genetics and Cell Biology, The University of Chicago, Chicago, Illinois 60637*

A. Oaks (85), *Department of Botany, University of Guelph, Guelph, Ontario, Canada N1G 2W1*

Leonard William Scheibel (165), *Department of Preventive Medicine, Uniformed Services University of the Health Sciences, School of Medicine, Bethesda, Maryland 20814*

K. A. Sechley (85), *Department of Botany, University of Guelph, Guelph, Ontario, Canada N1G 2W1*

K. G. Wagner (1), *Gesellschaft für Biotechnologische Forschung, D-3300 Braunschweig, Germany*

T. Yamaya (85), *Department of Agricultural Chemistry, Tokoku University, Sendai 981, Japan*

Dynamics of Nucleotides in Plants Studied on a Cellular Basis

K. G. Wagner and A. I. Backer
Gesellschaft für Biotechnologische Forschung
D-3300 Braunschweig, Germany

I. Introduction

The last general reviews on nucleotide metabolism in plants appeared in 1982 and earlier (Suzuki and Takahashi, 1977; Wasternack, 1982; Ross, 1981; Feingold and Avigad, 1980; Feingold, 1982), although there are more recent special reviews on nucleotide compartmentation (Wasternack and Benndorf, 1983) and on nucleotide regulation in energy metabolism (Pradet and Raymond, 1983; Raymond *et al.*, 1987). In the meantime new data have been elaborated in the field of nucleotide metabolism and also in several neighboring fields of plant metabolism that are interconnected with and dependent on nucleotide "dynamics." This justifies a new general review.

To write a review on nucleotide dynamics is a challenging task but also includes several risks. This title may stimulate one to write a review under a special aspect; whether this has been done successfully with the present article, our colleagues will have to judge and we are waiting for critical remarks. Most of the analytical methods are not suitable to study dynamic processes but give "static" information and hence dynamics must be derived from data composition and interpretation. Metabolic dynamics are like a giant "puzzle"; to compose it, one must handle many pieces. The topic of nucleotide dynamics is at a stage where only some of the pieces are available and many are missing; we therefore tried to add available pieces such as pool sizes and K_m values of relevant enzymes without being able to compose them in each case. But we hope that in future research more pieces and interconnections will be elucidated, in order to make the whole picture more apparent.

In designing this article, we realized that before discussing the main aspect, which is nucleotide dynamics in metabolic regulation and in growth, we had to discuss the current knowledge on the metabolic pathways (Section II), their localization within the cell and intracellular transport (Section III), as progress in these topics in the last 10 years has been large. The present article is devoted to nucleotide dynamics on a cellular basis only; Section IV deals with nucleotide dynamics in metabolic regulation and Section V with growth regulation of suspension cultured cells.

The topics concerning nucleotide dynamics in the whole plant could not be treated in this issue; thus topics such as germination, growth and senescence, the mature plant and its developmental aspects, nutrition, environmental stress, and plant–microbe interactions should be reserved for a further review.

Abbreviations

In the abbreviations of metabolites we do not use any hyphens. For the abbreviation of enzymes, in general two terms separated by a hyphen are used; the first term is the substrate and the second term describes the function of the enzyme, which always ends with the syllable "ase."

Metabolites

Ade	Adenine
Ado	Adenosine
AIR	Aminoimidazole ribotide
AICAR	Aminoimidazole-carboxamide ribotide
ALC	Allanotoic acid
ALN	Allantoin
AXP	Σ(ATP + ADP + AMP)
CAsp	Carbamylaspartate
CAIR	Carboxyaminoimidazole ribotide
Chl	Chlorophyll
CP	Carbamylphosphate
Cyd	Cytidine
Cyt	Cytosine
DHAP	Dihydroxyacetone phosphate
FAICAR	Formylaminoimidazole carboxamide ribotide
5FdUMP	5-Fluoro-dUMP
Fru6P	Fructose 6-phosphate
FGAM	Formylglycineamidine ribotide
FGAR	Formylglycineamide ribotide
GAR	Glycineamide ribotide
GlcNAc	*N*-Acetylglucosamine
Glc6P	Glucose 6-phosphate
GlcUA	Glucuronic acid
Gua	Guanine
Guo	Guanosine
Hyp	Hypoxanthine
Ino	Inosine
Msa	Monosaccharide
MTHF	Methylenetetrahydrofolate
NA	Nicotinic acid
NAAD	Nicotinic acid adenine dinucleotide
NAMN	Nicotinic acid mononucleotide
NDP	Nucleoside diphosphate
NMN	Nicotinamide mononucleotide
NMP	Nucleoside monophosphate
NTP	Nucleoside triphosphate
OA	Orotic acid (orotate)
OMP	Orotate monophosphate
PEP	Phosphoenolpyruvate
P5C	Pyrroline-5-carboxylate
P_i	Inorganic phosphate
PP_i	Inorganic pyrophosphate
PGA	3-Phosphoglycerate
PRA	Phosphoribosylamine
PRPP	Phosphoribosylpyrophosphate
QA	Quinolinic acid
SAICAR	Succinoaminoimidazole-carboxyamide ribotide

SAMP	Succinoadenosine monophosphate	UDPGlcNAc	UDP-*N*-acetylglucosamine
Thd	Thymidine	Ura	Uracil
Thy	Thymine	Urd	Uridine
UA	Uric acid	Xan	Xanthine
UDPGlc	UDP-glucose	Xao	Xanthosine
UDPGlcUA	UDP-glucuronic acid	XMP	Xanthosine monophosphate
		Xyl	Xylose

Enzymes

ATase	Adenyltransferase	Nase	Amidase/deaminase
Case	Carboxylase	NTase	Amidotransferase
CTase	Carbamyltransferase	Pase	Phosphatase
DCase	Decarboxylase	PPase	Pyrophosphatase
DHase	Dehydrogenase	PTase	Phosphotransferase
Ease	Epimerase	Pyase	Phosphorylase
GHase	Glycohydrolase	Rase	Reductase
Hase	Hydrolase	RTase	Phosphoribosyltransferase
Kase	Kinase	Sase	Synthase
Lase	Lyase	Tase	Transferase
MTase	Methyltransferase	UTase	Uridyltransferase

II. Nucleotide Synthesis, Utilization, and Degradation

A. Pathways of Synthesis and Degradation

1. *De Novo* Pathways

Figure 1 shows the pathways of *de novo* synthesis of ribonucleotides as developed from bacteria and animal cells. In plants there is no evidence of routes different from those described in this figure (Ross, 1981; Wasternack, 1982; Lovatt *et al.*, 1979; Lovatt, 1983), although this subject has not been studied in all details and not all the intermediates and enzymes have been identified up to now. In contrast to animal cells and in accord with bacteria, in plant tissue only one CP-Sase was described that is involved obviously in both pyrimidine nucleotide and Arg synthesis (Ong and Jackson, 1972a; O'Neal and Naylor, 1976). The catalytic activities for the two steps converting OA to UMP were shown to reside in one polypeptide (Walther *et al.*, 1984). Studies with bean seedlings suggested a conversion of UTP into 5-ribosyl-Ura (pseudo-Urd) without the prior incorporation into tRNA (Al-Baldawi and Brown, 1983).

Although not thoroughly investigated, one can conclude from pool measurements of different species (Neuhard and Nygaard, 1987; Sabina *et al.*, 1981) that the intermediates in *de novo* synthesis (Fig. 1) are present in concentrations considerably lower (with the exception of IMP) than those

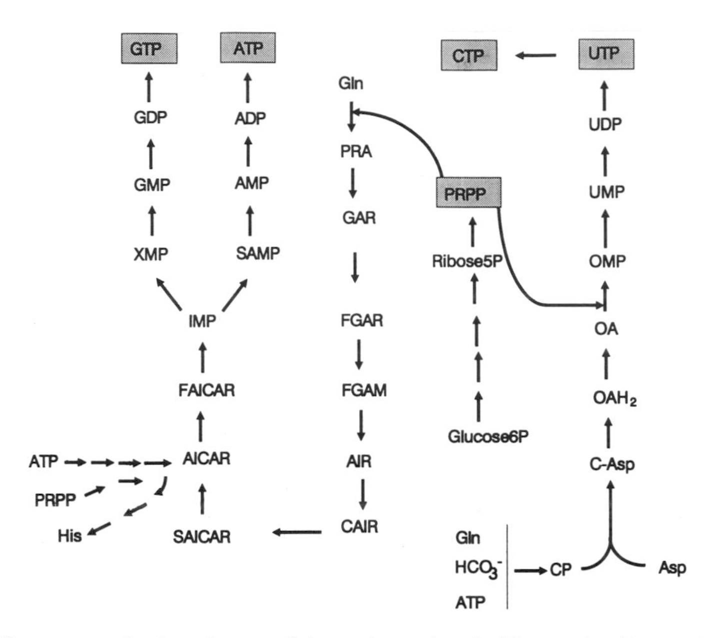

FIG. 1 *De novo* synthesis pathways of the purine and pyrimidine nucleotides as developed from bacteria and vertebrate cells. Only the intermediates are shown; for a more detailed description see textbooks and Ross (1981), Wasternack (1982), and Neuhard and Nygaard (1987).

found for the known nucleoside mono-, di-, and triphosphates, of which the triphosphates in general predominate. CMP and CDP are not produced by *de novo* synthesis; they are, however, obtained through recycling of phospholipid precursors such as CDPcholine and CDPdiacylglycerol. A total ribonucleotide pool of 20 μmol/g dry weight (DW) was determined for a *Salmonella* bacterium (Neuhard and Nygaard, 1987), which corresponds to a cellular concentration of about 8.6 mM. The total nucleotide content of tobacco suspension cultured cells (Table I) corresponds to a cytoplasmic concentration of about 8.8 mM (Meyer and Wagner, 1985c). Table I shows the distribution of the total ribonucleotide pool of different tissues into the main fractions. It is obvious that plant tissue is characterized by relatively large NDPsugar levels. Most of the nucleotides synthesized are incorporated, however, into the nucleic acids; with growing bacteria (*Escherichia coli* and *Salmonella typhimurium*) 95% is incorporated into RNA and DNA and only 5% into the soluble fraction, i.e., the nucleotides and nucleotide coenzymes. With *Datura* suspension cultured cells (day 11 of the growth cycle), of the total ribonucleotides about 83% were in the

RNA and 17% in the soluble fraction (Wylegalla *et al.*, 1985); the rather large soluble nucleotide pool is mainly due to the UDPsugars (Table I).

In both *de novo* and salvage pathways the ribose moiety is derived from PRPP, which is also the precursor for His, Trp, and NAD. With growing bacteria, 60–80% of the total PRPP pool is consumed for the synthesis of ribonucleotides, 10–15% for His and Trp and only 1–2% for NAD (Munch-Petersen, 1983). Regulation of plant PRPP synthesis is up to now only poorly elucidated, although inhibition of a partially purified PRPP-Sase by NMPs and NDPs was reported (Ashihara and Komamine, 1974; Ashihara, 1977a). With a bacterial PRPP-Sase, it was shown that ADP is an allosteric inhibitor and P_i concentrations of about 20 m*M* are required for both stability and activity of the enzyme (Neuhard and Nygaard, 1987); stimula-

TABLE I

Distribution of Nucleotides into the Main Fractions

Tissue	Total content (nmol)	Ribonucleotides (%)	NDPsugars (%)	Pyridinnucleotides (%)	Ref.[a]
Salmonella typhimurium (g DW)	20,000	74	14	12	1
Rat skeletal muscle[b] (g FW)	6,660	98	2		2
Rat sarcoma[b] (g FW)	3,800	86	14	—	2
Wheat flag leaf (g DW)	2,660	61	21	18	3
Strawberry leaves[b] (g FW)	450	65	35	—	4
Young tobacco leaf (g FW)	400	62	28	10	5
Tobacco suspension cultured cells (g FW)	440	42	53	5	6
Datura suspension cultured cells (g FW)	265	45	43	12	7
Sycamore suspension cultured cells[b] (g FW)	530	61	39	—	8

[a]1, Neuhard and Nygaard, 1987; 2, Weber *et al.*, 1983; 3, Sawert, 1988; 4, Isherwood and Selvendran, 1970; 5, Meyer and Wagner, 1986; 6, Meyer and Wagner, 1985c; 7, Meyer and Wagner, 1985a; 8, Brown and Short, 1969.

[b]Pyridine nucleotides were not determined; with the other plant tissues reduced pyridine nucleotides are not included.

tion through similar concentrations of P_i was also reported with mammalian enzymes (Becker *et al.*, 1979). This is in contrast to preliminary investigations with partially purified enzymes from spinach leaves (Ashihara, 1977b) and cultured *Catharanthus* cells (Ukaji and Ashihara, 1987a), which showed a 50% inhibition at 20–30 mM P_i. There are few data on cellular pools of PRPP; in bacteria a concentration of 0.5 mM was estimated (Munch-Petersen, 1983), whereas from suspension cultured cells (Hirose and Ashihara, 1983c) 2.2 nmol/g fresh weight (FW), which corresponds to a cytosolic concentration of about 40 μM, and from germinating peas a maximum value of 23 nmol/pair of cotyledons (Ross and Murray, 1971) was reported.

Regulation of the purine nucleotide *de novo* pathway is poorly understood in plants (Wasternack, 1982). In bacteria and in animal cells, the first enzyme, PRPP-NTase, is feedback inhibited by AMP, GMP, and IMP; furthermore, the levels of the first enzymes of the pathway have been shown in bacteria to be regulated simultaneously (Neuhard and Nygaard, 1987; Wyngaarden, 1976). A purified PRPP-NTase from soybean nodules revealed similar properties as bacterial or animal enzymes (Reynolds *et al.*, 1984), although the K_m values were slightly higher (Table II). The enzyme accepts both Gln and NH_4^+ as amino donor; it is inhibited by several purine NMPs and K_i values of 1.2 mM (XMP) and 2.0 mM (AMP) were determined. From feeding soybean embryonic axis with radiolabeled precursors, it was suggested that in the *de novo* route the flux from IMP to AMP (Fig. 1) is larger than that to GMP (Anderson, 1979; Ross, 1981). In

TABLE II

K_m Values of Enzymes from the Purine Nucleotide *de Novo* Routes

Enzyme	Substrate	K_m (μM)	Ref.[a]	Enzyme	Substrate	K_m (μM)	Ref.[a]
PRPP-Sase	ATP	36	1	SAMP-Lase	SAMP	30	4
	Rib5P	10					
	ATP	180	2				
	Rib5P	14					
PRPP-NTase	PRPP	400	3	AMP-Kase	AMP	69	5
	Gln	18,000			ADP	3	
	NH_4^+	16,000			MgADP	6	
					MgATP	25	
SAMP-Sase	IMP	90	4	GMP-Kase	GMP	16	6
	Asp	40			ATP	420	

[a] 1, Ashihara, 1977b (spinach leaves); 2, Ashihara and Komamine, 1974 (bean hypocotyl); 3, Reynolds *et al.*, 1984 (soybean root nodules); 4, Hatch, 1966 (wheat germ); 5, Kleczkowski *et al.*, 1990 (maize leaves); 6, Le Floc'h and Lafleuriel, 1990 (mitochondria of artichoke tubers).

the roots of 2-day-old squash plants (Lovatt, 1983), the rate of synthesis of adenine nucleotides was estimated as 80 nmol/(g FW · hr).

Feedback regulation of the pyrimidine nucleotide pathway (Fig. 1) in plants by pyrimidine nucleotides (most active are UMP and UDP) was claimed to occur solely at the Asp-CTase step (Lovatt and Cheng, 1984; Lin and Lovatt, 1986), although from earlier experiments inhibition of CP-Sase (O'Neal and Naylor, 1976; Ong and Jackson, 1972a) and of OMP-DCase was also reported (Ashihara, 1978). Stimulation of CP synthesis by Orn was suggested from studies on extracts of bean roots (Ong and Jackson, 1972a), pea leaves (O'Neal and Naylor, 1976), and summer squash callus (Lin and Lovatt, 1986); in extracts of summer squash roots, however, this could not be confirmed (Lovatt and Cheng, 1984). Whether these divergent findings reflect different properties of enzymes from different species remains to be elucidated. With a purified Asp-CTase from wheat germ, allosteric behavior and regulatory kinetics with the substrates and the feedback inhibitor UMP was thoroughly studied and a K_D of 2.5 μM for UMP was reported (Yon, 1984; Cole and Yon, 1984). In extracts of several plants, the activity of Asp-CTase was found to be 3- to 20-fold higher than that of CP-Sase; this would provide a fast transformation of the rather instable CP and further indicate that the formation of CP is a rate-limiting step (Shibata *et al.*, 1986). Low CP levels (2 nmol/g FW; Asp, 140 nmol/g FW) were determined in *Helianthus* tuber explants (Parker and Jackson, 1981). An estimation of fluxes of CP into Arg or UMP was possible with data from an extract of *Phaseolus aureus* seedlings; whereas the ratio of the V_{max} values was 3:1 in favor of Orn-CTase, the K_m values of CP were much lower for Asp-CTase than for Orn-CTase (Ong and Jackson, 1972). This was also found with extracts from *Helianthus* tubers (Parker and Jackson, 1981) (Table III) and indicates a predominant flow of CP toward UMP at the low CP levels found in the cells. In cultured *Catharanthus* cells, CP consumption for UMP synthesis was estimated as 8- to 10-fold higher than for Arg synthesis (Sasamoto and Ashihara, 1988). In roots of 2-day-old squash plants the rate of synthesis of Ura nucleotides was estimated as 50 nmol/(g FW · hr) (Lovatt, 1985).

Little is known on the synthesis of the deoxyribonucleotides in plants; Fig. 2 shows the possible routes. In higher plants the ribonucleotide reductase is a type I enzyme with NDPs as substrates (Follmann, 1974) and its activity in plant tissue was found to be rather low (Hovemann and Follmann, 1979). Regulation by dNTPs, as found with other species, was tentatively assumed from effects of dAdo on cell division (Fernandez-Gomez *et al.*, 1970). The route to dUMP is not clear; whereas with *Chlorella* a dCMP deaminase was described (Shen and Schmidt, 1966), this enzyme is not established in higher plants. Deamination of dCTP is a major route in bacteria, while 25% of the cellular dUMP is derived from

TABLE III

K_m Values of Enzymes from the Pyrimidine Nucleotide *de Novo* Routes

Enzyme	Substrate	K_m (μM)	Ref.[a]	Enzyme	Substrate	K_m (μM)	Ref.[a]
CP-Sase	Gln	170	1	OA-RTase	OA	4.5	7
		140	2		PRPP	1.6	
					OA	4.5	5
Asp-	CP	90	13		PRPP	5.4	
CTase	Asp	1700			OA	26	8
	CP	140	2		OA	17	6
		80	3		PRPP	10	
						75	9
Orn-	CP	1600	1				
CTase		9000	3	NDP-	CDP	200–300	10
				Kase	(dCDP)		
OAH_2-	CAsp	6200	4				
Sase				dTMP-	UMP	1.8	11
				Sase	MTHF	130	
OMP-	OMP	2	5				
DCase		4	6	dUTP-	dUTP	6	12
				Hase			

[a] 1, Ong and Jackson, 1972a, (bean seedlings); 2, O'Neal and Naylor, 1976 (pea leaves); 3, Parker and Jackson, 1981 (*Helianthus* tubers); 4, Ross, 1981 (pea leaves); 5, Walther *et al.*, 1984 (cultured tomato cells); 6, Walther *et al.*, 1981 (*Euglena gracilis*); 7, Ashihara, 1978 (black gram seedlings); 8, Kapoor and Waygood, 1965 (wheat embryos); 9, Wolcott and Ross, 1967 (bean leaves); 10, Klein and Follmann, 1988 (green algae); 11, Nielsen and Cella, 1988 (cultured carrot cells); 12, Pardo and Gutiérrez 1990 (onion roots); 13, Ong and Jackson, 1972b (bean seedlings).

phosphorylation of dUDP followed by hydrolysis of dUTP (Neuhard and Nygaard, 1987). In extracts from root meristem cells a dUTP-Hase was recently described (Pardo and Gutiérrez, 1990), which cleaves dUTP into dUMP and PP_i. The correlation between its activity and S-phase cells in proliferating meristem suggests a role of dUTP-Hase in the synthesis of dUMP and thus dTTP. dTMP-Sase was recently investigated after partial purification of the enzyme from cultured carrot cells; it revealed kinetic properties similar to those found in other species (Table III) and was highly sensitive toward 5FdUMP (Nielsen and Cella, 1988). Studies on cell synchronization with cultured root tips from pea and dThd feeding (Schvartzman *et al.*, 1984) suggested that control of DNA synthesis and propagation through the cell cycle may be exerted by the dTTP pool. The deoxyribonucleotide pools are usually rather small; in bacteria the total pool comprises about 5% of the total soluble nucleotide pool (Neuhard and Nygaard, 1987); in plant tissue the dNTP pools are obviously still smaller. A comparison of the dNTPs with NTPs in growing tissue (Nygaard, 1972) showed a ratio of 1:11 with bacteria, whereas with sycamore callus tissue and germinating pine pollen ratios of 1:205 and 1:270 were found. From tomato root tips levels of total dNDPs and dNTPs of 7.4 and 21 nmol/g FW were reported (Dutta *et al.*, 1991).

In contrast to the nucleobase-unselective NMP kinases and NDP kinases, AMP-Kase has been thoroughly studied in plants (Hatch, 1982; Kleczkowski *et al.*, 1990; Kleczkowski and Randall, 1986; Manetas *et al.*, 1986). The enzyme catalyzes the equilibration between the Ade nucleotides (2 ADP $\leftrightarrow$ ATP + AMP) and is usually present at high levels (Raymond *et al.*, 1987). From leaves of several C_3 and C_4 plants, two immunologically distinct forms were reported (Kleczkowski and Randall, 1987). With a highly purified enzyme from maize leaves, properties were reported (Kleczkowski *et al.*, 1990) similar to those already known from yeast and animal enzymes (Table II). It uses one molecule each of MgADP and free ADP in the forward and MgATP and AMP in the reverse reaction. Thus the enzyme actually governs the equilibrium between MgADP, free ADP, MgATP, and free AMP (Kleczkowski and Randall, 1991).

High activities of NMP and especially NDP kinases were reported from cultured *Catharanthus* cells (Hirose and Ashihara, 1984), while an NDP-Kase was described from pea seeds (Edlund, 1971) and also from a green alga (Klein and Follmann, 1988). From mitochondria of Jerusalem artichoke tubers a GMP-Kase was characterized (Le Floc'h and Lafleuriel, 1990) that showed product inhibition with GDP (K_i 95 μM) and GTP (K_i 40 μM).

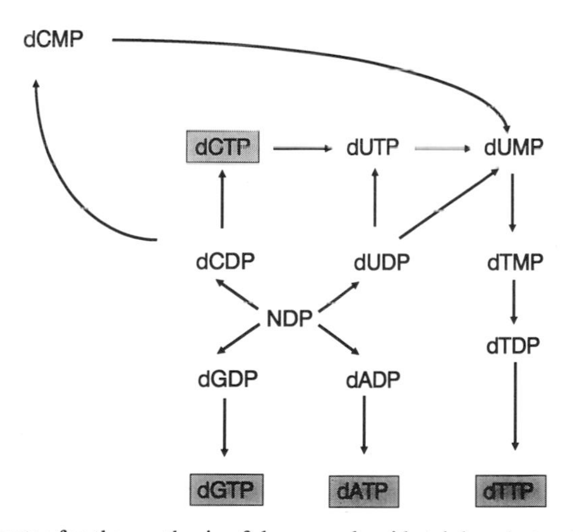

FIG. 2 Possible routes for the synthesis of deoxynucleoside triphosphates. The scheme is a modified proposal from bacterial metabolism (Neuhard and Nygaard, 1987).

2. Salvage Pathways

Salvage pathways fulfill different physiological functions (Fig. 3 shows a generalized scheme), although these are only poorly elucidated. They play a role in the utilization of exogenous nucleobases and nucleosides obtained after mobilization, e.g., in reserve tissue (germination) or during leaf senescence, and in the reutilization of endogenous nucleobases and nucleosides obtained from nucleic acid and nucleotide turnover. The latter process may also be important for cellular compartments lacking or with incomplete *de novo* pathways, as import of nucleosides or bases may be facilitated relative to the more polar nucleotides.

With bacteria (Neuhard and Nygaard, 1987) and animal cells, significant pools of nucleosides and nucleobases were not reported; obviously activities of salvage enzymes are high enough for fast transformation into nucleotides. With plants, however, several reports deal with considerable levels of both nucleobases and nucleosides. Earlier analyses excluded the presence of free bases in pea seeds (Brown, 1963, 1965), whereas low levels of free bases (Ade, Gua, Ura) were found in wheat grain (5–7 nmol/g) (Grzelczak and Buchowicz, 1975) and the rather high free Ade content found in soybeans (150–300 nmol/g) was suggested to interfere with food quality (Yokozawa and Oura, 1986). Free Ade was also found in leaves of cereals (Sawert *et al.*, 1987, 1988); young barley leaves contained up to 300 nmol/g DW, whereas young leaves of wheat and rye showed lower levels.

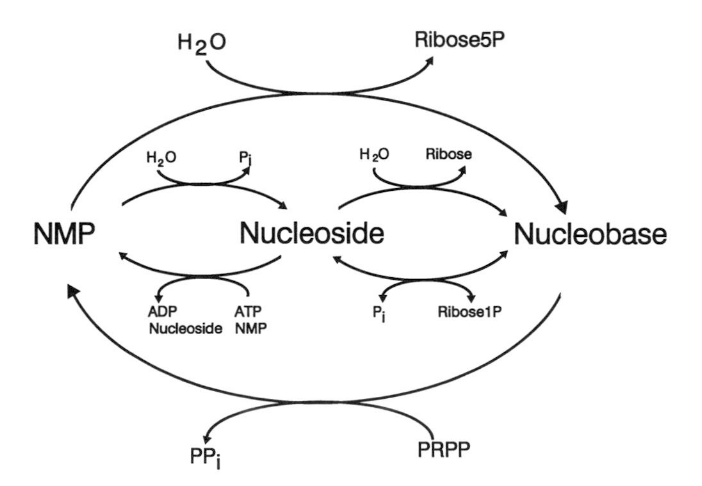

FIG. 3 Main enzymatic steps in pathways of nucleobase and nucleoside salvage.

Pools of nucleosides are more ubiquitous than free bases; in different tissues rather high levels of Ado and Urd were reported, lower levels of Guo, whereas significant levels of Cyd were not detected. Nucleosides are obviously stored in several seeds; while wheat grains (Grzelczak and Buchowicz, 1975) contain medium levels of Ado and Urd (30–50 nmol/g), pea seeds (Brown, 1963, 1965) have a rather high content (150–200 nmol/g); both seeds contain lower levels of Guo. Furthermore, young barley leaves contain high levels of nucleosides that may exceed the content of the respective nucleotides (Table IV), whereas the leaves of wheat and rye showed lower levels. There is preliminary evidence that large pools of nucleosides in growing leaves are most common to the monocotyledons, while dicotyledon leaves showed lower or insignificant pools (Sawert *et al.*, 1988). The question arises whether the high levels of nucleosides and bases in growing leaves are indicative of low activities of salvage enzymes or reflect compartmentation. With tomato suspension cultured cells, the presence of 13–35% of the total Urd content in the vacuole was reported (Leinhos *et al.*, 1986).

Salvage of bases occurs both with PRPP (phosphoribosyltransferases, RTases) and ribose1P (nucleoside phosphorylases, Pyases) (Fig. 3). Transferases using Ura (Kanamori *et al.*, 1980; Bressan, *et al.*, 1978), Gua, Hyp, and Ade (Guranowski and Barankiewicz, 1979; Doree, 1973; Le Floc'h and Lafleuriel, 1981b; Barankiewicz and Paszkowski, 1980) have been described, while there is no evidence of Cyt and Thy salvage, although

TABLE IV

Nucleoside and Adenine Contents of Plant Leaves and Cultured Cells[a]

Species	AXP nmol/g FW	Total precursor content nmol/g FW	Ado (%)	Guo (%)	Urd (%)	Ino (%)	Ade (%)	Ref.[b]
Leaves								
Barley	74	197	35	14	29	1	21	1
Wheat	98	93	45	24	17	—	14	1
Rye	147	58	70	8	7	—	15	1
Pea	278	35	54	14	31	—	—	2
Peanut	347	9	100	—	—	—	—	3
Tobacco	54	2	—	—	100	—	—	4
Cell culture								
Tobacco	102	8	47	—	53	—	—	5
Datura	72	8	12	29	59	—	—	6

[a]For comparison the content of total Ade nucleotides (AXP) was included; —, not detectable.

[b]1, Sawert *et al.*, 1987; 2, P. Grabo, unpublished observations; 3, Backer, 1990; 4, Meyer and Wagner, 1985b; 5, Meyer and Wagner, 1985c; 6, Meyer and Wagner, 1985a.

with *Euglena* cells the incorporation of labeled Thy into TMP was reported (Wasternack, 1975). The uracil transferases are obviously distinct from enzymes using OA in the *de novo* pathway (Ashihara, 1978), whereas Gua and Hyp are possibly salvaged by the same enzyme entity (Table V). A kinetic investigation with an enzyme from Jerusalem artichoke shoots is in favor of this suggestion; this enzyme showed product inhibition with GMP (K_i 6.3 μM) (Le Floc'h and Lafleuriel, 1981b). As tested with several tissues, salvage activity for Ade by RTases was considerably higher than that for Hyp (Ashihara and Nobusawa, 1981). Product inhibition by several purine nucleoside monophosphates was reported for both Ade-RTase and Hyp/Gua-RTase (Hirose and Ashihara, 1983a; Le Floc'h and Lafleuriel, 1978, 1981b). *Arabidopsis* mutants lacking Ade-RTase activity were isolated that showed reduced growth and male sterility (Moffatt and Somerville, 1988).

Purine nucleoside phosphorylases could not be detected in mesophyll protoplasts from tobacco leaves (Barankiewicz and Paszkowski, 1980) and cultured *Catharanthus* cells (Hirose and Ashihara, 1983b), whereas enzymes with specificities toward Ado and Ino or Guo were identified in wheat embryos; the Ado and Ino/Guo specificities obviously reside in different proteins (Chen and Petschow, 1978). Activity of Urd-Pyase in cultured *Catharanthus* cells, however, was 12-fold lower than that of the Ura-RTase (Kanamori *et al.*, 1980).

Salvage of nucleosides may occur by an indirect route, i.e., removal of the ribose moiety by hydrolysis or phosphorylysis (Fig. 3) and salvage of the base. Nucleoside hydrolases with specificity toward Ado, Ino, and Guo have been described (Guranowski, 1982; Guranowski and Schneider, 1977; Le Floc'h and Lafleuriel, 1981a) and varying activities, from undetectable levels to 80 μmol/(min · g FW) of Ado-Hase, were reported from different plants (Leszczynska *et al.*, 1984). In extracts of lupin seedlings activities toward Ino, Ado, and Guo were found in the ratio of 100 : 11 : 1 (Guranowski, 1982), while extracts from Jerusalem artichoke shoots revealed activities that suggest two distinct enzymes specific for Ado and Ino/Guo, respectively (Le Floc'h and Lafleuriel, 1981a). Product inhibition was shown for Ado-Hase from barley leaves with a K_i value for Ade of 4 μM (Guranowski and Schneider, 1977). From cowpea nodules a rather unspecific nucleoside hydrolase was highly purified that hydrolyzed both purine and pyrimidine nucleosides, although the enzyme prefered Xao. The V_{max} values for Xao, Ino, Ado, and Guo were in the ratio of 28:7:1:0.4 (Atkins *et al.*, 1989).

Direct salvage of nucleosides is performed by either a kinase or a phosphotransferase (Fig. 3). There are many data on kinases using dThd, while Urd-, Cyd-, and Ado-specific enzymes were also described (Ross, 1981; Guranowski, 1979a). A partially purified kinase from maize seeds

TABLE V

K_m Values of Enzymes from the Salvage Routes

Enzyme	Substrate	K_m (μM)	Ref.	Enzyme	Substrate	K_m (μM)	Ref.[b]
Ado-Hase	Ado	0.8–2.3	1	Ino/Guo-Hase	Ino	2.5	2
		17	2		Guo	8.5	
Ado-Kase	Ado	1.5	3	Xao/Ino-Hase	Xao	800	10
	ATP	300			Ino	830	
AMP-Nase[a]	AMP	400	4	dThd-Kase	dThd	0.15	11
					ATP	11	
Hyp/Gua RTase	Hyp	6.1	5	Ino/Guo-Kase	Ino	70	12
	Gua	5.5			Guo	14	
	PRPP	88			ATP	1450	
Ade-RTase	Ade	6.6	6	Urd/Cyd-Kase	Urd	53	13
	PRPP	9			Cyd	125	
	Ade	5.5	7		ATP	590	
	PRPP	64			GTP	61	
Nucleo-side-PTase	Ino	400	8	Nucleo-side-PTase	Urd	3500	14
	AMP	400			PhenylP	3500	
	Ado	540	9		dThd	840	15
	Guo	2130			Urd	4710	
	Urd	3930			Ado	1280	
	Cyd	6390			AMP	110	

[a]Determined in the presence of 1 mM ATP.

[b]1, Guranowski and Schneider, 1977 (barley leaves); 2, Le Floc'h and Lafleuriel, 1981a (artichoke shoots); 3, Guranowski, 1979a (lupin seeds); 4, Yoshino and Murakami, 1980 (spinach leaves); 5, Le Floc'h and Lafleuriel, 1981b (Jerusalem artichoke shoots); 6, Hirose and Ashihara, 1983a (cultured *Catharanthus* cells); 7, Le Floc'h and Lafleuriel, 1978 (artichoke shoots); 8, Guranowski, 1979b (lupin seedlings); 9, Prasher *et al.*, 1982 (barley seedlings); 10, Atkins *et al.*, 1989 (cowpea nodules); 11, Swinton and Chiang, 1979 (*Chlamydomonas*); 12, Combes *et al.*, 1989 (artichoke tuber mitochondria); 13, Deng and Ives, 1975 (maize seeds); 14, Brunngraber and Chargaff, 1970 (carrot roots); 15, Grivell and Jackson, 1976 (bird's nest fern).

gave evidence that the specificity for Urd and Cyd resides in one protein (Deng and Ives, 1975). Guo- and Ino-specific kinases, however, may be less abundant; in extracts from black gram seedlings low Ino-Kase activity was detected (Nobusawa and Ashihara, 1983). Only recently an Ino/Guo-Kase of low stability was described that was extracted from mitochondria of Jerusalem artichoke tubers (Combes *et al.*, 1989). It is inhibited by Gua nucleotides; GMP revealed a K_i value of 75 μM. Low activities of Guo/Ino-specific enzymes and the predominance of Ade-RTase activity over that of Ado-Kase are obviously the reasons why salvage of purine bases in several plant tissues is faster than that of purine nucleosides (Ross, 1981); this was shown, e.g., by feeding of soybean embryonic axes with different purine compounds (Anderson, 1979). In extracts from black gram embryonic axes the range of salvage enzyme activities was Gua-RTase > Ade-RTase >> Hyp-RTase > Ado-Kase > Ino-Kase, whereas in the cotyledons the range was different: Ade-RTase > Ado-Kase > Gua-RTase > Hyp-RTase >> Ino-Kase (Nobusawa and Ashihara, 1983). Also, *Catharanthus* stem tissue showed higher uptake rates with Ade than with Ado (Hirose and Ashihara, 1983b); cultured *Catharanthus* cells, however, showed slightly higher incorporation rates for Ado than for Ade (Hirose and Ashihara, 1983b, 1984), although the extracted activities of Ade-RTase were slightly larger than those of Ado–Kase. The same cells incorporated Urd into nucleotides and RNA with considerably higher efficiency than Ura (Kanamori-Fukuda and Ashihara, 1981; Sasamoto *et al.,* 1987) and also the extracted activity of Urd-Kase was about twice as large as that of Ura-RTase. In general, salvage of pyrimidine compounds may be governed by nucleoside kinases, whereas especially with differentiated tissue purine nucleobase salvage by RTases may predominate.

Phosphotransferases using NMPs as phosphoryl donor are complementary enzymes for the kinases and have a broad substrate specificity with respect to both the phosphoryl donor and the acceptor (Brawerman and Chargaff, 1955; Billich and Witzel, 1986; Billich *et al.,* 1986; Brunngraber and Chargaff, 1970; Brunngraber, 1978; Guranowski, 1979b; Prasher *et al.,* 1982; Richard *et al.,* 1979); there are enzymes that accept phenylP or similar phosphoryl compounds as phosphoryl donors. From barley seedlings an enzyme was purified that showed a slight preference for deoxyribonucleosides over ribonucleosides with respect to the phosphoryl acceptors (K_m values) (Prasher *et al.,* 1982), although this is probably not of physiological importance. With the same enzyme, the pyrimidine nucleoside monophosphates were slightly better phosphoryl donors than the purine nucleoside monophosphates (Richard *et al.,* 1979). Although their cellular function has not been elucidated, nucleoside-PTases may play a role in balancing the nucleoside–nucleotide pools or in the export/import of nucleosides within the plant. In cultured *Catharanthus* cells, however, phosphotransferase activity was found to be fivefold lower than that of Urd-kinase (Kanamori *et al.,* 1980).

The interconversion of the purine nucleotides via IMP includes the deamination of adenine compounds and overlaps with salvage routes. In Fig. 4 possible routes with enzymes known from plant tissues are shown; these routes also demonstrate possible entries of labeled purine bases and nucleosides added in the respective feeding experiments. With animal cells (muscle) a purine nucleotide cycle (IMP → SAMP → AMP → IMP) was proposed with a presumably regulatory role in balancing the level of Ade nucleotides (Aragón and Lowenstein, 1980). Turnover of AdoMet

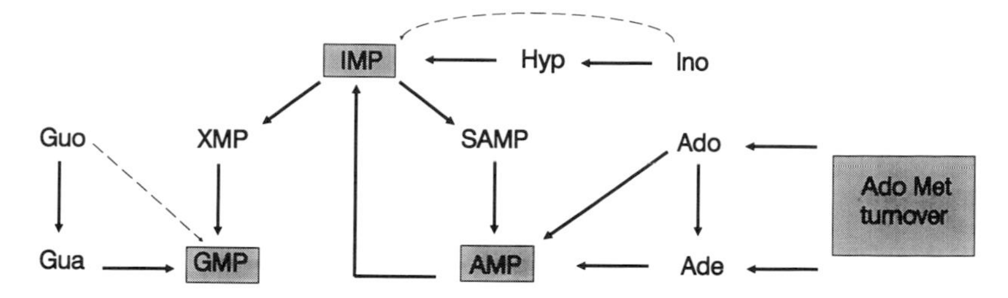

FIG. 4 Interconversion of purine nucleotides by enzymatic steps from salvage and *de novo* routes and connection to AdoMet turnover.

(SAM) is also connected with purine salvage pathways (Guranowski and Pawelkiewicz, 1977; Guranowski *et al.*, 1981).

With *E. coli* and *S. typhimurium* (Neuhard and Nygaard, 1987) deamination of Ade compounds is possible only with Ado, while deamination of Cyt compounds occurs with dCTP only (Fig. 2). In plants no activities for Ade and Ado deamination could be detected up to now (Le Floc'h *et al.*, 1982), whereas AMP deaminase (AMP-Nase) activity was reported from several species (Yoshino and Murakami, 1980; Le Floc'h *et al.*, 1982). AMP-Nase from spinach leaves revealed allosteric behavior with ATP as a positive and P_i as negative effector (Yoshino and Murakami, 1980). AMP deamination was suggested to regulate the pool size of Ade nucleotides with the effect of stabilizing the energy charge; it may also equilibrate the Ade and Gua nucleotide pools (Yoshino and Murakami, 1982; Doree, 1973; Le Floc'h *et al.*, 1982). Little is known on the interconversion of pyrimidine compounds by deamination; in higher plants the transformation of Cyd to Urd is apparently the only route (Ross, 1981), although from *Chlorella* deamination of dCMP is reported (Shen and Schmidt, 1966). Hence in plants, salvage of Cyd is possible by a kinase leading to CMP or by deamination to Urd and subsequent formation of UMP by probably the same kinase.

3. Degradation Pathways

The degradation of nucleic acids to nucleotides, a possible function of the vacuole (Abel and Glund, 1986; Abel *et al.*, 1990), will not be treated in this article. Enzymes that remove the phosphate from nucleotides, however, will be included, whereas those transforming nucleosides to nucleobases were discussed already in the preceding section. Nucleotidases with broad and narrow substrate specificities have been described. Examples of the former group are enzymes that cleave both 5′- and 3′-ribonucleotides; they were purified from wheat and maize seedlings (Polya, 1974; Carter and Tipton, 1986), whereas an enzyme with high specificity toward 5′-AMP was obtained from peanut cotyledons (Mittal *et al.*, 1988). Enzymes unselective toward the nucleobase but directed either to NMPs or NDPs were purified from soybean root nodules (Doremus and Blevins, 1988a,b).

Catabolism of pyrimidine bases (Fig. 5) in plants apparently follows the reductive pathway that is established for uracil, whereas Thy degradation has been poorly studied (Wasternack, 1980, 1982). A massive degradation to CO_2 of Urd, up to 75% of the amount absorbed, was reported with cultured sycamore cells (Cox *et al.*, 1973). Precursor feeding of protoplasts prepared from cultured *Catharanthus* cells (Sasamoto *et al.*, 1987) revealed high rates of degradation with Thy and Thd, 94 or 84% of the absorbed fraction, whereas the degradation of Ura (64%) and Urd (17%) was lower; the degradation product was mostly CO_2 but small amounts (2–5%) of ureidoisobutyrate and ureidopropionate (Fig. 5) were reported. The convertion of Cyt compounds for degradation in the uracil route obviously occurs by deamination of Cyd only and not of the free base (Ross, 1981). As salvage of Cyt is also not possible, as discussed above, plants apparently do not metabolize free Cyt (Wasternack, 1975).

Purine degradation follows the oxidative pathway (Fig. 5); this route was primarily studied in plants that form the ureides, allantoin and allantoate, during N_2 fixation (Wasternack, 1982; Reynolds *et al.*, 1982; Schubert, 1986), although degradation of purine compounds to ureides has also been described in seedlings, leaves, and cultured cells of various dicotyledon plants (Ashihara and Nobusawa, 1981; Nobusawa and Ashi-

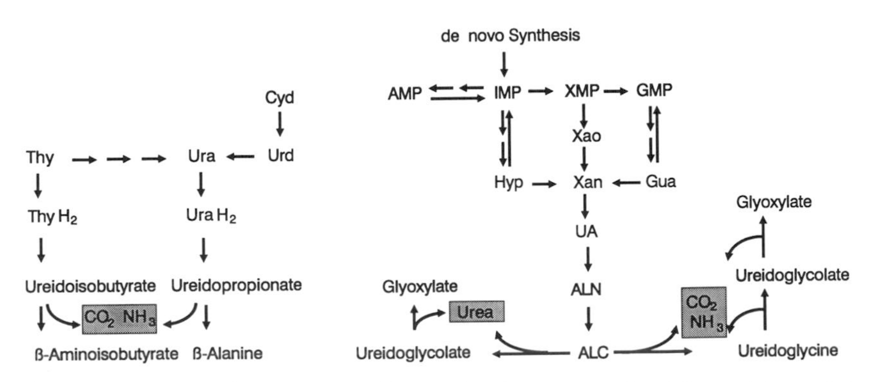

FIG. 5 Degradation pathways of pyrimidine and purine compounds. Only the intermediates are shown; for a more detailed description see textbooks and Wasternack (1982), Reynolds *et al.* (1982), and Winkler *et al.* (1988).

hara, 1983). Nevertheless little is known on purine catabolism in those plants that do not accumulate ureides (Reinbothe *et al.*, 1981) and the question arises as to which stage degradation of purine nucleotides proceeds and which recycle compounds are used in intraplant transport such as, e.g., in leaf senescence. This is especially interesting in plants that have high pools of nucleosides and Ade (Sawert *et al.*, 1987, 1988), as one could suggest that in these species nucleosides and Ade are used for intraplant transport.

As shown in Fig. 4, deamination of the base Ade in plants obviously occurs at the level of AMP only, which is in contrast to earlier suggestions from experiments with tobacco protoplasts (Barankiewicz and Paszkowski, 1980). Deamination of AMP (Fig. 5) is also consistent with results from several tissues that showed a faster transformation of exogenous Hyp or Gua into ureides than that of Ade (Ashihara and Nobusawa, 1981; Woo *et al.*, 1980). For the degradation of IMP to Xan there are two possible routes, although in nodules the route via XMP predominates (Schubert, 1986). The existence of IMP-DHase was shown in ureide-forming nodules (Shelp and Atkins, 1983) and thoroughly characterized (Atkins *et al.*, 1985); it is inhibited by AMP, NADH, and GMP (K_i of GMP is 60 μM). The Mo-containing Xan-DHase catalyzes the oxidation of both Hyp and Xan and is inhibited by allopurinol, as known from mammalian enzymes. With an enzyme from soybean nodules the V_{max} for Hyp oxidation was twice that of Xan oxidation (Boland *et al.*, 1983), although the K_m values were lower with Xan (Table VI). Uricase from soybean nodules, which transforms UA into ALN, was found to be strongly inhibited by Xan with a K_i value of 10 μM (Lucas *et al.*, 1983). The bases formed in the degradation pathway can be retransformed into nucleotides by salvage enzymes using PRPP (Fig. 5); besides the known transferase activities specific for Hyp and Gua, recycling of Xan, however, is not reported (Suzuki and Takahashi, 1975).

In several legumes of tropical origin, the ureides play an important role as transport compounds for nitrogen fixed in the nodules (Reynolds *et al.*, 1982; Winkler *et al.*, 1988; Schubert, 1986; Thomas and Schrader, 1981).They are synthesized in the nodules via purine nucleotide *de novo* synthesis and purine degradation. Catabolism of allantoate toward glyoxylate, occurring mainly in the leaf, is still not completely solved; although earlier reports stressed the urea-forming route, more recent data showed that the alternative pathway, with cleavage of CO_2 and NH_3 by amidohydrolases (Fig. 5), is more important (Winkler *et al.*, 1987, 1988).

In tea and coffee plants, considerable amounts of purine bases are transformed into caffeine (Suzuki and Takahashi, 1977); with certain *Camellia* species it was shown that caffeine synthesis (Fig. 6) occurs in the leaves and flowers (Ashihara and Kubota, 1986; Fujimori and Ashihara,

TABLE VI

K_m Values of Enzymes of the Degradation Routes

Enzyme	Substrate	K_m (μM)	Ref.[a]	Enzyme	Substrate	K_m (μM)	Ref.[a]
AMP-Hase	AMP	900	1	ALN-Hase	ALN	10,000	9
IMP-DHase	IMP	4	2	Xao-MTase	Xao	250	10
		9	3		AdoMet	3.3	
	NAD	18–35					
Uricase	UA	7.4	4	NDP-Hase	ADP	54	11
		10	5		GDP	31	
	O_2	31			UDP	143	
	UA	18	6		CDP	220	
	O_2	29					
Nucleotidase	5′-AMP	1.4	7	Xan-DHase	Xan	5	12
	3′-AMP	6.1			Hyp	52	
	AMP	58	8		NAD	12.5	
	GMP	57				20[b]	
	CMP	333			Xan	70	13
	UMP	200			Hyp	160	
					NAD	35	

[a]1, Mittal *et al.*, 1988 (peanut cotyledons); 2, Shelp and Atkins, 1983 (cowpea nodules); 3, Atkins *et al.*, 1985 (cowpea nodules); 4, Theimer and Beevers, 1971 (castor bean seedlings); 5, Lucas *et al.*, 1983 (soybean nodules); 6, Rainbird and Atkins, 1981 (cowpea nodules); 7, Polya, 1974 (wheat seedlings); 8, Carter and Tipton, 1986 (maize seedlings); 9, Thomas *et al.*, 1983 (soybean shoots); 10, Negishi *et al.*, 1985b (tea leaves); 11, Doremus and Blevins, 1988b (soybean root nodules); 12, Boland *et al.*, 1983 (soybean nodules); 13, Pérez-Vicente *et al.*, 1988 (*Chlamydomonas*).

[b]Value of NAD determined with Hyp as substrate; the other values of NAD were determined with Xan as substrate.

1990). Caffeine synthesis is connected to purine degradation and the branching metabolite is obviously the nucleoside Xao (Fig. 5); by feeding excised shoot tips of tea, it was shown that Xan is not a precursor of caffeine (Suzuki and Takahashi, 1976). Removal of the ribose occurs obviously after methylation of Xao at C-7 (Negishi *et al.*, 1988) (Fig. 6) followed by two further methylations to form caffeine (Negishi *et al.*, 1985a–c). The connection with the purine *de novo* and degradation pathways has not been fully elucidated; the fact that among the administered purine bases, Ade is the most efficient precursor for caffeine synthesis in tea and coffee plants (Ashihara and Kubota, 1986; Suzuki and Waller, 1988; Suzuki and Takahashi, 1976) is difficult to reconcile with the scheme of Fig. 5. One could speculate on separate compartments for salvage and degradation and the presence of high activities of Ade-RTase, AMP-Nase, and IMP-DHase in this tissue.

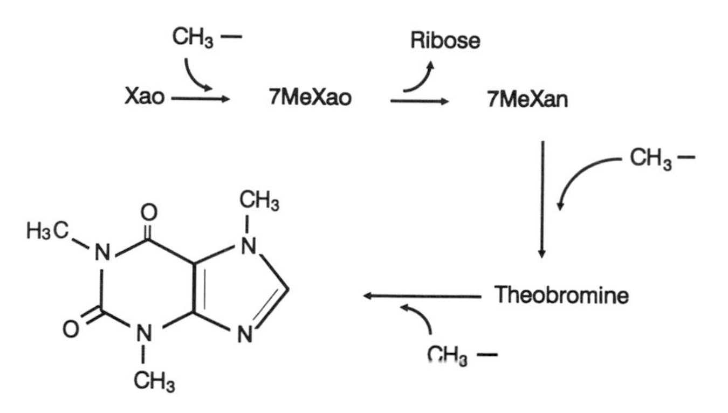

FIG. 6 Pathway for the biosynthesis of caffeine. For more details see Suzuki and Waller (1988).

B. Nucleotide Utilization in Carbohydrate and Lipid Metabolism

In growing cells nucleotides are consumed in the synthesis of nucleic acids, mainly RNA. Turnover of nucleotides, however, with conservation of the NDP or NMP moiety, occurs in the different routes of energy transformation and in the synthesis and utilization of various energy-rich anabolic intermediates such as the NDPsugars and the CDP-activated lipid precursors. Whereas the role of nucleotides in energy transformation, which involves mainly ATP and GTP, has been thoroughly treated in textbooks (Dennis, 1987), their role in the formation and consumption of energy-rich anabolic intermediates has been dealt with primarily in special reviews (Feingold and Avigad, 1980; Feingold, 1982).

1. Carbohydrate Metabolism

A major fraction of the total nucleotide pool in plants consists of nucleoside diphosphate sugars (Table I); they contribute to the synthesis of sucrose and other transport forms of carbon source, starch, the saccharide moieties of glycoproteins, and the various polysaccharides of the cell wall. Furthermore, several important sugar transformations occur at the level of NDPsugars (Feingold and Avigad, 1980; Feingold, 1982; Feingold and Barber, 1990). Figure 7 shows the main types of reactions in the synthesis and utilization of nucleoside diphosphate sugars.

Although many sugar nucleotide species with different nucleobases have been described (Feingold and Avigad, 1980; Feingold and Barber,

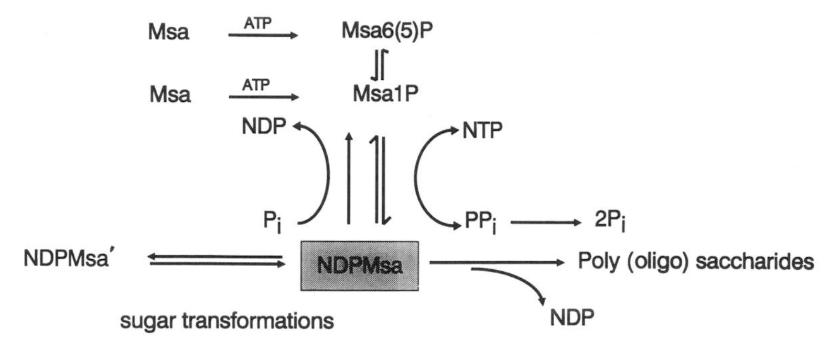

FIG. 7 The role of NDPsugars in oligo- and polysaccharide synthesis. Msa, Monosaccharide.

1990), UDPsugars extracted from different tissues prevail. In suspension cultured plant cells, the UDPsugar pools comprise up to 55% of the total nucleotide content, with low fractions of other NDPsugars (Table XI). Also in leaf and root tissue of tobacco (Meyer and Wagner, 1986), UDP-sugar fractions of about 25% of the total nucleotide content were found. In strawberry leaves the ratio of UDPsugars to GDPsugars was 100:1 (Isherwood and Selvendran, 1970), whereas the levels of UDPGlc and ADPGlc in soybean suspension cultured cells and in developing clubs of *Arum maculatum* were found in ratios of 14:1 and 10:1, respectively (ap Rees *et al.*, 1984). The ADPGlc concentrations found in pea chloroplasts in the light was about 0.4 m*M* (30 nmol/g FW, 8 nmol/mg Chl) and was very low in the dark (Kruger and ap Rees, 1983).

The predominance of the UDPsugar is also reflected in the activities of the NDPsugar-forming enzymes extracted from different tissues, which were much higher with UTP than with ATP or other NTPs (Feingold, 1982). As an example, in extracts of soybean cultured cells the ratio of the Glc1P-UTase and Glc1P-ATase activities was 15–50 : 1, while in developing clubs, active in starch production, the above ratio was 2.6 : 1 (ap Rees *et al.*, 1984). Glc1P-ATase activity is strongly regulated and is activated by PGA and inhibited by P_i (Preiss, 1988). Little is known about the regulation of Glc1P-UTase; with an enzyme from soybean nodules control through the availability of the substrates was proposed (Vella and Copeland, 1990). Product inhibition with a K_i value of 130 μM for UDPGlc was reported for Glc1P-UTase from pollen (Lucas and Madore, 1988).

It is generally accepted that ADPGlc is the main glycosyl donor in starch synthesis (Preiss, 1988); however, the discussion on the role of UDPGlc as substrate for a special form of starch synthase in nonphotosynthetic tissue, the granule-bound enzyme, is still contradictory (Preiss, 1988). There are reports on UDPGlc utilization for starch synthesis in maize kernel amyloplasts (Echeverria *et al.*, 1988), whereas a recent investigation with wheat grains identified the product of the UDPGlc-consuming activity as β-glucan (Kosegarten *et al.*, 1988). Synthesis of cellulose, (1,4)-β-glucan, has not been unequivocally demonstrated in cell-free extracts of higher plants; this is the reason why the respective synthase is poorly characterized, although it is established that UDPGlc is the substrate (Delmer and Stone, 1988). The related (1,3)-β-glucan (callose) synthase complex (Table VIII), which also accepts UDPGlc, has been described more thoroughly (Delmer and Stone, 1988; Hayashi *et al.*, 1987). An enzyme obtained from sugar beet tissue was shown to be activated by β-glucosides and Ca^{2+} and inhibited by UDP (Morrow and Lucas, 1987).

In plants UDPGlc is a very important metabolite not only in its role of supplying glucose moieties for polysaccharide synthesis; it plays a further key role in the synthesis and utilization of sucrose, the main carbon source for intraplant translocation (Fig. 8). SucroseP-Sase, the enzyme responsible for sucrose synthesis, is regulated by the cellular levels of Glc6P (activation) and P_i (inhibition) (Doehlert and Huber, 1984,; Stitt *et al.*, 1987). In the pathway of sucrose synthesis from Glc1P and Fru6P, one UTP is consumed and one UDP produced for every sucrose molecule synthesized, indicating a high turnover during active sucrose synthesis.

Sucrose utilization in sink tissue may proceed with the hydrolytic action of invertase leading to free glucose and fructose, which enter glycolysis by ATP-consuming kinase reactions (ap Rees, 1988). From several tissues, however, an energy-saving pathway was suggested that includes the reverse reaction of sucrose-Sase and the production of UDPGlc (Huber and Akazawa, 1986; Xu *et al.*, 1989; Stitt *et al.*, 1987; Black *et al.*, 1987) (Fig. 8). With cultured sycamore cells (Huber and Akazawa, 1986) the K_m value of sucrose with sucrose-Sase (15 mM) was lower than with the alkaline invertase (65 mM). UDPGlc obtained from sucrose may be consumed for cell wall synthesis or further converted to Glc1P and used for starch synthesis or consumed in glycolysis (ap Rees, 1988). In this "sucrose

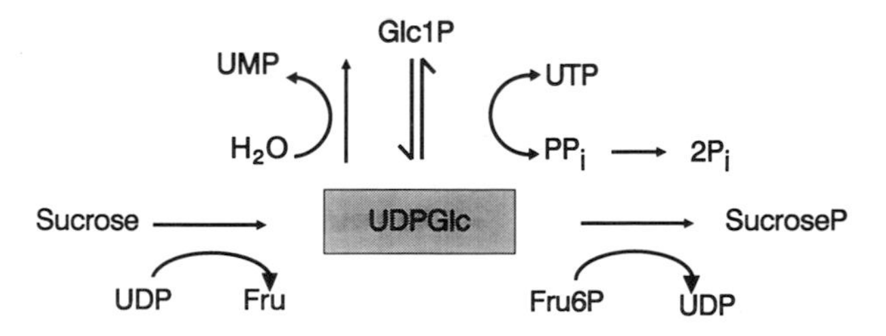

FIG. 8 The role of UDPGlc in sucrose synthesis and sucrose utilization.

synthase pathway'' one sucrose molecule is transformed to Fru and Glc1P with the formation of UTP and the consumption of UDP and PP_i (Fig. 8). It was further suggested that UTP may be recycled to UDP by an NTP-nonselective Fru-Kase using UTP (Xu *et al.,* 1989); hence both moieties of sucrose are channeled into glycolysis with the consumption of one PP_i. A Fru-Kase was described from developing maize kernels with a V_{max}/K_m value for UTP that was twice as large as that for ATP (Doehlert, 1990), although the K_m value for UTP was higher than for ATP. Similar properties revealed a further NTP-nonselective enzyme from suspension cultured cells (Yamashita and Ashihara, 1988), which showed, however, lower K_m values (Table VII). There is evidence that the sucrose synthase pathway plays a role in several sink tissues that also have elevated levels of PP_i dependent Fru6P-PTase. This enzyme may regulate the level of PP_i by its reverse reaction (reaction of Fru1,6P_2 with P_i); it is positively regulated by Fru2,6P_2 with a K_a of 0.2 μM (Xu *et al.,* 1989; Stitt, 1990).

Besides the pyrophosphorylase reaction there are further possible ways for the transformation of NDPsugars into hexose1P. They include a phosphorylytic cleavage resulting in NDP (Fig. 7); such an enzyme, specific for ADPGlc, was purified from potato tubers (Murata, 1977) and may be involved in the regulation of the ADPGlc level in the plastid (Feingold, 1982). In a third route NMP and hexoseP are formed by pyrophosphatase cleavage (Fig. 8); several enzymes with different specificities have been described (Feingold, 1982). A comparison of enzyme activities in different tissues (ap Rees, 1988) yielded the conclusion that, in the case of UDPGlc degradation, the pyrophosphorylytic route is the most important one. A highly NDPsugar-specific pyrophosphatase, however, was recently described from sugar cane vacuoles (Thom and Maretzki, 1989).

UDPGlcUA levels in plant tissues are low, about 5% of UDPGlc (Meyer and Wagner, 1985c, 1986); it is a direct precursor for cell wall synthesis or a substrate for the formation of UDPXyl by decarboxylation (Feingold, 1982; Delmer and Stone, 1988). There are two routes for its synthesis; the first route starts from UDPGlc and involves the action of a dehydrogenase. The second route (Fig. 9) starts with inositol and leads to GlcUA (oxidase), GlcUA1P (kinase), and finally UDPGlcUA (UTase). In the latter route the kinase step seems to be rate limiting; it is inhibited by its immediate product and UDPGlcUA (Feingold and Avigad, 1980; Feingold, 1982). With suspension cultured cells, specific enzyme activities were larger for UDPGlc-DHase than for GlcUA1P-UTase (Amino *et al.,* 1985a), indicating a possible higher rate for the route starting with UDPGlc than for the inositol pathway.

Synthesis of the oligosaccharide moiety of plant glycoproteins needs several NDPsugars and the transfer of the sugar moiety first onto a lipid membrane anchor (dolichol) before the attachment to the proteins. With

TABLE VII

K_m Values of Enzymes from Routes of NDPsugar Formation and Transformation

Enzyme	Substrate	K_m (μM)	Ref.[a]	Enzyme	Substrate	K_m (μM)	Ref.[a]
Glc1P-	Glc1P	40	1	ADPGlc-	ADPGlc	50	11
ATase	ATP	110		Pyase	P_i	2,500	
UDPGlcUA-	UDPGlcUA	240	2	GlcUA1P-	GlcUA1P	200	12
DCase		180	3	UTase	UTP	390	
		530			UDPGlcUA	250	
					PP_i	410	
Sucrose-	UDP	23	4	UDPGlc-	UDPGlc	500	13
Sase	Sucrose	40,000		Pyase	P_i	1,300	
	UDP	5	5				
	ADP	130		Fru-Kase	ATP	100	6
	Sucrose	31,000			UTP	500	
	UDPGlc	12			ATP	460	14
	Fru	3,700			UTP	1,010	
	Sucrose	15,000	6		ATP	50	15
	UDPGlc	76	7		UTP	150	
	Fru	7,800			Fru	130	
	UDP	190	8				
	ADP	190					
	Sucrose	17,000					
Glc1P-	Glc1P	940	9	Glc1P-	Glc1P	230	16
UTase	UTP	300		UTase	UTP	110	
	PP_i	230			PP_i	190	
	UDPGlc	445			UDPGlc	70	
	Glc1P	48	10		Glc1P	460	17
	UTP	30			UTP	140	
	PP_i	54			PP_i	25	6
	UDPGlc	56				29	4

[a] 1, Preiss, 1988 (maize endosperm, at saturating glycerateP levels); 2, Hayashi *et al.*, 1988 (soybean seedlings); 3, John *et al.*, 1977 (wheat germ); 4, Xu *et al.*, 1989 (developing potato tubers); 5, Morell and Copeland, 1985 (soybean nodules); 6, Huber and Akazawa, 1986 (cultured sycamore cells); 7, Doehlert, 1987 (maize endosperm); 8, Delmer, 1972 (*Phaseolus aureus* seedlings); 9, Otozai *et al.*, 1973 (artichoke tubers); 10, Gustafson and Gander, 1972 (sorghum seedlings); 11, Murata, 1977 (potato tubers); 12, Hondo *et al.*, 1983 (pollen); 13, Salerno, 1986 (developing maize endosperm); 14, Doehlert, 1990 (developing maize kernels); 15, Yamashita and Ashihara, 1988 (cultured *Catharanthus* cells); 16, Vella and Copeland, 1990 (soybean nodules); 17, Hopper and Dickinson, 1972 (*Lilium* pollen).

N-linked glycoproteins the main sugars are GlcNAc and Man, with the respective precursors UDPGlcNAc and GDPMan, and several minor sugar compounds (Kaushal *et al.*, 1988). UDPGlcNAc levels were determined in suspension cultured plant cells (Meyer and Wagner, 1985a, c) and amounted to about 10% of the UDPGlc pool. GlcNAc is formed from

TABLE VIII
K_m Values of Enzymes for NDPsugar Utilization

Enzyme	Substrate	$K_m (\mu M)$	Ref.[a]	Enzyme	Substrate	$K_m (\mu M)$	Ref.[a]
Sucrose P-Sase	UDPGlc	3000	1	GlcNAc1P-Tase	UDPGlcNAc	0.42	5
	Fru6P	3000			DolicholP	6.2	
Starch-Sase	ADPGlc	100–140	2	Xyloglucan-Sase	UDPXyl	28	6
βMan-Tase	GDPMan	1.7	3	Callose-Sase	UDPGlc	200–300	7
	Dolichol-GlcNAc-GlcNAc	9				260	8
						1–150	9
Galacturonan-Sase	UDPGalUA	770	4				

[a] 1, Doehlert and Huber, 1985 (spinach leaves); 2, Preiss, 1988 (maize endosperm); 3, Kaushal and Elbein, 1987 (cultured soybean cells); 4, Bolwell *et al.*, 1985 (sycamore trees); 5, Kaushal and Elbein, 1986 (cultured soybean cells); 6, Hayashi *et al.*, 1988 (soybean seedlings); 7, Hayashi *et al.*, 1987 (mung bean seedlings, cotton fibers); 8, Morrow and Lucas, 1987 (sugar beet); 9, Ingold and Seitz, 1986 (cultured carrot cells).

Fru6P (Fig. 9); the first enzyme of this pathway, the amidotransferase, is feedback inhibited by UDPGlcNAc (Beevers, 1982).

2. Lipid Metabolism

Cytosine nucleotides play a pivotal role in the synthesis and turnover of phospholipids. However, little is known about the pools of NDP-activated lipid precursors such as CDPcholine, CDPethanolamine, or CMPphospha-

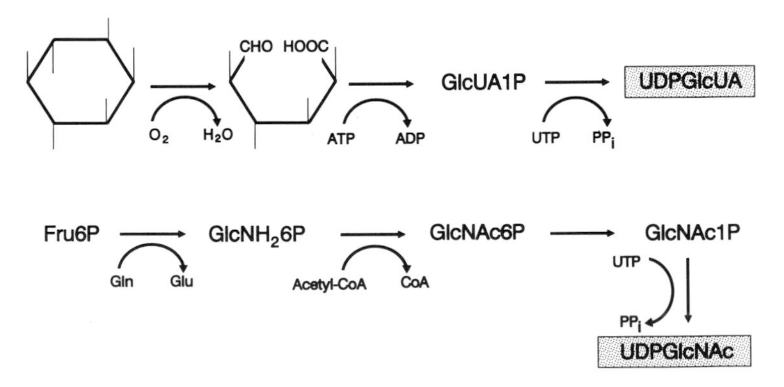

FIG. 9 Routes of synthesis of UDPGlcUA from inositol and of UDPGlcNAc from Fru6P.

tidic acid or about their respective enzymes. Levels of CDPcholine between 0.2 and 4.2 μg/g FW, cholineP between 0.6 and 6.5 μg/g FW, and choline between 11 and 28 μg/g FW were reported from pea stems (Price-Jones and Harwood, 1983). From spinach leaves and cultured sycamore cells a very efficient level of cholineP production in response to administered choline was observed, which indicated high activities of endogenous (cytoplasmic) choline-Kase (Bligny *et al.*, 1989, 1990). CholineP is a major phosphorylated compound in xylem exudate and is probably used for transport of phosphate and nitrogen (Martin and Tolbert, 1983).

C. The Pyridine Nucleotides

As found with all investigated species, *de novo* synthesis of pyridine nucleotides involves quinolinic acid (QA), which is transformed into NAD via NAMN and NAAD (Preiss-Handler pathway, Fig. 10). In higher species the amino donor for NAD synthesis is Gln, although the enzyme from tobacco roots accepts both Gln and Asn (Wagner and Wagner, 1985). Production of QA is more complex and starts with Trp as found in mammalian cells and some fungi; the alternative route, starting with Asp oxidation and reaction with dihydroxyacetoneP, is established in bacteria (Tritz, 1987); precursor feeding in maize seedlings suggested that QA is produced from Trp, although the final confirmation for the higher plants remains to

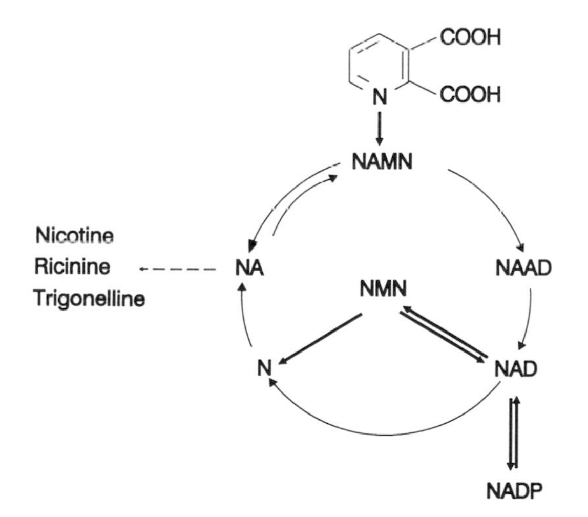

FIG. 10 Pyridine nucleotide cycle and connection to nicotinic acid-derived alkaloids; for more details see Wagner *et al.* (1986c), and Wagner (1987).

be established (Wilder *et al.*, 1984; Arditti and Tarr, 1979; Tarr and Arditti, 1982).

The total pyridine nucleotide pool is rather high; in growing bacteria it constitutes about 11% of the total nucleotides (Munch-Petersen, 1983), whereas in wheat flag leaves and in cultured *Datura* cells the sums of NAD and NADP were 18 and 12%, respectively, of the total nucleotide pool (Table I). On the other hand, the rate of NAD synthesis is usually lower than that of the ribonucleotides, as NAD is not consumed in the formation of a biopolymer, in contrast to the nucleotides. In growing bacteria only 1–2% of the total PRPP pool is consumed for NAD synthesis, relative to 60–80% for the ribonucleotides. NAD is turned over, however, with full conservation of the pyridine ring (Fig. 10). In *E. coli* a half-life for NAD of about 2 hr was found, whereas in HeLa cells the half-life was only 1 hr (Hillyard *et al.*, 1981). The function of the pyridine nucleotide cycle is not only to salvage NAD from pyridine moieties but also to supply intermediates for other pathways such as ADP-ribosylation, biosynthesis of vitamin B_{12}, and the formation of secondary plant products such as nicotine, ricinine, trigonelline, and others.

In different species different pyridine nucleotide cycles exist (Fig. 10), which are termed according to the number of enzymatic steps required for the reformation of NAD. For tobacco roots a six-membered cycle, with the formation of NMN by NAD pyrophosphatase, was suggested after characterizing the respective enzymes (Wagner and Wagner, 1985; Wagner *et al.*, 1986b; Wagner, 1987). However, part of the NAD may also be recycled by a five-membered cyclus bypassing NMN, as ADP-ribosylation with the production of nicotinamide (Fig. 10) also occurs in plant cells (Willmitzer and Wagner, 1982). A considerable activity of NAD glycohydrolase was recently reported from *Lemna* (Taguchi *et al.*, 1989). There is further evidence that nicotinic acid (NA) can be directly produced from NAMN by a glycohydrolase (Wagner *et al.*, 1986b). On the other hand, salvage of nicotinamide with direct formation of NMN, found in mammalian cells, seems to be excluded (Wagner *et al.*, 1986b) as well as deamination of NMN to form NAMN, which occurs in bacteria (Hillyard *et al.*, 1981).

Although not proven in every case, the most important metabolite of the pyridine nucleotide cycle for secondary product formation is NA (Fig. 10). There is indirect evidence from tobacco roots and callus (Wagner *et al.*, 1986a; Wagner, 1987) that NA is the immediate precursor for nicotine formation. It was further shown that nicotine production in tobacco is based on the stimulation of QA-RTase activity, stimulation of NAD degradation via the pyrophosphatase, and on NAMN glycohydrolase activity, which opens a direct route to nicotinic acid (Wagner *et al.*, 1986a, c). The latter enzyme has a rather large K_m value for NAMN (4 mM) (Table IX), obviously to prohibit complete depletion of the NAD pool.

TABLE IX

K_m Values of Enzymes from Pyridine Nucleotide Pathways

Enzyme	Substrate	K_m (μM)	Ref.[a]	Enzyme	Substrate	K_m (μM)	Ref.[a]
QA-RTase	QA	5	1	NMN-GHase	NMN	220	3
	PRPP	21					
				NAMN-GHase	NAMN	4000	3
NAD-Sase	NAAD	83	2				
	ATP	180					
	Gln	2000		NA-RTase	NA	150	3
					PRPP	330	
NAMN-ATase	NAMN	1300	2	Nicotanimide-Nase	Nicotanimide	150	3
	ATP	4000					
NMN-ATase	NMN	500	2	NA-MTase	NA	78	4
	ATP	190			AdoMet	55	
NAD-PPase	NAD	110	3	NAD-Kase	NAD	110	5
					ATP	180	
					NAD	550	6

[a] 1, Wagner and Wagner, 1984 (tobacco roots); 2, Wagner and Wagner, 1985 (tobacco roots); 3, Wagner *et al.*, 1986b (tobacco roots); 4, Upmeier *et al.*, 1988b (cultured soybean cells); 5, Muto, 1983 (pea seedlings, calmodulin dependent); 6, Laval-Martin *et al.*, 1990 (*Euglena gracilis*, calmodulin-independent).

Trigonelline (*N*-methylnicotinic acid) and ricinine formation starts with the methylation of NA (Tramontano *et al.*, 1983a; Johnson and Waller, 1974; Hiles and Byerrum, 1969; Taguchi *et al.*, 1989); furthermore, NA is also transformed to the *N*-glucoside (Upmeier *et al.*, 1988a,b). A more thorough study with different plant tissues showed that *N*-glucosylation occurs in several differentiated tissues and suspension cultured cells, whereas *N*-methylation is obviously restricted to differentiated plant tissues (Köster *et al.*, 1989). Results on the degradation of NA (Fig. 11) in parsley suspension cultured cells showed hydroxylation at C-6 as the first step and glutaric acid as a possible product (Komossa and Barz, 1988).

NAD is transformed to NADP by a kinase and NADP can be transformed back to NAD by a phosphatase (Fig. 10). The production of NADP and the regulation of the NAD(H)/NADP(H) ratio are very important for plant metabolism. There exist in plants calmodulin/Ca^{2+}-dependent and -independent NAD kinases (Raymond *et al.*, 1987 and references therein). There is controversial discussion about whether photoregulation of NAD kinase activity is mediated by phytochrome and/or cellular Ca^{2+}-levels (Dieter and Marmé, 1986; Kansara *et al.*, 1989; Muto *et al.*, 1981). NADP phosphatase activities have been reported from mammalian tissue, but also from *Euglena gracilis* (Laval-Martin *et al.*, 1990).

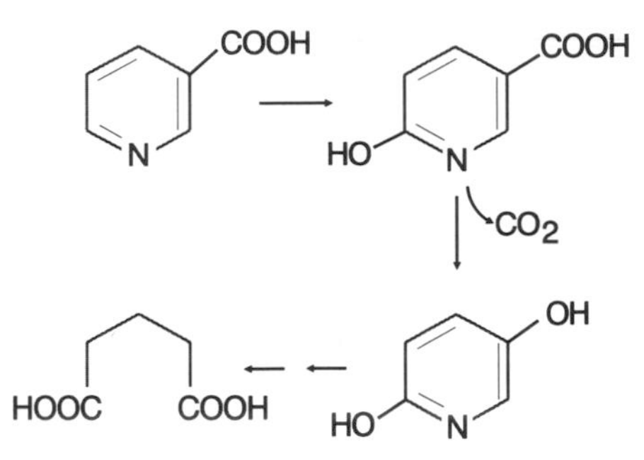

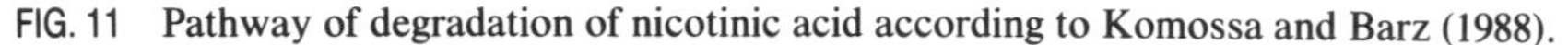

FIG. 11 Pathway of degradation of nicotinic acid according to Komossa and Barz (1988).

III. Localization of Pathways and Pools

A. The Cellular Compartments

Discussing the organization of metabolism on the cellular level, one must consider that plant cells have the most sophisticated intracellular anatomy. Whereas several compartments may play a role similar to those known in other eukaryotic cells, the plastid is a unique, plant-specific compartment and the vacuole has properties that differ from those of related organelles of other eukaryotic cell types. In many differentiated mature cells, but also in suspension cultured cells, the cytoplasmic volume is rather small relative to the vacuole. The cytoplasm, plastids, mitochondria, and vacuoles are the most frequently studied compartments; little is known of peroxisomes (glyoxisomes) and the lumen of the endoplasmic reticulum (ER) and Golgi apparatus. The cytoplasmic space may be extended to the nuclear compartment and also to the outer intermembrane space of plastids and mitochondria, as the respective membranes are provided with pores that allow the free diffusion of most of the metabolic intermediates. However, in these spaces specific enzymes and metabolic steps may be located. Enzymes located in the extracellular space, i.e., in the cell wall, may be important for import of solutes.

Methods for the determination of metabolite contents of cellular compartments often include the lysis of protoplasts and subsequent filtration techniques; in this case the cytosolic fraction is obtained from the total amount of metabolites, subtracting the corresponding amounts of the mitochondria and chloroplasts. This is correct, however, only if the vacuolar content is negligible, which holds for the main nucleotides. There is evidence from animal cells that the lumen of the Golgi apparatus has a different nucleotide composition than the cytoplasm; the lumen space contains significant NDPsugar pools that are used for synthesis of the glyco-moieties of proteins and of polysaccharides. Although not thoroughly investigated, lack of ATP in this compartment may be concluded from the fact that phosphorylation of the Man moieties of glycoproteins, the marker for glycoprotein addressed to the lysosome, is performed with UDPGlcNAc and not with ATP (Lang *et al.*, 1984).

For leaf mesophyll protoplasts from oat (Hampp *et al.*, 1982) fractional volumes of cellular compartments were estimated (total space 23 μl/10^6 protoplasts) as 77% for the vacuole, 11% for the chloroplasts, 5% for the mitochondria, and 7% for the cytosol. For root tips (length 1.5 mm) volume fractions for the cytoplasm, vacuole, and intercellular spaces including cell walls were estimated as 50, 40, and 10%, respectively (Lee and Ratcliffe, 1983). With protoplasts from suspension cultured tomato cells, for the vacuolar space a fraction of 80% was estimated by staining (Leinhos *et al.*, 1986).

B. Localization of Pathways

1. *De Novo* Routes

a. Synthesis of Phosphoribosylpyrophosphate and Pyrimidine Nucleotides The ribose moiety (ribose5P) for nucleotide synthesis is provided by the oxidative pentoseP pathway, the enzymes of which are located in the cytoplasm and the plastids (ap Rees, 1987), although from different species differences in the plastidial enzymatic equipment, e.g., absence or presence of the first enzyme, Glc6P-DHase, were reported. Conversion of ribose5P to PRPP, which is required for both *de novo* and salvage pathways, obviously also occurs in the cytoplasm and in organelles. With suspension cultured cells (Kanamori *et al.*, 1980), 60% of the PRPP-Sase activity was found in the cytosol (100,000 *g* supernatant) and 30% were in a particulate fraction (12,000 *g* pellet). From an earlier investigation with spinach leaves, 95% of the total activity was reported to be in the cytosol and only a little in the chloroplasts and mitochondria (Ashihara, 1977b). PRPP-Sase activity was, however, reported from mitochondria of Jerusalem artichoke tubers (Le Floc'h and Lafleuriel, 1983) and its location in the outer intermembrane space was suggested. Recent data from a green alga (Ashihara, 1990) showed 55% of the PRPP-Sase activity in the cytoplasm and 45% in the chloroplast, whereas negligible amounts were detected in the mitochondrial fraction. These data, although preliminary, suggest that both the cytoplasm and the plastids possess activities for PRPP synthesis; there is, however, no respective evidence for the mitochondrial matrix, as enzymes located in the outer membrane space may not necessarily supply the matrix. Hence one could speculate that the mitochondrial matrix may depend on import of nucleosides or nucleotides.

De novo synthesis of pyrimidine nucleotides and Arg begins with CP and there is obviously only one CP pool produced by the same enzyme (Ong and Jackson, 1972b). As synthesis of Arg and many other amino acids occurs in the plastid (ap Rees, 1987), one should also have expected that the pyrimidine nucleotide *de novo* pathway starts in this compartment. Recent investigations with several species (Shibata *et al.*, 1986) showed that CP-Sase and Asp-CTase (Fig. 1) are both located in the chloroplast, although earlier results from cultured *Catharanthus* cells (Kanamori *et al.*, 1980) suggested also a cytoplasmic location; this could, however, have been a consequence of breakage of organelles during homogenization (Doremus and Jagendorf, 1985). Further work on pea leaves gave evidence that the first three enzymes, leading to OAH_2, and the last two enzymes, transforming OA to UMP, of the pathway (Fig. 1) are stromal proteins (Doremus, 1986; Doremus and Jagendorf, 1985). The enzyme, OAH_2-DHase, is apparently associated with the outer surface of the inner mitochondrial membrane (Doremus and Jagendorf, 1985; Miersch *et al.*, 1986), and directly participates with the electron transport chain of the mitochondrion, which is a common feature of other eukaryotic species (Doremus and Jagendorf, 1985). The chloroplast-located enzymes are obviously coded by nuclear genes and synthesized on cytoplasmic ribosomes, as shown with oat leaves (Doremus and Jagendorf, 1987). Contradictory results, however, were reported recently from cytochemical studies with cultured tomato cells; after fixation of the cells and activity assays in the presence of lead acetate, Asp-CTase was assigned to the nucleus and Orn-CTase to the vacuole (Walther *et al.*, 1989). Obviously additional investigations are necessary to confirm the location of the enzymes for pyrimidine nucleotide synthesis.

Synthesis of OAH_2 in the chloroplast, its subsequent oxidation to OA in the mitochondria, and the conversion of the latter into UMP in the plastid would suggest an efficient metabolite traffic between these two compartments. Furthermore, UMP or probably Urd, as a less polar compound, must leave the stroma as the main uracil nucleotide pools are in the cytoplasm. On the other hand, for feedback regulation of Asp-CTase, the pools of UMP inside and outside the chloroplast should be in close equilibrium.

b. Purine Nucleotide* de Novo *Routes and Nucleotide Kinases Localization of enzymes of the purine nucleotide *de novo* pathway is only poorly studied, with the exception of investigations on N_2-fixing nodules of certain ureide-forming legumes such as cowpea and soybean (Schubert, 1986). Nodule fractionation and labeled precursor incorporation nicely showed that the enzyme activities for the pathway from PRPP to IMP (Fig. 12) are quantitatively associated with the plastids (Boland and Schubert, 1983; Shelp *et al.*, 1983; Atkins *et al.*, 1982). Whether this holds also for non-ureide-forming tissues remains to be elucidated. Little is known about the location of the routes (Fig. 1) from IMP to AMP and GMP, respectively. Purification of a GMP-Kase from isolated mitochondria of artichoke and potato tubers suggested its location in the inner mitochondrial membrane or the matrix (Le Floc'h and Lafleuriel, 1990).

A major fraction of AMP-Kase is apparently associated with the chloroplasts; 40–90% of the total activity was extracted from the plastid fraction of both C_3 and C_4 plants (Hampp *et al.*, 1982; Birkenhead *et al.*, 1982; Moore *et al.*, 1984). With various C_4 plants, considerable fractions were also found in the cytoplasm (Moore *et al.*, 1984), whereas with several C_3 plants little if any activity was found in this compartment, but significant activity was associated with the mitochondria (Birkenhead *et al.*, 1982; Day *et al.*, 1979). These findings are consistent with a report on immunologically distinct forms of the enzyme in leaf extracts (Kleczkowski and Randall, 1987). With protoplasts of wheat leaves, 56% of the AMP-Kase activity was localized in the chloroplasts, 41% in the mitochondria, and only 3% in the cytosol fraction (Stitt *et al.*, 1982). With several preparations of isolated mitochondria, AMP-Kase activity was reported to be located in the outer intermembrane space (Birkenhead *et al.*, 1982; Day *et al.*, 1979; Stitt *et al.*, 1982). Location of a major fraction of activity in the

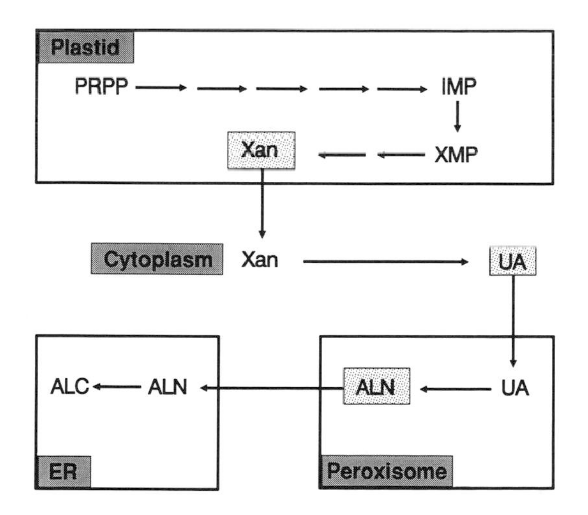

FIG. 12 Localization of purine *de novo* synthesis and ureide formation as elaborated from legume nodules. This is a modified scheme derived from Fig. 4 of Schubert (1986).

outer intermembrane space was also found with chloroplasts from spinach leaves (Rodionova *et al.*, 1978). As this space is accessible for Ade nucleotides from the cytoplasm, equilibration of the adenylates in the cytoplasm should be efficient with both C_4 and C_3 plants. Determination of Ade nucleotide levels and calculation of the mass action ratio of the AMP-Kase reaction suggested equilibration in the cytoplasm and the stroma, but not in the mitochondrial matrix (Stitt *et al.*, 1982). Taken together, these results show that AMP-Kase, which is usually present in the cell at abundant levels (Raymond *et al.*, 1987), is able to equilibrate the Ade nucleotide pool in the cytoplasm and the stroma, whereas the mitochondrial matrix obviously does not depend on its action. In a recent report the presence of AMP-Kase on the tonoplast membrane of suspension cultured cells was suggested (Hill *et al.*, 1988).

The cytoplasm also contains considerable activities of NDP-Kase for equilibration of the NTP and NDP pools, whereas these enzymes are obviously absent from the mitochondrial matrix; this was concluded from the determination of nucleotide levels (Dancer *et al.*, 1990a,b) and further supported by reports on ^{31}P NMR measurements of corn root tips, which revealed an NDP pool that was obviously inaccessible for NDP-Kase (Hooks *et al.*, 1989). In suspension cultured cells 50% of the NDP-Kase activity, which was assayed by the conversion of UDP into UTP, was located in the cytoplasm and the rest in the chloroplasts, with negligible activity in the mitochondria. The activities were high enough to recycle the UDP formed in sucroseP production in the leaf cytoplasm (Dancer *et al.*, 1990b).

2. Salvage Pathways

Localization of enzymes of the salvage pathway is particularly interesting, as cell compartments without *de novo* pathways will depend on import of respective precursors. Export or import of nucleosides or nucleobases should be conceivable as these are better candidates to cross membranes than the more polar nucleotides; however, little is known about this subject. With the assumption that both the purine and pyrimidine *de novo* synthesis pathways (Fig. 1) are located in the plastids, the cytosplasm and the mitochondrial matrix will depend on the salvage of imported precursors.

Enzymes involved in salvage of pyrimidine nucleotide precursors were studied by fractionation of cultured *Catharanthus* cells (Kanamori *et al.*, 1980); Urd-Kase activity was almost totally associated with the cytoplasm, whereas only 30% of the Ura-RTase activity was found in the cytosol and 60% in a particulate fraction. As the total activity of Urd-Kase was about threefold larger than that of Ura-RTase, one may conclude that Ura nucleotide formation in the cytoplasm depends mainly on the import of Urd and its phosphorylation. However, further studies with different plant tissues and improved cell fractionation techniques should be performed to confirm these preliminary results. Cell fractionation of rye seedlings showed that nucleoside-PTase activity was mainly found in the cytosol fraction, while 70% of Thd-Kase activity was located in the cytoplasm and about 30% in the chloroplast (Golaszewski *et al.,* 1975). Thd incorporation in *Chlamydomonas* cells suggested that Thd-Kase activities are located in the chloroplasts and the cytoplasm (Chiang *et al.,* 1975). One may conclude from these results that both the cytoplasm and the plastids contain the routes for dNTP synthesis, although more information, especially on the NDP-Rase, is urgently required.

Earlier results from labeled precursor feeding (Ross, 1981; Le Floc'h *et al.,* 1982; Anderson, 1979), were interpreted to suggested that purine salvage is locally separated from the *de novo* pathway. As *de novo* synthesis of purine nucleotides is most likely located in plastids, then especially the cytoplasm but also the mitochondria will depend on salvage enzymes. Cell fractionation studies with suspension cultured cells showed that several enzymes of purine salvage (Ade-RTase, Hyp/Gua-RTase, and Ado-Kase) had their main activity in the cytosol (Ukaji *et al.,* 1986), whereas about 8 to 13% of the activity was in the mitochondrial fraction. There are, however, also reports that chloroplasts contain enzymes for purine salvage, such as Ade-RTase and Ado-Kase (Ashihara and Ukaji, 1985).

Studies with Jerusalem artichoke mitochondria suggested that Ade-RTase activity is located in the outer intermembrane space and bound to the outer membrane (Le Floc'h and Lafleuriel, 1983). As activities of PRPP-Sase and AMP-Kase were also present, it was further suggested that complete salvage of Ade into AMP and ADP is possible in this space. Furthermore, considerable activities of AMP-Nase were also found in the mitochondrial intermembrane space; this enzyme may function in the adjustment of the Ade and Gua nucleotide pools, provided the enzymes of the IMP → GMP route would be available (Fig. 4). Also, a Guo/Ino-Kase activity was recently ascribed to the outer intermembrane space of mitochondria (Combes *et al.,* 1989). The importance of enzymes located in the outer intermembrane space of mitochondria is obvious for the cytoplasmic compartment; it is, however, not clear whether these enzymes play a role in supplying the matrix space with nucleotides. The same may hold for the outer intermembrane space of plastids.

3. Degradation Pathways

Vacuoles prepared from suspension cultured cells have the capacity to degrade RNA; they obviously contain RNases, phosphodiesterase, and acid phosphatases (Abel and Glund, 1986; Abel *et al.,* 1990). Vacuoles from cultured tomato cells contained about 25% of the total cellular Urd pool; the vacuolar concentration was estimated as 20 μM and that of the cytoplasm as 235 μM (Leinhos *et al.*, 1986). These findings suggest that Urd is supplied from RNA degradation in the vacuole and transported to the cytoplasm for the salvage route (Leinhos *et al.*, 1986).

An AMP-specific 5′-nucleotidase was isolated from the plasma membrane of peanut cotyledons and its obvious precursor, located in the Golgi apparatus, was also purified (Sharma *et al.*, 1986; Mittal *et al.*, 1988). Also in pea cotyledons, a 5′-nucleotidase was assigned to the outer surface of the plasma membrane and to the plasmadesmata (Nougarède *et al.*, 1985). However, a further nucleobase-nonselective 5′-nucleotidase was purified from the microsomal fraction of maize seedlings (Carter and Tipton, 1986) and may also be located in the plasma membrane. The generation of nucleosides for intraplant transport may be a possible role of these enzymes.

Nucleosides may also be used in intracellular transport; with the location of the *de novo* routes in the plastid, one should expect nucleotide-hydrolyzing activities in this compartment. In the stroma of *Beta vulgaris* chloroplasts, a 5′-nucleotidase was found that was, however, absent in spinach chloroplasts (Eastwell and Stumpf, 1982). NDP- and NMP-hydrolyzing activities from soybean nodules were found to be soluble proteins and neither appeared to be associated with nodule organelles (Doremus and Blevins, 1988a), although activities for degradation of XMP into Xan (Fig. 12) would be expected in the plastid (Schubert, 1986). A base- and ribose-nonselective nucleoside triphosphatase from the chromatin fraction of pea nuclei was described (Chen *et al.*, 1987) and stimulation by Ca^{2+}/calmodulin was reported.

AMP deaminase, the enzyme starting degradation of the Ade nucleotide pool (Fig. 5), was studied in spinach leaves (Yoshino and Murakami, 1980) and its main activity was found in the cytosolic fraction and in a particulate fraction that could be extracted with a mild detergent (Triton X-100). Considerable AMP-Nase activities were also reported from the outer intermembrane space of Jerusalem artichoke mitochondria (Le Floc'h and Lafleuriel, 1983). Location of the enzymes of the purine degradation route leading to the ureides (Fig. 5) is fairly established in species with N_2-fixing nodules (Schubert, 1986, and references therein; Wasternack, 1982), although some details remain to be clarified. The enzymes were assigned to different cellular compartments (Fig. 12); IMP-DHase is most likely located in the plastid, where the riboseP moiety is probably removed also. Xan-DHase is a cytoplasmic enzyme, while uricase, as in animal tissue, is peroxisomal. The intracellular location of allantoinase is not unambiguously solved (Webb and Newcomb, 1987); there is evidence for its location in the ER; however, difficulties in the preparation of pure ER fractions still must be overcome. It must be confirmed whether it is also so that, in those species that do not couple purine degradation and ureide biogenesis to N_2 fixation, the enzymes of the purine degradation route are located in the cellular compartment as shown in Fig. 12. Thus with pea leaves, Xan-DH was assigned to the cytoplasm (Nguyen and Feierabend, 1978), and with maize root tips, which did not contain allantoinase, uricase was found to be located in the peroxisomes (Parish, 1972). In the fat-degrading tissues of seedlings from several species, uricase activity was also determined in glyoxisomes (Theimer and Beevers, 1971), while allantoinase seemed to be located both in the glyoxisome and proplastid fractions.

4. Pathways for Nucleotide Utilization in Carbohydrate Metabolism

The enzymes involved in the formation of NDPsugars and their utilization also operate in different cellular compartments. It is generally accepted that in leaf tissue Glc1P-ATase and starch synthase are quantitatively associated with the chloroplast (Preiss, 1988; Okita *et al.*, 1979) and in nonphotosynthetic tissue with the plastid species such as amyloplasts. Location in amyloplasts was shown in developing potato tubers or wheat grains, but also in cultured soybean cells (Kim *et al.*, 1989; Entwistle and ap Rees, 1988; Macdonald and ap Rees, 1983). Location of the UDPGlc-forming enzyme is less unambiguous, although in leaf tissue this enzyme is confined to the cytosplasm. This holds also for amyloplasts of developing wheat grains (Entwistle and ap Rees, 1988) and cultured soybean cells (Macdonald and ap Rees, 1983); in amyloplasts of developing maize endosperm, however, higher activities of Glc1P-UTase were reported than of Glc1P-ATase (Echeverria *et al.*, 1988). In earlier work with onion roots, the major fraction of Glc1P-UTase activity was attributed to the Golgi apparatus and a minor fraction to the smooth ER (Feingold, 1982, and references therein). The distribution of Glc1P-UTase activity between the soluble part of the cytoplasm and the endomembrane system should be reinvestigated, however, with the more recently developed methods that improve the preparation of cellular compartments (Fleischer, 1989; Hirschberg and Snider, 1987).

Earlier results from several authors suggested distinct UDPGlc pools in the cell available to glycosyltransferases associated with different membrane structures (Feingold, 1982, and references therein), which may also hold for other NDPsugars. Synthesis and processing of glycoproteins and polysaccharides occur within the rough ER and Golgi apparatus, which require pools of NDPsugars in the lumens of these compartments. Specific carriers for the translocation of NDPsugars through Golgi membranes have been reported from animal cells (Fleischer, 1989; Milla *et al.*, 1989). Studies on xyloglucan synthesis in soybean seedlings revealed the location of the UDPGlcUA-DCase at the Golgi membranes, where the xyloglucan-forming activity was also associated (Hayashi *et al.*, 1988).

Location of callose synthase at the plasma membrane is fairly established (Frost *et al.*, 1990), and enzyme activities for glycoprotein synthesis, using dolicholP as carrier, were also reported to be membrane associated (Kaushal *et al.*, 1988; Torossian and Maclachlan, 1987; Riedell and Miernyk, 1988). Fractionation of membranes from different sources indicated that the latter glycosyltransferases were located in the rough ER and the Golgi apparatus (Kaushal *et al.*, 1988; Riedell and Miernyk, 1988). With developing bean cotyledons a GlcNAc-specific transferase was shown to be located in the Golgi apparatus (Chrispeels, 1985). Although more elaborated in animal cells, there is ample evidence that plant glycoprotein synthesis also proceeds by cotranslational glycosylation at the rough ER, with further processing in the Golgi apparatus (Kaushal *et al.*, 1988). A similar strategy also holds for cell wall polysaccharide synthesis, which is initiated in the lumen of the ER and moved to the Golgi apparatus; the subsequent budding of secretory vesicles from the Golgi and endocytosis with the plasma membrane releases the polysaccharide and to some degree also precursors into the apoplastic space (Delmer and Stone, 1988). Transferase activity may be small in the ER and increase in the Golgi apparatus, as shown with arabinan and xylan synthesis (Bolwell and Northcote, 1983). Studies with cultured carrot cells revealed transferase activity toward β1,4-glucan synthesis associated with Golgi membranes, whereas callose synthesis was associated with the plasma membrane (Ingold and Seitz, 1986). The enzymes for cellulose and callose synthesis seem to be latent up to the moment when they are integrated into the plasma membrane (Delmer and Stone, 1988).

Enzymes for sucrose synthesis including Glc1P-UTase and sucroseP-Sase, are confined to the cytoplasm (Stitt *et al.*, 1987). In addition, the enzymes for sucrose degradation, including alkaline invertase, or those of the reverse sucrose synthase pathway (Fig. 7) including sucrose-Sase, Glc1P-UTase, NTP-nonselective Fru6P-Kase, and PP_i-dependent Fru6P-PTase, are obviously located in the cytoplasm (Entwistle and ap Rees, 1988; Xu *et al.*, 1989; Macdonald and ap Rees, 1983; Wendler *et al.*, 1990). Due to low PPase activity, the cytoplasm has a significant PP_i pool, whereas the main (alkaline) PPase activity is obviously confined to the plastid (Weiner *et al.*, 1987). From studies with cotyledons of cucumber, however, a plasma membrane-associated PPase was described with properties different from those of the plastidial enzyme (Kirsch *et al.*, 1989).

The previously suggested group translocator concept, investigated with vacuoles from cultured sugar cane and red beet cells, which suggested a transport-linked vectorial sucrose synthesis at the tonoplast membrane

(Lucas and Madore, 1988; Thom and Maretzki, 1985, 1987; Thom *et al.*, 1986; Maretzki and Thom, 1987; Voss and Weidner, 1988), gave UDPGlc a pivotal role in transport of sucrose into the vacuoles. The experimental basis of this concept has been, however, severely criticized recently; hence the participation of UDPGlc in sucrose transport remains an open question (Preisser and Komor, 1988; Maretzki and Thom, 1988). Interestingly, vacuoles prepared from cultured sugar cane cells were shown to contain activities of both Glc1P-UTase and UDPGlc-PPase; the latter enzyme was highly specific for NDPsugars and did not cleave NAD (Thom and Maretzki, 1989).

5. Routes of Pyridine Nucleotide Synthesis and Turnover

Little is known about the localization of enzymes involved in the synthesis and turnover of the pyridine nucleotides (Fig. 8). As NAD(P)(H) is required in at least four compartments, i.e., in the cytoplasm, plastid, mitochondrion, and peroxisome, it is of special interest how *de novo* synthesis, transport of intermediates, and salvage routes may be interrelated. Unfortunately, almost nothing is known with the exception of the localization of NAD-Kases (see below). From enzyme activity measurements, it was suggested that the main fraction of NAD-PPase is separated, probably located in the vacuole, from the main cellular NAD pools (Wagner *et al.*, 1986b,c). From animal cells there is evidence for a nuclear localization of the enzyme that converts NMN into NAD (Jackson and Atkinson, 1966).

Activity of NAD-Kase, which forms NADP from NAD, was found in the cytosolic fraction and in the outer membrane of mitochondria and chloroplasts; the membrane-associated activity was suggested to face the cytoplasm (Dieter and Marmé, 1984, 1986; Simon *et al.*, 1982, 1984). These enzyme species are thought to supply the cytoplasm with NADP; furthermore, there is evidence for the localization of a calmodulin-independent enzyme in the stroma of chloroplasts obtained from pea seedlings (Simon *et al.*, 1984). In addition, soluble and particle-bound enzyme activities were reported from *E. gracilis* cells (Laval-Martin *et al.*, 1990). Disruption of leaf protoplasts, performed with several species, revealed over 85% of the activity of a calmodulin-independent NAD-Kase in the chloroplast fraction; only in the case of pea leaves and maize mesophyll cells were 30% and 95%, respectively, found in the cytosolic fraction (Muto *et al.*, 1981).

C. The Subcellular Pools

Subcellular pools of inorganic phosphate are usually estimated by ^{31}P NMR spectroscopy under *in vivo* conditions. An investigation with leaves and protoplasts of different species in the dark (Stitt *et al.*, 1985) revealed total P_i contents of 5–8 μmol/mg Chl; the main fraction hereof, 60–89%, was in the vacuole. For the combined compartments, cytoplasm and chloroplast, P_i concentrations of 16–33 mM were estimated. Recent studies on spinach leaves and isolated chloroplasts revealed a detailed picture of the P_i pools within the cell (Bligny *et al.*, 1990). Whereas in the chloroplast the P_i levels were rather high, about 12 mM, rather low levels were adjusted in the cytosol, in the range of 0.6–1.2 mM and lower. The P_i level in the vacuole fluctuated, however, strongly depending on the total supply and also on the metabolic state of the leaf. It was suggested that the cytosolic level is controlled by active transport of P_i into the vacuole and passive efflux from the vacuole into the cytoplasm. The vacuolar buffer level may increase to rather high values, exceeding 50 mM, depending on P_i availability.

Due to the low pyrophosphatase activity in the cytoplasm, plant cells possess significant PP_i levels. With different plant tissues, contents of 4–40 nmol/g FW were determined and, as a consequence of the cellular distribution of the PPases, one can assume that PP_i is mainly in the cytoplasm. With a cytoplasmic volume of about 10% of the total tissue volume, from the above PP_i levels cytoplasmic concentrations of 0.04–0.4 mM were estimated (Black *et al.*, 1987). The lowest value was obtained with suspension cultured cells (0.04 mM), but this value may be about twofold higher, as cultured cells usually have very large vacuoles and a small cytoplasmic volume.

In explants from *Helianthus* tubers, CP levels of 2.3 nmol/g FW and Asp levels of 143 nmol/g FW were determined; assuming the plastid volume to be about 10%, a plastidial CP concentration of 23 μM would result, whereas that of Asp would be 1.4 mM. Thus the concentration of CP is below the K_m value for Asp-CTase (Table III), while that of Asp is higher, suggesting that CP production is a rate-limiting step in pyrimidine nucleotide *de novo* synthesis (Parker and Jackson, 1981).

The main nucleotides for energy metabolism, the adenine nucleotides and pyridine nucleotides, have significant pools in at least three cellular compartments (Raymond *et al.*, 1987). For leaf mesophyll cells, Table X shows an equal distribution of the Ade nucleotides into the cytoplasm and the plastids, while the mitochondrial fraction was only about 9% of the total amount (Stitt *et al.*, 1982). With protoplasts obtained from *Vicia faba* guard cells, a different distribution of ATP was reported; 67% of the total amount of 92 nmol/mg protein was in the cytosol fraction, 31% in the mitochondria, and only 3% in the plastids (Michalke and Schnabl, 1987). This was, however, a consequence of the fact that the plants were in the dark for 24 hr before isolation; in the plastids 70% of the Ade nucleotides consisted of AMP.

With the pyridine nucleotides, the total NAD(H) level is usually higher than that of NADP(H). The leaf (Hampp *et al.,* 1985) and endosperm (Donaldson, 1982) cells (Table X) have large cytoplasmic fractions of both pyridine nucleotide species, while leaf chloroplasts have larger pyridine nucleotides pools than the plastids of the endosperm. Mitochondria of both leaf and endosperm have higher levels of NAD(H) than NADP(H); the glyoxisomes of the endosperm have also a considerable NADP(H) pool. Mitochondria have low, although significant, NADP(H) pools; this is obviously not an artifact, i.e., cytosolic or plastid impurities; there are reports on NADP(H)-dependent enzymes in mitochondria (Rasmusson and Moller, 1990).

Only a few data are available on the subcellular distribution of the other nucleotide pools such as, e.g., GTP, UTP, and CTP. These nucleotides are required both for nucleic acid synthesis in the nucleus and protein synthesis in the cytoplasm, but also for the same processes in the partially autonomous organelles, the chloroplasts and mitochondria. With cultured tomato cells, evidence for separate UTP pools in nuclear and mitochondrial rRNA synthesis was suggested (Hause and Wasternack, 1988). With *in vivo* ^{31}P NMR spectroscopy on root tips from mung bean, complexation of Mg^{2+} with the ATP pool was studied (Yazaki *et al.,* 1988). The ratio of MgATP to free ATP was 9.45:1, and the free Mg^{2+} concentration in the cytoplasm was estimated to 0.4 m*M*.

Concerning the distribution of the various NDPsugars, studies have been performed more thoroughly with UDPGlc only. With spinach leaves, it was shown that UDPGlc is located mainly in the cytoplasm, at least to 90%, with levels of 30–50 nmol/mg Chl (Stitt *et al.,* 1987). With barley leaf

TABLE X
Distribution of Ade Nucleotides and Pyridine Nucleotides into the Different Cellular Compartments[a]

	Total amount (nmol/mg Chl)	Cytoplasm (%)	Plastid (%)	Mitochondria (%)	Glyoxisome (%)	Ref.[b]
Leaf protoplasts						
AXP	103	46	45	9	—	1
NAD(H)	20	50	29	21	—	2
NADP(H)	9	54	35	11	—	2
Castor bean endosperm						
NAD(H)	—	48	2	47	3	3
NADP(H)	—	64	11	9	16	3

[a]Leaf mesophyll protoplasts in the light.
[b]1, Stitt *et al.,* 1982; 2, Hampp *et al.,* 1985; 3, Donaldson, 1982.

protoplasts the following cytosolic concentrations were estimated (the percentage of the metabolite present in the cytoplasm is in parentheses): UDPGlc, 4.5 m*M* (90%); UTP, 0.7 m*M* (70%); UDP, 0.35 m*M* (70%); and ATP, 2 m*M*. The rest of the uracil nucleotides not present in the cytoplasm may be in the mitochondria. Total uracil (adenine) nucleotide levels were 120 (45) nmol/mg Chl (Dancer *et al.*, 1990b).

D. Intracellular Transport

1. The Plastid Compartment

In the light the P_i/trioseP translocator at the inner envelope membrane of the chloroplast plays the pivotal role in supplying the cytoplasm with energy for ATP regeneration and with carbon skeletons; the K_m values for P_i and trioseP are in the range of 0.1–0.4 m*M* (Heldt and Flügge, 1987; Flügge and Heldt, 1984). Chloroplasts possess a further translocator, which counterexchanges ATP and ADP and is obviously highly specific for cytoplasmic ATP; in the case of spinach leaf chloroplasts, however, its rate was one to two orders of magnitude lower than those of the P_i/trioseP translocator (Heldt and Flügge, 1987). This plastidial ATP/ADP translocator, the properties of which have not been fully elucidated, may possibly function in the energy import in the dark or during plastid development. In the case of pea plants, chloroplasts of young leaves were reported to have higher transport rates than those of mature leaves (Robinson and Wiskich, 1977). In general the rate of plastidial ATP/ADP counterexchange varies with different tissues and species; with mesophyll chloroplasts from the C_4 plant *Digitaria sanguinales*, a very active ATP/ADP translocator was described (Huber and Edwards, 1976).

In nonphotosynthetic plastids, the ATP/ADP translocator apparently plays an important role in the energy import from the cytoplasm. This was shown with chromoplasts from daffodil flowers (Liedvogel and Kleinig, 1980), and by immunochemical methods the presence of the translocator in the inner envelope membranes of amyloplasts from cultured sycamore cells was shown (Ngernprasirtsiri *et al.*, 1989). The importance of energy import was further stressed by showing that in these amyloplasts the enzyme sets for glycolysis, gluconeogenesis, and the oxidative pentoseP pathway were not complete (Frehner *et al.*, 1990). Thus import of ATP would be required, e.g., for ADPGlc formation and starch production. In the chloroplasts of the green-mutant sycamore cell line, the antibodies, prepared from the mitochondrial ADP/ATP translocator from *Neurospora crassa*, did not react with a putative ATP/ADP translocator (Ngernprasirtsiri *et al.*, 1989).

Pure ATP/ADP translocation would not function in the net exchange of Ade nucleotides between different compartments, which is, however, required during the growth phase of tissues. With pea chloroplasts a net export of ATP was reported, as the ATP/ADP translocator obviously also exchanges PP_i, although with low rates, and in suspended chloroplasts depletion of the stromal ATP pool by the PP_i of the medium was shown (Robinson and Wiskich, 1977). Furthermore, counterexchange of ATP with pyruvate was reported for pea and maize chloroplasts, suggesting an interaction of the ATP/ADP and the P_i/trioseP translocators (Woldegiorgis *et al.*, 1983; Heldt and Flügge, 1987). However, it remains to be elucidated whether a net transport of Ade nucleotides from the site of their production in the chloroplasts into the cytoplasm will occur *in vivo* by the action of the ATP/ADP translocator.

Transport of nucleotide precursors across the plastid envelope have not been studied thoroughly. In earlier work, uptake of nucleobases such as Ade and Cyt by isolated pea chloroplasts was shown to occur with high initial rates, especially for Ade (Barber and Thurman, 1978). The nucleobases accumulated in the sucrose-impermeable space of the plastids; also, evidence of active transport was reported, although the latter may be the consequence of salvage of the nucleobases by the enzymes within the stroma. Transformation of the bases into nucleotides would drastically decrease their passive permeation.

2. The Mitochondrial Compartment

The major gate for energy export from the mitochondria is the ADP/ATP translocator, which shows a strict counterexchange for ADP and ATP with no activities for AMP and nucleotides with nucleobases other than Ade. As shown with mammalian mitochondria, it is the most abundant single protein of the inner membrane, with two identical subunits, and operates in a gated pore mechanism. With energized mitochondria the translocation is asymmetric with export of ATP and import of ADP; thus energy generated by electron transport in mitochondria is used both for ATP production and for its transfer to the higher phosphorylation potential of the cytoplasm (Heldt and Flügge, 1987, and references therein). Plant mitochondria obviously have an ADP/ATP translocator with structural and functional properties similar to those described in animal cells (Heldt and Flügge, 1987, and references therein). A cDNA encoding the translocator from maize mitochondria was cloned recently (Baker and Leaver, 1985); inhibition of the counterexchange by low concentrations of anionic detergents was also shown (Konstantinov *et al.*, 1988). Transport of P_i via the P_i/OH^- translocator (identical to P_i/H^+ symport described in animal mitochondria) is also present in plant mitochondria and has similar properties (Heldt and Flügge, 1987, and references therein).

Little is known about the net transfer of nucleotides or precursors into mitochondria with the exception of NAD. With isolated mitochondria from potato tubers, slow passive efflux and influx of NAD was reported (Neuburger and Douce, 1983); furthermore, evidence of carrier-mediated and energy-dependent import was elaborated (Neuburger *et al.,* 1985). Hence, NAD is usually added to the incubation medium in studies with isolated mitochondria.

3. Endoplasmic Reticulum and Golgi Apparatus

Plant cells have large UDPsugar pools that are synthesized in the cytoplasm, and a large fraction of them is consumed within the lumen of the ER and the Golgi apparatus for polysaccharide and glycoprotein synthesis; thus transport of UDPsugars into the lumen space of the cellular endomembrane system is very important. With higher animal cells, some details have been elaborated and the existence of specific transporters for NDPsugars at the Golgi apparatus is established. Apparent K_m values for UDPGlc, UDPGal, and UDPGlcNAc of 37, 420, and 400 μM respectively, were determined (Fleischer, 1989). There is also evidence for hydrolysis of liberated UDP to UMP in the Golgi apparatus, with the effect that the inhibition of sugar transferases by liberated UDP would be abolished (Delmer and Stone, 1988). On the other hand, for the rough ER, transfer of sugar residues via transmembrane movement of lipid-linked intermediates is discussed (Hirschberg and Snider, 1987), as is ATP transport into the Golgi apparatus (Milla *et al.,* 1989).

Very little is known about similar transport systems in the plant cell (Delmer and Stone, 1988). With soybean seedlings, transport of UDP-GlcUA into the lumen of the Golgi apparatus and its subsequent decarboxylation to UDPXyl were recently suggested, as was synthesis of xyloglucan in this space (Hayashi *et al.,* 1988). Finally, in ureide-forming legume nodules, transport of ALN into the ER (Fig. 12) before deliverance of the ureides to the xylem is discussed (Schubert, 1986).

IV. Nucleotide Dynamics and Metabolic Regulation

The discussion of nucleotide pool changes within the cell can be divided into two main parts: (1) net pool changes as a consequence of new synthesis or degradation and consumption of nucleotides and (2) changes in the individual components without changes in the total nucleotide content.

Whereas the former is mainly a subject of growth, development, and senescence, the latter occur in the energy metabolism, i.e., the production of ATP from ADP/AMP or the formation of NAD(P)H from the oxidized compounds; a further example is the generation of NDPsugars from NTPs and the utilization of NDPsugars with the formation of NDPs. In the following sections, we discuss these aspects separately; however, one should keep in mind that often these two aspects overlap.

A. General Nucleotide Metabolism

The chemical composition of nucleotides indicates that nucleotide synthesis is dependent on the phosphate, carbon, and nitrogen source. Their availabilities should influence the rate of nucleotide synthesis and the size of the nucleotide pools. Concerning first the phosphate source, the main phosphorylated metabolic intermediates are the nucleotides, the sugar phosphates, and lipid precurors such as cholineP. In the consumption of these intermediates for the synthesis of polysaccharides, nucleic acids, and lipids in the respective anabolic pathways, only with the nucleic acids, and the phospholipids is the phosphate retained in the final product. Hence, the main phosphate flux through the nucleotide pools is toward the nucleic acids (about 45% of the total P_i with cultured plant cells) and the phospholipds (up to 30%) (Ashihara and Tokoro, 1985).

Inorganic phosphate is in general a limiting nutrient in plant growth; however, control of nucleotide synthesis by P_i availability has been studied mainly with suspension cultured plant cells (see also Section V,D). As the precursor for the ribose moiety, PRPP, carries three phosphate groups, regulation of its synthesis by P_i or its usage as a regulatory metabolite is conceivable. As mentioned above, PRPP-Sase from bacteria and animal cells is stimulated by millimolar concentrations of P_i, whereas crude preparations from plant origin revealed an inhibition by P_i (Ashihara, 1977b; Ukaji and Ashihara, 1987a). From a further crude extract of suspension cultured cells, a slight stimulation of CP-Sase, the first enzyme in pyrimidine nucleotide synthesis, by PRPP was reported (Kanamori *et al.*, 1980). In view of these poor results, further studies with purified enzymes should be performed. Changes in amounts of PRPP in pea cotyledons showed a steep maximum at a very early phase of germination (Ross and Murray, 1971) and may suggest a trigger function of this metabolite. Furthermore, P_i addition to P_i-starved cultured cells (Ukaji and Ashihara, 1987a) showed a strong increase of the PRPP pool and its turnover (see Section V,D).

Regulation of nucleotide levels by P_i was more thoroughly studied; the results from cultured cells, however, are presented in Section V. Changes of Ade nucleotides due to P_i deficiency were investigated in isolated soybean leaf cells, wheat seedlings, and wheat leaf fragments (Miginiac-Maslow and Hoarau, 1982; Miginiac-Maslow *et al.,* 1986). Cultivation in P_i-deprived media or incubation with nonmetabolizable sugars such as glucosamine or mannose, which leads to P_i deficiency, only slightly reduced the energy charge (EC), while the Ade nucleotide pools were strongly reduced. This decrease was suggested to be caused by reduced synthesis and enhanced degradation. Pulse feeding experiments with labeled adenine showed a strong increase of the labeled oxidized products (Xan, UA, ALN, and ALC) on P_i depletion; in the case of wheat, P_i deficiency caused a strong reduction of the rate of labeling of AMP and IMP, whereas those of Ado, Ino, and Hyp were less affected (Miginiac-Maslow *et al.,* 1986). Interestingly, AMP-Nase, the first enzyme in the degradation of Ade nucleotides (Figs. 4 and 5), was shown to be allosterically inhibited by P_i; with an enzyme from spinach leaves the K_m value of AMP increased in the presence of P_i (Yoshino and Murakami, 1980).

The interconnection of nucleotide synthesis with carbohydrate metabolism is evident in the biosynthetic route leading to the ribose moiety (Fig. 1), i.e., the pathway from Glc6P to ribose5P and PRPP (Becker *et al.,* 1979). Regulation of ribose5P production in the oxidative pentoseP pathway has been thoroughly studied in plants and some details on the coarse and fine control are known (Copeland and Turner, 1987; Turner and Turner, 1980). The two dehydrogenases transforming Glc6P into ribulose5P are mainly regulated by the NADP:NADPH ratio; both enzymes are inhibited by NADPH and inhibition of gluconate6P-DHase by Fru2,6P_2 is also reported. The chloroplast Glc6P-DHase activity is also inhibited by light-mediated reduction. Due to a low ratio of NADP:NADPH in the light, in chloroplasts the rate of the oxidative pentoseP pathway may be low in the light but increased in the dark due to an increase in the NADP:NADPH ratio (Copeland and Turner, 1987, and references therein). In human erythrocytes, it was further shown that ribose5P levels stimulate PRPP-Sase activity and hence nucleotide salvage synthesis (Yeh and Phang, 1988). The rate of synthesis of Ura nucleotides in the roots of squash plant seedlings strongly depended on the glucose content of the roots (Lovatt, 1985), which indicates a tuning of pyrimidine nucleotide synthesis, probably via PRPP formation, to the availability of the carbon source.

As activity of the pentoseP pathway and hence ribose5P production is controlled by the NADP:NADPH ratio, dehydrogenase reactions that consume NADPH and produce NADP would positively interfere in nucleotide synthesis. Transformation of Glu into Pro via pyrroline-5-carboxylate (P5C), with the help of two NADPH-specific dehydrogenases, P5C-Sase and P5C-Rase (Fig. 13), is considered in erythrocytes to be

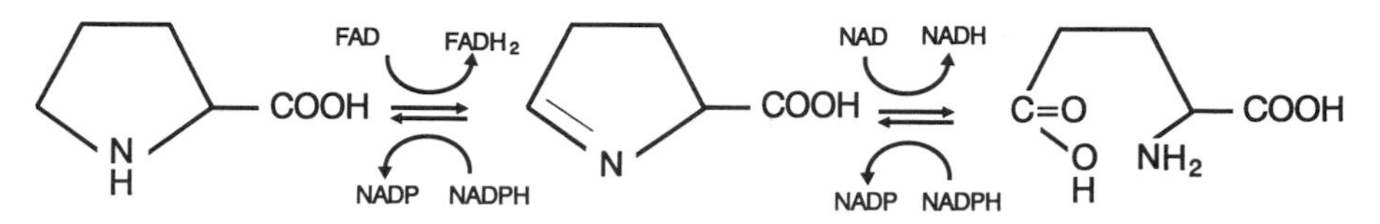

FIG. 13 Routes of transformation of proline into pyrroline-5-carboxylate (P5C) and glutamate and vice versa; this is a modified scheme derived from Fig. 5 of Kohl *et al.* (1988).

coupled to ribose5P production and it was shown that addition of P5C stimulates PRPP and purine nucleotide synthesis (Yeh *et al.*, 1984; Yeh and Phang, 1988). Similar suggestions were recently reported from studies on N_2-fixing root nodules of ureide-producing legumes; high activities of P5C-Rase were found in the cytoplasm of the nodules as well as evidence of P5C-Sase activity (Kohl *et al.*, 1988). The further findings of high activities of Pro-DHase (Fig. 13) in the bacteroid lead to the suggestion that Pro is generated in the cytoplasm and used in the bacteroid for fueling the energy-intensive N_2 fixation. The role of Pro production in the generation of NADP, in order to stimulate the pentoseP pathway and hence PRPP synthesis, has not been fully elucidated up to now, as a comparison between ureide- and amide-exporting nodules revealed similar high P5C-Rase activities, although only the ureide-forming type would require stimulated purine nucleotide synthesis (Kohl *et al.*, 1990). The only significant difference was the considerably higher activities of Pro-DHase in the bacteroids of the ureide-exporting species; but the importance of this findings is not clear.

Little is known about the regulatory interconnections between the assimilation and utilization of the nitrogen source and nucleotide metabolism. The first step in nitrate assimilation is its reduction in the cytoplasm followed by nitrite reduction in the chloroplast in the light. It is still not fully elucidated which metabolites mediate the adjustment of cytoplasmic nitrate reduction to the photosynthetic electron transport (Sanchez and Heldt, 1990), although from *in vitro* studies with nitrate reductase inhibition by ADP was suggested (Watt *et al.*, 1987) and stimulation of nitrate reduction in the dark on absorption of exogenously supplied ATP was reported from studies with leaf disks of different plants (Prakash *et al.*, 1986). Furthermore, control of nitrate reduction by the cytoplasmic level of NADH is unlikely, as the K_m value of NADH is very low (1.4 μM) and the malate/oxalacetate shuttle should be able to import enough redox equivalents from the mitochondria to the cytoplasm in the dark (Sanchez and Heldt, 1989; Krömer and Heldt, 1991a). Control of key enzymes of amino acid synthesis by nucleotide levels of the chloroplast such as Ade nucleotides or NADPH, reported recently (Schmidt *et al.*, 1990; Bryan,

1990), is connected to this subject, although this field cannot be treated comprehensively in this article.

B. Energy Metabolism and Photosynthesis

This is a very thoroughly studied field and it is not possible to discuss it critically in all details; there are, however, excellent recent reviews (Stitt *et al.*, 1987; Stitt, 1990; Walker and Sivak, 1986; Dennis, 1987; Champigny, 1977). This section will be confined to data concerned with nucleotide pools and pool changes connected to energy metabolism; we apologize, however, for not being able to consider all the relevant results in this field.

Green plant cells have two separate, electron transport-coupled, ATP-generating systems, the chloroplasts and mitochondria. Whereas in darkness and in nonphotosynthetic tissue only the mitochondrial system is operative, in the light (in source leaves) the light-dependent reactions of the chloroplasts provide the main amount of ATP and NADPH. Whether and how much mitochondria also contribute to ATP production in the light phase is a controversial subject (see below).

For the presentation and interpretation of pool changes due to dark–light transitions, several terms are used (Raymond *et al.*, 1987; Pradet and Raymond, 1983; Erecinska *et al.*, 1977; Heber *et al.*, 1987): (1) the ATP:ADP ratio; due to difficulties in the determination of AMP, especially when using enzymatic methods, the ATP:ADP ratio is often more reliable than the EC; (2) the Ade nucleotide-derived EC (ATP + $\frac{1}{2}$ADP):(ATP + ADP + AMP); (3) the ATP phosphorylation potential ATP/ADP × P_i; (4) the mass action ratio of AMP-Kase K = ATP × AMP/$(ADP)^2$; (5) pyridine nucleotides are usually presented as redox ratios such as, e.g., NADPH:NADP; and (6) the assimilatory force F_A = ATP × NADPH/ADP × P_i × NADP. The nucleotide levels determined by extraction, however, do not necessarily reflect the thermodynamically active *in situ* concentrations, as several enzymes bind nucleotides with sufficient activity and the concentrations of these enzymes may be on the same order as that of the nucleotides (Dry *et al.*, 1987; Usuda, 1988, 1990).

1. Pools in the Chloroplast

The Ade nucleotide content (AXP) in the chloroplasts of mesophyll cells from C_3 and C_4 plants ranges from 30 to 50 nmol/mg Chl (Usuda, 1988; Wirtz *et al.*, 1980; Gardeström and Wigge, 1988; Stitt *et al.*, 1982; Hampp *et al.*, 1982); in maize mesophyll cells a stroma concentration of about 1.4 m*M* for the AXP pool was estimated. In general the total AXP pool is the same in the light and the dark phases. The mass ratio for AMP-Kase was found in maize mesophyll chloroplasts to be 0.6 and invariable during the dark–light transition, which indicates full equilibration between the Ade nucleotides (Usuda, 1988). With spinach chloroplasts a total NADP(H) level of 14–24 nmol/mg Chl was reported (Takahama *et al.*, 1981), whereas with maize mesophyll chloroplasts a level of 6 nmol/mg Chl (0.5 m*M*) was determined (Rébeillé and Hatch, 1986). Using a nonaqueous method for chloroplast isolation, the total NADP(H) pool of maize mesophyll chloroplasts was determined to 7.5 nmol/mg Chl (0.3 m*M*) in the dark and 12 nmol/mg Chl (0.5 m*M*) in the light (Usuda, 1988). The increase in the light phase is obviously caused by the light-dependent NAD-kinase action. NAD(H) levels in the chloroplasts are low (Table X).

Transition from dark to light usually causes an increase in the ATP and NADPH pools and a decrease in the ADP, AMP, and NADP pools. The following *in situ* changes were reported in the literature: (1) The ATP:ADP ratio in the chloroplasts of several C_3 plants was between 0.4 and 0.1 (dark) and 1.6 and 3.1 (light), with maize mesophyll chloroplasts about 1 (dark) and 4 (light). (2) EC values of chloroplasts of C_3 plants were between 0.55 and 0.61 (dark) and 0.71 and 0.81 (light) (Dietz and Heber, 1984, 1986; Kobahashi *et al.*, 1979; Takahama *et al.*, 1981; Goller *et al.*, 1982; Hampp *et al.*, 1982; Stitt *et al.*, 1982), with maize mesophyll chloroplasts about 0.6 (dark) and 0.74–0.81 (light) (Usuda, 1988). (3) NADPH:NADP ratios in chloroplasts of several C_3 plants were 0.3–0.5 (dark) and 0.3–1.1 (light) (Heber and Santarius, 1965), recent values from spinach leaves were 1.1–1.8 in the light (Dietz and Heber, 1984), with maize mesophyll chloroplasts between 0.1 and 0.18 (dark) and 0.23 and 0.48 (light) (Usuda, 1988). (4) DHAP:PGA ratios for chloroplasts of C_3 plants were very low in the dark and between 0.05 and 0.25 in the light (Dietz and Heber, 1984, 1986; Gerhardt *et al.*, 1987), with maize mesophyll chloroplasts near zero in the dark and 0.4–1.8 in the light (Usuda, 1988). (5) The assimilatory force (F_A) of maize mesophyll chloroplasts was 8 M^{-1} (dark) and between 40 and 170 M^{-1} (light); it increased with light intensity. With isolated chloroplasts F_A values of 30,000–80,000 M^{-1} were reported; this is a maximum force when chloroplasts were not used for work (Usuda, 1988; Heber *et al.*, 1987). The F_A of chloroplasts in spinach leaves was between 60 and 150 M^{-1} in the light and decreased with light intensity (Dietz and Heber, 1986). (6) P_i concentrations in spinach chloroplasts were about 15 m*M* (dark) and 5 m*M* (light) (Lilley *et al.*, 1977), and in maize mesophyll chloroplasts 20 m*M* (dark) and 10–17 m*M* (light) (Usuda, 1988). (7) ADPGlc levels in pea leaves were not detectable in the dark and 4.2 nmol/mg Chl in the light (Smith *et al.*, 1990).

2. Pools in the Cytoplasm

The total AXP pool in the cytoplasm has been found in several cases to be near that of the chloroplasts; from wheat protoplasts 47 nmol/mg Chl was reported, which remained constant during the light–dark transition (Stitt *et al.*, 1982). A total cytoplasmic NADP(H) pool of 8.4 nmol/mg Chl was determined in oat mesophyll protoplasts, which also remained constant during the dark–light transition (Hampp *et al.*, 1984). The UDPGlc pool was found to range from 20 to 50 nmol/mg Chl and increases or decreases were observed from dark to light transition (Gerhardt *et al.*, 1987; Stitt *et al.*, 1985; Weiner *et al.*, 1987). The total NAD(H) levels were in the range of those of mitochondria (Table X). To calculate cytoplasmic concentrations, in the case of spinach and wheat leaves a 20-μl cytoplasmic volume was assumed per milligram chlorophyll (Weiner *et al.*, 1987).

Results on changes in the cytoplasmic energy status due to photosynthesis were controversial (Stitt *et al.*, 1987); they obviously depend on the applied CO_2 concentration, as high CO_2 levels suppress photorespiration (see below). The following changes were reported: (1) The ATP:ADP ratio in the cytoplasm of oat leaf protoplast was 9.1 (dark) and 11 (light) (Goller *et al.*, 1982), with wheat protoplasts (high CO_2) 9.2 (dark) and 6.4 (light) (Stitt *et al.*, 1982), in oat protoplasts 3.0 in the dark and in the light 2.8 (high CO_2) and 7.5 (low CO_2) (Gardeström, 1987). (2) The NADH:NAD ratio was 0.0012 (light) with spinach leaves, with levels of NAD of 0.5 m*M* and NADH 0.6 μM (Krömer and Heldt, 1991a); the cytoplasmic NAD(H) system obviously operates at very low NADH:NAD ratios, which are much lower than the NADPH:NADP ratio in the stroma and also lower than the NADH:NAD ratio in the mitochondrial matrix (Sanchez and Heldt, 1990). (3) The NADPH:NADP ratio was 3.6 (dark) and 6.1 (light) with oat mesophyll protoplasts (Hampp *et al.*, 1984). (4) P_i levels in the cytoplasm assumed to be in the range of 10–25 m*M* were criticized recently as being too high (Bligny *et al.*, 1990). In general P_i levels were assumed to decrease in the light by 20–50% (Stitt *et al.*, 1985) and PP_i levels of 0.2–0.3 m*M* did not change during the light–dark transition (Weiner *et al.*, 1987). (5) Sugar phosphate levels with spinach leaves were for Fru1,6P_2 0.02 m*M* (dark) and 0.2 m*M* (light), for Fru6P 0.9 m*M* (dark) and 3.0 m*M* (light), and for Glc1P 0.6 m*M* (dark) and 1.2 m*M* (light); DHAP:PGA ratios were near zero (dark) and 0.13 (light) (Weiner *et al.*, 1987; Gerhardt *et al.*, 1987).

3. Mitochondrial Contribution in the Light and Photorespiration

The contribution of mitochondria to cellular ATP generation during the light phase is still not fully elucidated (Stitt *et al.*, 1987; Krömer and Heldt, 1991b; Kelly, 1983; Singh and Naik, 1984; Krömer *et al.*, 1988; Gans and

Rébeillé, 1988; Avelange *et al.*, 1991; Hampp, 1985; Goller *et al.*, 1982). From earlier work an inhibition of oxidative phosphorylation in the light was suggested (Santarius and Heber, 1965; Heber *et al.*, 1982; Sawhney *et al.*, 1978; Graham, 1980) to be caused by a high ATP:ADP ratio in the cytoplasm, which in turn is generated by efficient cytoplasmic ATP production via the trioseP/PGA shuttle (Fig. 14) (Flügge and Heldt, 1984). This hypothesis suggests that the ATP:ADP ratio in the cytoplasm is higher in the light than in the dark. With isolated mitochondria, it was shown that very high extramitochondrial ATP:ADP ratios (>20) were required to restrict respiration (Dry and Wiskich, 1982). As shown above, ATP:ADP ratios in the cytoplasm, determined at low CO_2 concentrations, i.e. under natural atmospheric conditions, were only slightly higher in the light than in the dark (Gardeström, 1987).

In leaf mitochondria a major part of the matrix protein consists of two enzymes that convert Gly, a product of photorespiration, to Ser (Fig. 14); NADH produced during Gly decarboxylation is a potential source for ATP production in the light, although a stoichiometric amount of NADH is required for hydroxypyruvate reduction in the peroxisome (Dry *et al.*, 1987; Douce and Neuburger, 1989; Gardeström and Wigge, 1988; Oliver *et al.*, 1990). NADH is, however, also provided through trioseP oxidation in the cytoplasm, which generates one NADH per ATP formed (Stitt *et al.*, 1987). Different experimental strategies were applied to probe the contribution of the mitochondrial energy metabolism in the light. Aminoacetonitrile, an inhibitor of the photorespiratory conversion of Gly to Ser, had

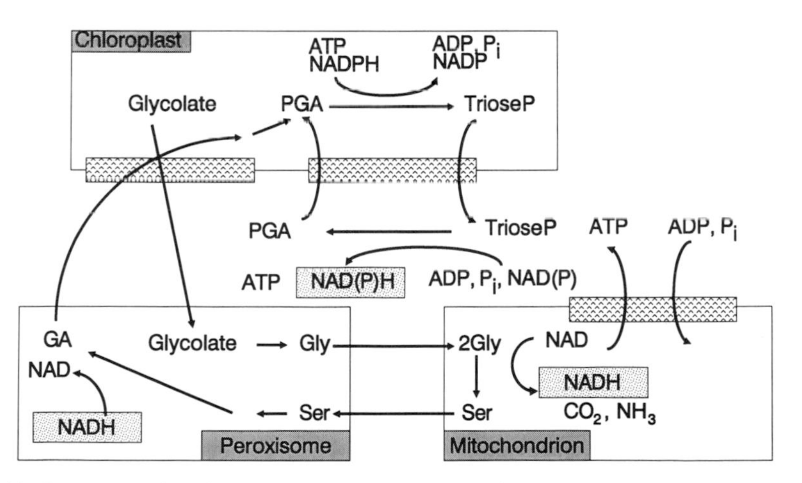

FIG. 14 Interconnections in the energy metabolism of chloroplast, mitochondrion, and peroxisome. The scheme was designed according to models published by Heldt and Flügge (1987) and by Stitt *et al.* (1987).

no effect on the ATP:ADP ratio at saturating CO_2 concentration; under photorespiratory conditions (limiting CO_2), however, the ATP:ADP ratio in the mitochondria and the cytoplasm decreased and that of the chloroplasts increased in the presence of the drug. Suppression of photorespiration by high CO_2 or glycolaldehyde also decreased the cytosolic ATP:ADP ratio, whereas the ratio in the chloroplast remained high. These results were interpreted to indicate that oxidative phosphorylation provides ATP for the cytoplasm due to photorespiratory Gly oxidation (Gardeström and Wigge, 1988).

Low concentrations of the antibiotic oligomycin were shown to inhibit selectively oxidative phosphorylation in mitochondria of barley leaf protoplasts and barley leaves, but had no direct significant effect on chloroplast photophosphorylation as shown with lysed protoplasts. Oligomycin administered to leaves in the light, however, reduced photosynthesis by 60%, decreased the ATP:ADP ratio in the cytoplasm more strongly than that in the stroma, and was accompanied by a decrease in sucrose synthesis (Krömer *et al.,* 1988; Krömer and Heldt, 1991b). These findings are consistent with the hypothesis that mitochondria contribute to the ATP production in the cytoplasm and play a role in the oxidation of surplus redox equivalents formed by the photosynthetic electron transport and delivered to the cytoplasm (Fig. 14). Recent investigations of isolated mitochondria from pea leaves suggested a route for the oxidation of excess cytoplasmic redox equivalents; the high concentration of NADPH in the cytoplasm relative to NADH makes the NADPH-DHase located at the outside of the inner mitochondrial membrane (Palmer and Moller, 1982) a possible candidate for the import of electrons into mitochondria; NADPH could be delivered from the chloroplasts to the cytoplasm via the trioseP/PGA shuttle, coupled in the cytoplasm to the action of the nonphosphorylating, NADP-dependent GAP-DHase (Krömer and Heldt, 1991a).

The relative contribution of mitochondrial oxidative phosphorylation in the light, however, may vary and its quantity is not exactly known. Furthermore, an excess of cytoplasmic redox equivalents may also be consumed by nitrate reduction or by the alternative cyanide-insensitive respiratory pathway (Singh and Naik, 1984). There are further recent results from light-dependent pH changes in leaves of C_3 plants (Yin *et al.,* 1990) and the determination of O_2 and CO_2 gas exchange (Avelange *et al.,* 1991; Avelange and Rébeillé, 1991), which stress the trioseP/PGA shuttle as the main supply of energy for the cytosol in the light.

Total NAD(H) levels of 5.4 and 4.2 nmol/mg Chl were reported from mitochondria of oat mesophyll protoplasts; the total level did not change during the dark–light transition, nor did the total NADP(H) level of about 0.8 nmol/mg Chl (Hampp *et al.,* 1984, 1985). With isolated mitochondria from pea leaves, the following levels of NAD and NADH (NADH:NAD ratio) were reported: state 3 (presence of ADP), 4.7 and 0.3 nmol/mg protein (0.065); state 4 (controlled state), 3.8 and 1.2 nmol/mg protein (0.33) (Krömer and Heldt, 1991a). From these data (state 4) and an NADH:NAD ratio in the cytoplasm smaller than 0.003 (see above), the redox gradient from the mitochondrial matrix to the cytoplasm is at least 100. Thus in both animal tissue and plant cells the NAD(H) pool in the matrix is more reduced than in the cytoplasm; this must be considered in discussing the redox transfer between the cytoplasm and the matrix (Krömer and Heldt, 1991a; Rustin *et al.*, 1987).

Pool changes in mitochondria during the dark–light transition are as follow: (1) The ATP:ADP ratio in oat leaf protoplasts was 1.3 (dark) and 0.7 (light) (Goller *et al.*, 1982), in wheat leaf protoplasts (high CO_2) 0.6 (dark) and 2.6 (light) (Stitt *et al.*, 1982), in barley leaf protoplasts 0.8 in the dark and in the light 1.5 (high CO_2) or 4.2 (low CO_2) (Gardeström, 1987). (2) The NADH:NAD ratio in oat leaf protoplasts was 0.35 (dark) and 0.54 (light) (Hampp *et al.*, 1984), 0.14 (dark) and 0.34 (light) (Hampp *et al.*, 1985). (3) The NADPH:NADP ratio in oat leaf protoplasts was 0.75 (dark) and 1.0 (light) (Hampp *et al.*, 1985).

4. Total Cellular Pools

With several investigations it is not possible to produce data for separate cellular compartments, because it would be too time consuming; hence often total metabolite pools are determined, which are of course more difficult to interpret. Therefore available data related to the total cellular pools are also listed. (1) The ATP:ADP ratio in wheat mesophyll protoplasts was 1.1 (dark) and 2.1 (light) (Stitt *et al.*, 1983a), in oat protoplasts 1.3 (dark) and 2.4 (light) (Goller *et al.*, 1982), in barley protoplasts about 1.0 in the dark and in the light 1.4 (high CO_2) and 4.0 (low CO_2) (Gardeström, 1987), in excised maize leaves 1.8 (dark) and 7.9 (light) (Roeske and Chollet, 1989); ATP levels from *Asparagus* mesophyll cells, suspended in liquid medium, showed a 30% reduction from dark to light that was interpreted to indicate photosynthetic consumption depending on CO_2 availability (Bown and Nicholls, 1985; Ciccarelli and Bown, 1988). (2) Total pyridine nucleotide pools, NAD(H) plus NADP(H), in general did not change significantly during the dark–light transition; the NADP(H) levels, however, increased in most cases, while the NAD(H) levels decreased due to light stimulation of NAD–Kase. The following data (values for the light phase in parentheses) were reported: bean leaves, NAD(H) 39 (33) and NADP(H) 16 (25) nmol/mg Chl (Orgen and Krogmann, 1965); mung bean leaves, NAD(H) 50 (37) and NADP(H) 14 (26) nmol/g FW (Graham and Cooper, 1967), pea seedlings, NAD(H) 438 (329) and NADP(H) 151 (264) nmol/g DW (Muto *et al.*, 1981); oat leaf protoplasts, NAD(H) 16.4

and NADP(H) 12.7 nmol/mg Chl, with no significant differences between the dark and light phases (Hampp *et al.,* 1985). (3) Redox ratios: the data from the literature revealed no consistent trend for the total NADPH:NADP ratios, whereas the total NADH:NAD ratios in most cases slightly increased from dark to light. (4) The NAD:NADP ratio is an interesting value, because with the acidic extraction procedure only the oxidized pyridine nucleotides can be determined. With the exception of data from protoplasts (Hampp *et al.*, 1984, 1985) values obtained from leaf tissues decreased markedly from dark (3.0–5.1) to light (1.3–1.8) (Muto *et al.*, 1981; Orgen and Krogmann, 1965; Graham and Cooper, 1967).

C. Respiration and Nonphotosynthetic Tissue

The properties of plant mitochondria and their contribution to energy metabolism have been thoroughly studied (Douce *et al.*, 1987; Douce and Neuburger, 1989; Dry *et al.*, 1987). Control of mitochondrial activity is in general exerted both through the supply of respiratory substrates, such as sucrose and O_2, and the rate of energy consumption by synthetic events. This control is mediated by the Ade nucleotide pools; when respiratory substrates are readily available, the rate of respiration matches the cellular rate of ATP consumption. The latter is sensed by the supply of ADP and with isolated mitochondria it was shown that it is more the absolute concentration of ADP than the ATP:ADP ratio that determines the rate of respiration. Escape of strict control by the Ade nucleotides is manifested by the cyanide-insensitive respiration; this contribution, which varies from tissue to tissue, is controlled mainly by the supply of substrates (Dry *et al.*, 1987, and references therein). From studies with isolated mitochondria from pea leaves, the maximum capacity of mitochondrial ATP synthesis was estimated as about 25% of the rate of noncyclic photophosphorylation at the maximum rate of photosynthesis. Furthermore, respiration measurements with the leaves showed an average rate of only 35% of the maximum rate [67 μmol O/(mg Chl · hr)], indicating a high reserve available for metabolic emergencies (Krömer and Heldt, 1991a).

From nonphotosynthetic tissues fewer data on nucleotide pools are available and the interaction of the different cellular compartments in energy metabolism and in transformations of the carbon skeletons has not been thoroughly elucidated. In these cells the mitochondria are the most important cellular compartments, with their ADP/ATP translocators and malate/oxalacetate shuttles for export of energy and redox equivalents. Whereas mitochondria have been more thoroughly investigated, the energy metabolism in plastids from nongreen cells is little elaborated. The enzyme sets for carbon catabolism and anabolism are not fully known and whether amyloplasts import trioseP or hexoseP for starch production is not fully elucidated. In Section III,D,1 a putative plastidial ATP/ADP translocator is discussed that may serve for energy import. NAD concentrations in the matrix of potato tuber mitochondria were reported to vary strongly and values of 0.25–4 mM were estimated (Tobin *et al.*, 1980).

In general the Ade nucleotide-derived EC is stabilized at higher values in nongreen tissues than in leaves (Raymond *et al.*, 1987); this is caused by the EC values in the chloroplasts, which are lower than in the cytosol (see above). Several investigations on young roots of maize reported ATP:ADP ratios in the range of 3.3–5.3 and EC values of 0.74–0.91 (Drew *et al.*, 1985; Gronewald and Hanson, 1982), whereas *in vivo* ^{31}P NMR studies on excised maize roots revealed ATP:ADP ratios higher than 25 (Roberts *et al.*, 1985); in the extracts, however, the values were between 1.2 and 3.8; these discrepancies may reflect ATP hydrolysis during extraction or NMR-invisible nucleotide pools. A recent *in vivo* NMR study on excised mung bean roots showed an ATP:ADP ratio of 3.8 and a phosphorylation potential of 960 M^{-1} (Yazaki *et al.*, 1988).

D. Control of Anabolic Precursor Synthesis and Utilization

In the cytoplasm there are at least three partially autonomous energy systems (Dancer *et al.*, 1990a): the Ade nucleotides, the Ura nucleotides, and the PP$_i$ system; the latter seems to be a unique, although not fully elucidated, energy donor system in plants (Black *et al.*, 1987; Weiner *et al.*, 1987; Dancer and ap Rees, 1989). Whereas in the cytoplasm activation of precursors for sucrose, glycoprotein, and cell wall synthesis is separated from the general phosphoryl donor system (ATP) by using mainly Ura nucleotides and minor quantities of Gua nucleotides, in the plastids the Ade nucleotides are used both for precursor activation in starch synthesis and general energy donation.

The key regulatory enzyme in starch synthesis, Glc1P-ATase, is activated by PGA and inhibited by P$_i$; thus the PGA:P$_i$ ratio in the plastids is an important regulatory value (Preiss, 1982, 1988). PP$_i$ obtained from ADPGlc synthesis is hydrolized to P$_i$, as plastids contain high activities of alkaline PPase, and PP$_i$ concentration in plastids is very low (Gross and ap Rees, 1986; Weiner *et al.*, 1987). Starch synthesis in chloroplasts is stimulated by the reduced supply of P$_i$, leading to an accumulation of PGA and an increase in the PGA:P$_i$ ratio. This activates Glc1P-ATase and hence starch synthesis, which in turn enhances P$_i$ recycling in the plastid. With a mutant defective in the starch branching enzyme, about three- to fivefold higher levels of ADPGlc were observed than with the wild type (Edwards *et al.*, 1988; Smith *et al.*, 1990). A defect in the branching enzyme obviously reduces the rate of ADPGlc consumption and leads to a depletion of the Ade nucleotide pool; reduced ATP levels will in turn decrease the rate of PGA reduction and hence photosynthesis. The effective regulation of starch synthesis by the PGA:Pi ratio obviously depends on the presence of adequate levels of the subsequent enzymes (Smith *et al.*, 1990).

Whereas starch synthesis is regulated by the ADPGlc-forming enzyme, sucrose synthesis is not controlled by the UDPGlc-forming enzyme, but at steps before and after this position in the pathway, i.e., at Fru1,6P_2-Pase and sucroseP-Sase (Neuhaus *et al.*, 1990a). Furthermore, the cytoplasm contains a considerable PP_i pool (0.2–0.3 mM) and little or no alkaline PPase (Weiner *et al.*, 1987), hence Glc1P-UTase catalyzes a near-equilibrium reaction. Precursor synthesis reactions for starch production (ADPGlc in plastids) and for sucrose production or mobilization (UDPGlc in the cytoplasm) are separated for these reasons. The cytoplasmic PP_i participates in two further near-equilibrium reactions: phosphorylation of Fru6P (Fru6P-PTase) and the tonoplast PP_i-dependent proton pump (Weiner *et al.*, 1987). The PP_i pool in the cytoplasm is obviously not in equilibrium with the adenine and uracil nucleotide pools; the energy status of PP_i can vary independently (Dancer *et al.*, 1990a) and the PP_i pool is remarkedly constant (Neuhaus *et al.*, 1990a). PP_i has a compensatory role with respect to ATP; Fru6P-PTase saves ATP by consumption of PP_i. During P_i limitation, maintenance of high PP_i levels allows the Fru6P-PTase to replace the ATP-consuming Fru6P-Kase. In a P_i-depleted black mustard suspension culture (Duff *et al.*, 1989) a selective increase in Fru6P-PTase and a decrease in Fru6P-Kase were reported. In leaf disks, fluoride ions were found to interfere with the PP_i turnover, as a four- to fivefold increase in the PP_i level was observed concomitant with a marked decrease in the UDPGlc pool and the photosynthetic sucrose production (Quick *et al.*, 1989); inhibition of the PP_i-dependent H^+ pump of the tonoplast by $F^-$was suggested.

In the leaves of spinach and barley, maximum rates of sucrose production were estimated as 10–12 μmol/(mg Chl × hr) (Quick *et al.*, 1989; Stitt *et al.*, 1988); this also requires high rates for recycling of UDP to UTP (Fig. 8). In extracts of these leaves, the activity of NDP kinase was found to be 2.5- to 6-fold higher than that required for maximum sucrose synthesis (Dancer *et al.*, 1990b). The K_m of UDPGlc for sucroseP-Sase is rather high (3–8 mM) (Stitt *et al.*, 1987); hence the rate of this enzyme may be limited by the UDPGlc level. On the other hand, the size of the total Ura nucleotide pool and also the PP_i concentration may influence the UDPGlc levels (Fig. 8) and thus the rate of sucrose synthesis (Dancer *et al.*, 1990a; Quick *et al.*, 1989). Removal of PP_i from sucrose synthesis may occur through Fru6P-PTase action; in spinach leaves its activity was found to be twice that of sucrose synthesis (Neuhaus *et al.*, 1990a). However, activity of this enzyme depends on the cytoplasmic P_i level, as P_i is a product in this near-equilibrium reaction. For full activity of Fru6P-PTase the P_i level should not exceed 4 m*M* (Neuhaus *et al.*, 1990a). Recent NMR studies in spinach leaves reported cytoplasmic P_i levels in the range of 0.6–1.2 m*M*, which are below this limit (Bligny *et al.*, 1990).

Control of sucrose synthesis is further exerted by Fru2,6P_2, a metabolite that is restricted to the cytoplasm (Stitt, 1990; Stitt *et al.*, 1983b). The key regulatory enzyme, cytoplasmic Fru1,6$_2$-Pase, which has a low K_m value for its substrate (1.5 μM for the enzyme from spinach leaves), is activated by a threshold level of trioseP and inhibited by Fru2,6P_2 (K_i 0.07 μM, spinach leaves) (Ladror *et al.*, 1990). In the presence of Fru2,6P_2 higher substrate concentrations are required for full activity. The enzyme is also inhibited by UDP (Foyer *et al.*, 1982) and AMP is a very potent inhibitor (K_i 120 μM, spinach leaves) (Ladror *et al.*, 1990): it enhances the sensitivity for Fru2,6P_2, thus increasing the level of substrate required for full activity (Stitt *et al.*, 1987; Foyer *et al.*, 1982). The susceptibility of this enzyme toward Fru2,6P_2 and AMP is higher at lower temperatures (Stitt and Grosse, 1988). The plastidial Fru1,6P_2-Pase, however, which regulates photosynthetic starch synthesis, is not controlled by Fru2,6P_2 and AMP, but is light activated (Scheibe, 1987). Furthermore, sucroseP-Sase is inhibited by UDP (Stitt *et al.*, 1987). Hence a high energy status for both the Ade and Ura nucleotides in the cytoplasm is required for active sucrose synthesis. Coarse control of sucroseP-Sase through light is performed by phosphorylation/dephosphorylation; light activation occurs by dephosphorylation, which may in turn be controlled by the P_i level (Huber and Huber, 1990).

The second enzyme that is controlled by Fru2,6P_2 is the PP_i-dependent Fru6P-PTase; the role of Fru2,6P_2 activation of this enzyme is unclear (Stitt, 1990). The reaction it catalyzes, equilibration between Fru6P, Fru1,6P_2, PP_i, and P_i, is fully reversible (Weiner *et al.*, 1987), hence it may function in glycolysis, gluconeogenesis, the regulation of the PP_i concentration, or in the sucrose synthase cycle for the utilization of sucrose in sink tissue (Black *et al.*, 1987). Its reversible reaction may reflect the adaptive role of plant metabolism in the development of response to varying external conditions (Stitt, 1990).

NDPsugar utilization for cell wall synthesis in growing and differentiating cells and its regulation is a very important topic (Northcote, 1982; Delmer and Stone, 1988). There is evidence that the main control is exerted by regulation at the level of the polysaccharide synthases. Transition from primary to secondary wall synthesis, studied with growing sycamore trees, showed large changes in the amount of polysaccharide synthases, whereas the level of enzymes involved in the interconversion of nucleotide sugars did not change strongly. With growing callus and hypocotyl tissue of beans, induction of differentiation was also accompanied by increases in the synthase levels, which were shown to arise from *de novo* synthesis. In addition, however, to changes in the enzyme levels, alterations in the nucleotide sugar pools and direct effects of nucleotide intermediates, e.g., inhibition of xylan synthase by UMP and UDP, may also contribute (Northcote, 1982; Delmer and Stone, 1988, and references therein).

V. Nucleotide Dynamics in Suspension Cultured Cells

Plant cells can be adapted to grow *in vitro* in suspension culture as single cells or cell aggregates. Cultured cells constitute a unique physiological state; the cells, selected for high growth rates, are in a more or less dedifferentiated state and offer the advantage of dealing with a large cell population with rather uniform properties, although extrapolation to *in vivo* growing plant tissues should be made cautiously. Cultured cells are grown mainly heterotrophically, although cell lines with different degrees of photosynthetic capacity have been established. Suspension cultured cells have in general large vacuoles and mitochondria with presumably the same properties as differentiated nongreen plant cells. Little is known about the physiology of the plastid equivalent of those cells, although it should be comparable to plastids of normal nongreen plant cells. A recent investigation of purified amyloplasts from suspension cultured sycamore cells gave evidence of incomplete glycolytic and gluconeogenic pathways (Frehner *et al.,* 1990), thus revealing dependency on the cytoplasmic and mitochondrial energy metabolism.

A. The Growth Cycle and Its Phases

Suspension cultured cells can be grown in a discontinuous batch or in a continuous chemostat culture. Both techniques have their unique advantages; discontinuous batch culture, however, has been more frequently studied. In a batch culture, inoculation with starved cells leads to a certain degree of synchrony in the uptake and utilization of nutrients and thus in growth physiology (King and Street, 1977; De Gunst *et al.,* 1990; Wylegalla *et al.,* 1985). In most cases four different phases can be discriminated that show unique features and are more or less separated in the time scale of the growth cycle, which usually has a length of 7 to 14 days; the phases are as follow: lag phase, cell division phase, cell elongation phase, and starvation. The length of the lag phase usually depends on the duration of the preceding starvation phase; however, extended starvation is not always possible due to cell deterioration or death. An extended lag phase was described, e.g., with cultured sycamore (Brown and Short, 1969) and *Datura* cells (Meyer and Wagner, 1985a).

With cultured *Catharanthus* cells (Kanamori *et al.*, 1979), determination of fresh weight per cell showed a decrease that started at about 24 hr after inoculation; this suggests that onset of cell division was accompanied by a decrease of the vacuolar space. Fresh weight per cell increased again at day 4 after inoculation, indicating the onset of cell elongation exerted by an enlargement of the vacuole. Furthermore, cultured *Datura* cells indicated that transition from cell division to cell elongation may be characterized by the halt of net increases of RNA and protein (Fig. 15), suggesting that cell elongation occurs without a significant net increase of the cytoplasmic volume (Wylegalla *et al.*, 1985). Usually the border between cell division and cell elongation is assigned to about the middle of the total fresh weight increase of the growth phase. The sharpness of this border, however, and its location may vary with the different cultured species and with the conditions of growth, suggesting that the capacity of cells for elongation may vary.

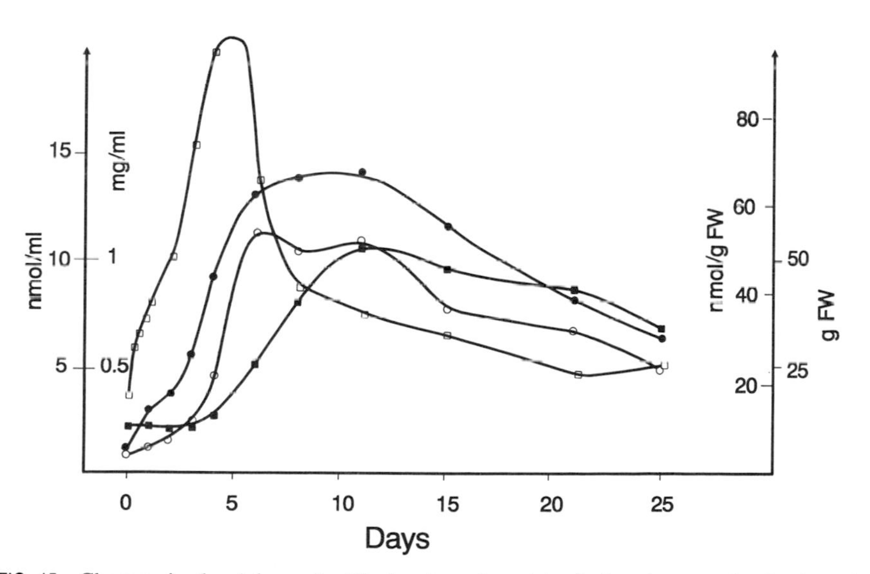

FIG. 15 Changes in the Ade nucleotide levels and protein during the growth of cultured *Datura* cells (Meyer and Wagner, 1985a; Wylegalla *et al.*, 1985). (■) Growth cycle in grams fresh weight (FW) per 100-ml culture; (○), amount of protein in milligrams per milliliter culture; (●), total Ade nucleotides (AXP) in nanomoles per milliliter; (□), AXP levels in nanomoles per gram FW.

B. Lag Phase Induction

1. Nucleotide Pools

Starved cells have low levels of PRPP and nucleotides; inoculation into fresh medium results in a tremendous increase in these pools. There are many reports (Kanabus *et al.*, 1986; Shimazaki *et al.*, 1982; Shimizu *et al.*, 1977; Okamura *et al.*, 1975; de Klerk-Kiebert and van der Plas, 1984; Brown and Short, 1969; Miginiac-Maslow *et al.*, 1981; Meyer and Wagner, 1985a,c) that show that the onset of increase in pool size occurs very early in the lag phase, preceding cell division. With cultured *Catharanthus* cells the level of PRPP was 0.4 nmol/g FW at the starvation phase and increased to about 2.2 nmol/g FW at day 3 of the growth cycle (Hirose and Ashihara, 1983c). PRRP availability, measured as utilization of exogenous Ade for nucleotide and nucleic acid synthesis, increased from 33 (starvation phase) to 125 nmol/(g FW · hr) 24 hr after inoculation, indicating a high turnover of the PRPP pool. Also, the individual nucleotide pools increase very early, as shown in Fig. 15 for the AXP pool of cultured *Datura* cells (Meyer and Wagner, 1985a). The high rate of nucleotide synthesis beginning in the lag phase correlates with a high respiratory capacity starting in the lag phase, as shown with cultured tobacco cells (Miginiac-Maslow *et al.*, 1981; Kanamori *et al.*, 1979).

2. Activities of *de Novo* and Salvage Enzymes

Enzyme activity measurements showed a very early induction of activities involved in the synthesis of precursors such as ribose5P and PRPP; enzymes of the oxidative pentoseP pathway and of PRPP-Sase had their activity maximum at day 2 of the growth cycle of cultured *Catharanthus* cells (Kanamori *et al.*, 1979). The enzyme activities for both salvage and *de novo* nucleotide synthesis and the capacity to incorporate exogenous nucleosides and nucleobases into cellular nucleotide pools and RNA had early maxima in the lag phase and/or beginning of the cell division phase (Kanamori-Fukuda and Ashihara, 1981; Hirose and Ashihara, 1984; Shimazaki and Ashihara, 1982). With cultured *Catharanthus* cells, the activities of enzymes involved in pyrimidine nucleotide salvage and *de novo* synthesis decreased in the following order: OA-RTase/OMP-DCase > Urd-Kase > Ura-RTase >> Urd-Pyase; the latter activity was rather low. This range correlated with the incorporation efficiency of exogenous precursors (OA > Urd > Ura), although labeling efficiency of RNA was highest with Urd. It is difficult to assess whether salvage or *de novo* synthesis of nucleotides contributes more to the enlargement of the pools in this early phase.

Activity measurements of purine salvage enzymes showed rather high levels at the beginning of the growth cycle, although about 24 hr after inoculation an activity maximum was observed (Shimazaki *et al.*, 1982; Hirose and Ashihara, 1984), which may, however, reflect a general increase in enzyme synthesis due to ongoing protein synthesis. Its position in the time scale and probably also the starting levels of enzyme activity may depend on the duration of the preceding starvation phase. A more thorough study with *Catharanthus* cells revealed the following enzyme activities decreasing in the order: NDP-Kase >> NMP-Kase >> Ade-RTase = Gua-RTase = Ado-Kase >> Hyp-RTase >> nucleoside-PTase; the latter two enzymes had very low activities (Hirose and Ashihara, 1984). Apparently salvage of Ura nucleotides is most efficient with Urd by Urd-Kase, whereas salvage for Ade nucleotides is equally efficient with both Ade and Ado. This is supported by studies of cultured carrot cells, which also showed similar activities of Ade-RTase and Ado-Kase, which did not change significantly during the growth cycle, whereas enzyme levels of *de novo* synthesis such as PRPP-ATase and GAR-Sase were lower and had their maximum activity at the onset of the cell division phase (Ashihara and Nygaard, 1989). At later stages of the growth cycle purine salvage is obviously more important than *de novo* synthesis.

3. Role of Cytoplasmic Nucleotide Concentrations

Nucleotide levels and enzyme activities are usually related to the fresh weight; this term, however, will respond both to changes in net synthesis (or degradation) of nucleotides and to increases in the fresh weight during growth. To indicate the duration of net nucleotide synthesis in a cell population one must relate it to the volume (ml), which showed in the case of cultured *Datura* (Meyer and Wagner, 1985a; Wylegalla *et al.*, 1985) (Fig. 15), tobacco (Meyer and Wagner, 1985c), and carrot cells (Okamura *et al.*, 1975) (Fig. 16) that total nucleotide synthesis actually stops at the end of the cell division phase, concomitantly with the end of net RNA and protein synthesis (Fig. 15). In the cell elongation phase, there is obviously no significant net increase in the nucleotide content as the cytoplasmic volume also does not increase; this suggests mechanisms that adjust constant nucleotide concentrations in the cytoplasm and probably also in the organelles. A similar conclusion was deduced from cultured sycamore cells after sucrose deprivation; the decrease in the NTP pools correlated well with a decrease in the cytoplasmic volume (Roby *et al.*, 1987).

A crude measure of the amount of cytoplasm is the amount of extractable protein. When nucleotide contents were related to 1 mg protein, with cultured *Datura* cells fairly constant levels were obtained for the cell elongation and starvation phase (20 days), which were about 13 (26) nmol/

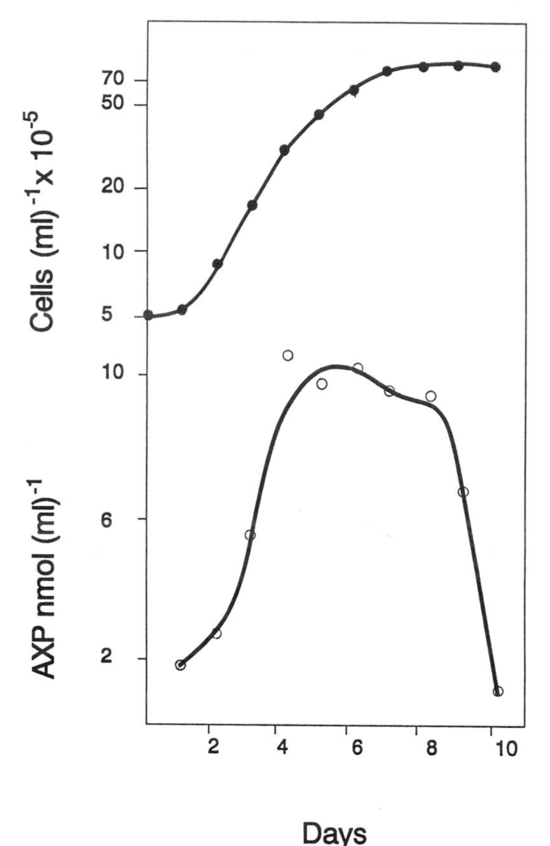

FIG. 16 Growth cycle and Ade nucleotide levels of cultured carrot cells. The data were redrawn from Okamura *et al.* (1975). (●), Number of cells per milliliter; (○), AXP levels in nanomoles per milliliter culture.

mg for the total adenine (uracil) nucleotide pools (Wylegalla *et al.*, 1985). During the lag phase, however, these values were twice as high and declined in the cell division phase to the above values, which suggests that high nucleotide concentrations in the cytoplasm at the onset of proliferation may constitute a regulatory signal. In the same culture the total NTP pool had a peak value slightly before (about 6 hr) the maximum of the cellular RNA content; this may also be indicative of a trigger function of nucleotide pools, as it is discussed to be in animal cell cultures (Grummt and Grummt, 1976; Grummt *et al.*, 1977; Chou *et al.*, 1984; Rapaport *et al.*, 1979). Nucleotide analyses 24 hr after inoculation of *Catharanthus* cells into P_i-deficient medium revealed a three- to fivefold reduction in the NTP pools (relative to the control, which obtained P_i), whereas the other nucleotide pools were only slightly affected (Ashihara *et al.*, 1988); this also suggests a role of the triphosphates in growth promotion. It is, furthermore, interesting that in two different cell cultures the level of UDPGlcNAc had a maximum about 2 days earlier than the other nucleotide pools, such as ATP or UDPGlc (Meyer and Wagner, 1985a,c). UDPGlcNAc is a precursor for glycoprotein synthesis, and it was shown for *Chlorella* (Quigley *et al.*, 1984) that new synthesis of *N*-glycosylated proteins is essential for the G_1/S phase transition in the cell cycle. With cultured *Catharanthus* cells an early sharp maximum (day 1) of Fru2,6P_2 (450 pmol/g FW) was reported (Ashihara, 1986); a peak of this regulatory metabolite in the lag phase may indicate a signal for utilization of the exogenous carbon source (see Section IV,C).

4. Early Induction Events

In the rather extended lag phase of cultured *Datura* cells (Fig. 3 of Meyer and Wagner, 1985a), the initial increase in the nucleotide pools consisted of two phases, a very early increase immediately after inoculation and a second phase starting about 2 days later (Fig. 15). There were also two phases of uptake of inorganic phosphate, with the second phase preceding the second phase of nucleotide synthesis (Fig. 2 of Wylegalla *et al.*, 1985). One is tempted to suggest an early phase of nucleotide synthesis by existing enzymes mainly governed by salvage of precursors (which may be obtained from digestion of cellular constituents such as RNA) and synthesis of enzymes for nucleotide *de novo* synthesis at a slightly later stage. This is supported by the above-mentioned report on purine salvage enzymes, which had a rather constant activity throughout the cell cycle, whereas enzymes of the *de novo* route showed an induction at the beginning of the cell division phase (Ashihara and Nygaard, 1989). Cultured *Catharanthus* cells also showed early maxima of enzyme activities for some of the salvage enzymes, which precede (by about 1 day) that of *de novo* enzymes. Furthermore, NMP- and NDP-kinases had high initial values and early activity maxima and also precursor feeding experiments showed early maxima of incorporation (Kanamori-Fukuda and Ashihara, 1981; Kanamori *et al.*, 1979; Cox *et al.*, 1973; Hirose and Ashihara, 1984; Shimazaki and Ashihara, 1982). However, not all data are in favor of this explanation; uridine kinase, e.g., was in phase with *de novo* enzymes (Kanamori-Fukuda and Ashihara, 1981) and nucleic acid- or nucleotide-degrading enzymes had their maxima only at day 5/6 of the growth cycle (Hirose and Ashihara, 1984). Furthermore, in several cell cultures no significant pools of precursors such as bases or nucleosides were detected (Meyer and Wagner, 1985a,c).

C. Individual Nucleotide Pools

1. C:U and U:A Nucleotide Ratios

With several cultured plant species comprehensive nucleotide analyses were performed that allowed a comparison of the individual nucleotide classes (Table XI). The results are rather uniform, showing very low levels of Cyt nucleotides, small fractions of GXP and UXP, medium fractions of AXP, and large fractions of the UDP sugars. Thus the ratio of the total Ura to Cyt nucleotides is very high, in the extreme case of cultured tobacco cells it is 70:1. There may also be levels of CDPcholine or CDPethanolamine, but not of considerable size, as they have not been reported up to now by ^{31}P NMR studies. Thus the interesting question is how the CTPSase activity and nucleotide turnover is regulated in the cell to ensure such a high U:C ratio.

With cultured tobacco and *Datura* cells the U:A nucleotide ratio was followed throughout the growth cycle (Meyer and Wagner, 1985a,c). Although the individual pools changed up to sixfold, the ratio was fairly constant and very near to 2 with *Datura* cells and 2.6 with tobacco cells. Also with cultured sugar cane cells, removal of P_i resulted in a synchronous decrease of the ATP and ADP as well as UTP and UDPGlc pools (Dancer *et al.*, 1990a). The ratio between Ade nucleotides and Ura nucleotides, and probably also between the total purine and pyrimidine nucleotides, seems to be tightly adjusted. Interestingly, with cultured mammalian cells (Chou *et al.*, 1984) and *E. coli* (Shimosaka *et al.*, 1984) such ratios were also reported to be critical for growth and initiation of DNA synthesis. An auxotrophic mutant cell line, which required Ade, Ado, or Ino for growth, had an approximately fivefold reduced capacity of purine nucleo-

TABLE XI

A Comparison of Nucleotide Pools of Suspension Cultured Cells

	Total pool (nmol/g FW)	CXP (%)	GXP (%)	UXP (%)	AXP (%)	UDPsugars (%)	Ref.[a]
Sycamore	530	1	12	16	32	39	1
Tobacco	413	1	4	14	25	56	2
Carrot[b]	1570	—	10	16	37	37	3
Datura	227	1	7	13	31	48	4
Datura	400	3	15	11	19	52	5

[a] 1, Brown and Short, 1969; 2, Meyer and Wagner, 1985c; 3, Kanabus *et al.*, 1986; 4, Meyer and Wagner, 1985a; 5, Mitsui and Ashihara, 1988.

[b] Cyt nucleotides were not determined as well as the NMPs.

tide *de novo* synthesis, whereas the salvage capacity was unaffected; the cells revealed slightly enlarged Ade nucleotide pools and UDPGlc pools, but much higher levels of ADPGlc and UMP than the wild type, when grown in the presence of 0.5 m*M* Ade (Mitsui and Ashihara, 1988). The data published reveal that the U:A ratio of the wild type (3.2) remained the same in the mutant, although grown with Ade. This was, however, correct only when the ADPGlc pool, which must be assigned to the plastid compartment, was not considered; one might speculate that adjustment of the Ade and Ura nucleotide pools occurs in the cytoplasm and the mutant is able to sequester an excess of salvaged Ade nucleotides into the plastid.

2. Energy Charge

Although there are reports of significantly lower values (Shimazaki *et al.*, 1982; Shimizu *et al.*, 1977), in general the Ade nucleotide-derived EC is rather high in suspension cultured cells under optimal growth conditions, i.e., in the range of 0.8 to 0.9 (Meyer and Wagner, 1985c, and references therein; Kanabus *et al.*, 1986; Okamura *et al.*, 1975). These values are usually higher than those reported for green plant tissue (Meyer and Wagner, 1986; Raymond *et al.*, 1987; Pradet and Raymond, 1983). In the case of cultured tobacco cells the mass action ratio for the AMP-Kase reaction was slightly higher (1.8–3.6) than the values reported for full equilibration (0.5–1.2) (Pradet and Raymond, 1983) and also during the initial growth phase of cultured *Datura* cells they were rather high (4.5–9) and declined to the expected values at the later growth stages (Meyer and Wagner, 1985a,b); unfortunately no enzyme activitiy measurements are available that could show whether equilibration is incomplete, especially at the early stages of growth.

3. Estimation of Cellular Concentrations

To calculate cellular concentrations, knowledge is required about the location of the nucleotide pools within the cells and the relative volume of the respective compartments. Nongreen cultured cells have in general large vacuoles and few amyloplasts; there is no evidence that the vacuole contains nucleotide pools of significant magnitude. Although data on nucleotide pools in plastids of cultured cells are not available, one should assume from the incomplete catabolic capacities found in amyloplasts of cultured sycamore cells (Frehner *et al.*, 1990) that the Ade nucleotide levels in these plastids may be low. As the fraction of mitochondrial pools is also low (Table X), one can assume that nucleotide levels determined in cultured cells roughly correspond to cytoplasmic pools; this holds certainly for the total Ura nucleotides, but approximately also for the other nucleotide classes. For growing plant cells it was assumed that the cytoplasm constitutes a fraction of about 10% of the total cell volume (Edwards *et al.,* 1984); with the large vacuole of suspension cultured cells a relative volume of 5% may be more realistic. With this assumption and relating 1 g fresh weight to 1 ml, cytoplasmic concentrations for UDPGlc/Gal and ATP were estimated as 2.1 and 1.1 m*M* for cultured *Datura* cells and 4.3 and 1.7 m*M,* respectively, for cultured tobacco cells (Meyer and Wagner, 1985a,c). With cultured *Catharanthus* cells concentrations of 3.2 and 0.71 m*M* for ATP and ADP were assumed (Kubota and Ashihara, 1990).

4. Deoxyribonucleotide Pools

As known from other species, the deoxynucleotide pools are much smaller than those of the ribonucleotides, although especially with plant tissues only a few data are available. Tobacco and sycamore callus tissue revealed dNTP levels (1–10 nmol/g DW) that were about 100-fold lower than the NTP levels. Although with sycamore callus dCTP was the smallest pool, with tobacco dCTP, dTTP, and dGTP pools were similar in size; with both tissues, however, the dATP pools were dominant (Nygaard, 1972). In extracts of cultured carrot cells dTMP-Sase activity was found to be rather low, much lower than reported for bacteria, and it did not change significantly during the growth cycle (Nielsen and Cella, 1988). This would help to explain why DNA synthesis in cultured *Datura* cells seemed to be less strictly regulated during the growth cycle than RNA synthesis (Wylegalla *et al.,* 1985). On the other hand, the low activity of NDP-Rase, found in cultured soybean cells, could be detected for only a short period at the beginning of the cell division phase (Hovemann and Follmann, 1979).

D. Cycle Phases and Growth Regulation

1. Role of Inorganic Phosphate

For several culture media, such as Murashige–Skoog and Linsmaier–Skoog media, inorganic phosphate is the growth-limiting component (Amino *et al.,* 1983). Although the various cell lines reveal different kinetics in P_i uptake (Rébeillé *et al.,* 1982; Wylegalla *et al.,* 1985; Ashihara and Nygaard, 1989), P_i absorption usually precedes the uptake of the nitrogen source and of sucrose. With cultured *Datura* cells P_i uptake was complete within 2 days (Wylegalla *et al.,* 1985) and with P_i-starved *Catharanthus* cells even within 10 hr (Ashihara and Ukaji, 1986). Excess phosphate is stored in the vacuole, which provides the cytoplasm for P_i, although at a low rate (Rébeillé *et al.,* 1983).

P_i availability strongly influences nucleotide synthesis and pool sizes; this was shown by inoculation into P_i-deficient medium (Li and Ashihara, 1990; Ukaji and Ashihara, 1987a; Ashihara *et al.*, 1988) and by P_i removal from a chemostat culture (Dancer *et al.*, 1990a). P_i addition to P_i-starved cells showed a more than 10-fold increase of the PRPP pool (0.8 nmol/g FW) and a still higher increase in PRPP availability [65 nmol/(g FW · hr)], as measured by salvage of added Ade (Ukaji and Ashihara, 1987a). The increase in the PRPP pool did not correlate with the extractable activity of PRPP-Sase, which was also rather high in the P_i-starved cells. It is interesting that the maximum PRPP level was obtained already at 10 hr after P_i addition, whereas PRPP availability was at a maximum only after 24 hr, when the ATP pool had also reached a high value (Ukaji and Ashihara, 1987a). One could speculate that at the very early phase the PRPP pool is not accessible for salvage of exogenous Ade, as it is first consumed in the plastid for *de novo* synthesis, and the cytoplasmic PRPP pool, which is mainly available for salvage activity, is established only later. Another explanation is that a high rate of PRPP turnover depends on a high ATP pool.

Cellular ATP levels increased very fast, within 3 hr, after P_i addition to P_i-starved *Catharanthus* cells, up to 180 nmol/g FW, which suggests a very early net synthesis of Ade nucleotides (Ashihara and Ukaji, 1986). In P_i-deficient medium the activity of both salvage and *de novo* nucleotide synthesis, estimated from the incorporation rates of labeled precursors, was strongly reduced relative to the P_i-containing medium (Ashihara *et al.*, 1988), which is obviously a consequence of the availability of both PRPP and ATP. Degradation of Ade, Ado, and Ura to CO_2, however, was similar in both media; only degradation of Urd was about fourfold enhanced in P_i-deficient medium. As salvage of Ura nucleotides is most efficient with Urd through Urd-Kase and the latter enzyme was shown to have its activity maximum 1 day later in the growth cycle than the other salvage enzymes (Kanamori-Fukuda and Ashihara, 1981), Urd salvage may partially depend on enzyme *de novo* synthesis, which is prohibited in P_i deficiency. On the other hand, the low ATP levels in P_i-deficient medium may also interfere with the ATP-dependent Urd-Kase reaction.

2. P_i Starvation

Sustained P_i deficiency caused markedly lower levels of respiration as measured by O_2 uptake and CO_2 release from labeled glucose (Li and Ashihara, 1990). In cultured sugarcane cells the total Ade and Ura nucleotides decreased three- and sixfold after P_i removal from the medium (Dancer *et al.*, 1990a); reduction was most pronounced in the ATP and UDPGlc pools, but also the sugarP pools showed strongly reduced levels (six- to sevenfold), as did Fru2,6P_2. The ATP:ADP and UTP:UDP ratios decreased synchronously about fourfold, indicating efficient coupling by NDP–Kase. P_i-starved cultured *Brassica* cells also showed low levels of Ade nucleotides and Fru2,6P_2, but no change in the PP_i pool (Duff *et al.*, 1989). Measurements of the respective enzyme activities suggested that ATP-dependent and P_i-consuming glycolytic reactions may be bypassed by the PP_i-dependent Fru6P-PTase and nonphosphorylating, NADP-dependent GAP-DHase and PEP–Hase. These suggestions could not, however, be confirmed by enzyme activity measurements in P_i-starved *Catharanthus* cells (Li and Ashihara, 1990). With cultured sugar cane cells, P_i removal caused a decrease in the PP_i level, but only after an extended (5 day) delay, also indicating that PP_i is an autonomous energy donor in the cytosol of higher plants (Dancer *et al.*, 1990a).

A further effect of P_i starvation is a considerable elevation of the amino acid pools correlated with increased protease activities (Duff *et al.*, 1989); increase in the amino acid pools is caused by *de novo* synthesis, inhibited protein synthesis, and protein degradation (Ukaji and Ashihara, 1987b). P_i deficiency further results in a strong decrease in RNA and protein levels and in rate of DNA synthesis (Ukaji and Ashihara, 1986). P_i-starved *Catharanthus* cells showed a retarded rate of extracellular sucrose hydrolysis, and of the uptake and incorporation of radioactivity from labeled sucrose into both ethanol-soluble and -insoluble cellular fractions (Li and Ashihara, 1989). Secretion of an extracellular RNase, probably to capture exogenous P_i, was reported at the onset of starvation of cultured tomato cells; secretion was strongly enhanced when the cells were transferred to P_i-deficient medium (Nürnberger *et al.*, 1990). Sycamore cells hydrolyze cholineP through a cell wall-located phosphatase, the activity of which is also enhanced with P_i-deficiency (Gout *et al.*, 1990).

3. Regulation of Cell Division

Under conditions when P_i is the growth-limiting component, one should expect that the available amount of P_i will determine the length of the cell division phase. This was shown for cultured *Datura* cells; the cellular phosphate pool was consumed during the cell division phase and it decreased to a constant low level at a point where the other biochemical data indicated the start of cell elongation (Wylegalla *et al.*, 1985). Deficiency of cellular phosphate is a signal for the end of cell proliferation; this signal may be exerted by a reduction of the nucleotide pools, especially the NTPs. With the same cultured cells it was shown that the decrease of the NTP pools nicely paralleled the decrease of the cellular P_i pool, which suggests that NTP pools may regulate RNA and protein synthesis (Fig. 2 of Wylegalla *et al.*, 1985). With cultured tobacco cells a correlation of the cellular P_i levels and the Ade nucleotide content was also observed and the decrease of the P_i level correlated well with the halt in protein net increase (Miginiac-Maslow *et al.*, 1981).

4. Regulation of Cell Elongation

Whereas with several cultured species P_i uptake occurred very early and often was already completed in the lag phase, the main phase of sucrose uptake occurred during the cell division phase, as shown with cultured *Datura* and *Catharanthus* cells (Wylegalla *et al.*, 1985; C. Helmbrecht, unpublished observations), although other cell lines showed a different behavior (Ashihara and Nygaard, 1989). With cultured *Catharanthus* cells, uptake of ammonium ions was also completed at the end of cell division, whereas nitrate uptake extended into the cell elongation phase. With cultured *Datura* cells the cellular sugar content increased slightly in the lag phase and strongly in the cell division phase; starch synthesis started at a threshold level of the cellular sugar pool and stopped when the medium sucrose was depleted. Cellular sugar and starch content was consumed during the cell elongation phase; these findings suggest that the end of the cell elongation phase and hence of total growth is signaled by a decrease of the cellular carbon source below a threshold value (Fig. 3 of Wylegalla *et al.*, 1985). Cultured soybean, however, showed a different behavior: while the specific activity of starch synthase was at a maximum in the middle of the cell elongation phase, the amount of starch was at a maximum at the onset of the stationary phase (Miyamoto *et al.*, 1989).

5. Stationary Phase

With cultured *Datura* cells an extended stationary phase (14 days) could be followed with only a slight decrease in cell mass, DNA content, and viability (Wylegalla *et al.*, 1985). The RNA level declined only at the onset of the stationary phase, whereas cultured carrot cells revealed RNA degradation already at the beginning of the cell elongation phase (Ashihara and Nygaard, 1989). In the extended stationary phase of the cultured *Datura* cells the EC was established at a high level, although the nucleotide pools declined to constant low, but significant, levels. This suggests a stabilized metabolism in the stationary phase, ensuring viability. It is suggested that metabolism is stabilized by metabolite "maintenance levels," which were 1 μmol/g FW P_i and about 14 μmol/g FW sucrose; protein levels and hence cytoplasm declined slowly in this phase, but the nucleotide levels related to 1 mg protein and hence approximately to the cytoplasmic volume remained constant (Fig. 4 of Wylegalla *et al.*, 1985). During extended sucrose deprivation with cultured sycamore cells, constant high values of the NTP:NDP ratio were also observed (Roby *et al.,* 1987). These are conditions for metabolic survival in the stationary phase and further imply that regulatory devices must exist that prohibit a complete consumption of these pools. With cultured cells exhibiting cell death during the stationary phase, low EC values approaching 0.5 were reported (Okamura *et al.,* 1975). The rather strong decline of nucleotide levels at this stage may also be indicative of cell deterioration (Fig. 16).

E. Sugar Assimilation and Nucleoside Diphosphate Sugars

1. Assimilation of Sucrose

Sugar assimilation occurs from external or vacuolar stores and has also been studied in cultured cells. Several species, such as carrot, tobacco, proso millet, and maize, efficiently hydrolyzed sucrose by extracellular invertases; carrot cells cultured in sucrose-containing medium were shown to synthesize sucrose within the cell from UDPGlc and Fru6P and to synthesize UDPGlc from Glc1P and UTP (Fig. 8) (Kanabus *et al.,* 1986). Cultured *Catharanthus* cells hydrolyzed medium sucrose through a cell wall-associated invertase within 96 hr; at this stage the cellular fructose and glucose concentrations were 230 and 210 m*M*, respectively (Li and Ashihara, 1989; Kubota and Ashihara, 1990). With cultured tobacco cells, transfer into new medium induced *de novo* synthesis of cell wall invertase independent of the presence of sucrose in the medium (Weil and Rausch, 1990). On the other hand, cultured cells from other species, such as cultured *Datura* cells (Wylegalla *et al.,* 1985), obviously had low or no cell wall invertase activities and absorbed sucrose intact.

In the case of sucrose uptake, UDPGlc could be formed directly from sucrose by sucrose-Sase and UDP; K_m values for UDP with different enzymes range from 5 to 190 μM (Table VII); the UDP concentrations estimated in cultured tobacco cells, 120–240 μM, or those found in *Datura* cells, which were slightly lower, are in this range (Wylegalla *et al.,* 1985; Meyer and Wagner, 1985a,c). The percentage of UDPGlc related to the total Ura nucleotide content found in these cultured cells showed a smooth maximum of about 80% at the phase of sucrose uptake, which could be indicative of such a transformation. UDPGlc would be converted into Glc1P with PP_i (Fig. 8) and used for energy metabolism or used for cell wall synthesis in the cell division and cell elongation phase.

2. The Individual NDPsugar Pools

Some analyses of the sugar moieties from mixtures of UDPsugars have been obtained from suspension cultured cells. With soybean cells the ratio found for Glc, Gal, Xyl, Ara, and Rha was 42:14:1:1:0.8; the levels of UDPGlc and UDPXyl was 102 and 2.4 nmol/g FW, respectively; the rather low value of the latter suggests a high turnover, as synthesis of xylose-containing polysaccharides was high (Hayashi and Matsuda, 1981). Cultured rice cells contained relative high levels of GDPsugars; the total amount of the nucleotide sugars, about 20% of the total soluble nucleotide pool, had fractions of UDPsugars, GDPsugars, and ADPsugars of 61, 35, and 16 nmol/g FW, respectively; UDPGlc, the main component, was 33% of the total NDPsugars (Mitsui *et al.*, 1989). Cultured *Catharanthus* cells showed rather high levels of UDPGlc, 75–90% of the total NDPsugar pool, and also considerable levels of UDPGlcUA, 6–16%, whereas the levels of UDPGal, UDPAra, and UDPXyl were smaller (Amino *et al.*, 1985b). The pools of UDPGlc and UDPGal are far from thermodynamic equilibrium. The specific activities of the related enzymes were in the range of 2 nmol/(min · mg) to 2.5 μmol/(min · mg); the order of activities was Glc1P-UTase >> sucrose-Sase > UDPGlc-Ease >> UDPGlc-DHase > GlcUA1P-UTase >> UDPGlcUA-DCase = UDPXyl-Ease (Amino *et al.*, 1985a). Glc1P-UTase played a more important role in UDPGlc formation than sucrose-Sase. On the other hand, formation of UDPGlcUA seemed to be governed by oxidation of UDPGlc and not by the inositol pathway (Fig. 9), which includes the GlcUA1P-UTase step. The cells were, however, cultured in medium without inositol.

Starch synthesis in the amyloplast depends on the uptake of carbon skeletons from the cytoplasm and it is still unclear whether trioseP is transported via the known trioseP/P_i translocator or if it is hexoseP, which is imported by a different translocator (Macdonald and ap Rees, 1983). Cultured *Glycine max* cells contained UDPGlc and ADPGlc levels in the ratio of 14:1, the respective enzyme activities, Glc1P-UTase and Glc1P-ATase, were in about the same ratio, and starch synthase activity was about 20% of that of Glc1P-ATase (ap Rees *et al.*, 1984). In spite of the high pools and synthetic activities for UDPGlc, which are ascribed to the cytoplasm, starch synthesis in the amyloplasts is performed with ADPGlc, which was concluded from a comparison of the rates of starch accumulation with the activities of Glc1P-ATase and starch synthase and from the localization of these enzymes in the amyloplasts (ap Rees *et al.*, 1984). From sugar cane suspension cultured cells, a high sucrose turnover, i.e., simultaneous synthesis and degradation, was reported (Wendler *et al.*, 1990); sucrose storage in the vacuole started after depletion of the nitrogen source and correlated with high activities of sucroseP-Sase, sucrose-Sase (which does not contribute to sucrose synthesis), and invertase in the cytoplasm. It is suggested that sucrose storage is regulated through this cycle of synthesis and degradation.

F. The Cell Cycle

Studies on the cell cycle of plant cells connected to nucleotide dynamics have been performed mainly with suspension cultured cells; there are, however, also investigations on algae that may be of interest in this respect.

1. UDPsugars and Cell Wall Synthesis

Although different techniques have been tested, in order to synchronize dividing cells in suspension culture (King, 1980), a double-phosphate starvation method has been successfully applied with cultured *Catharanthus* cells to study cell wall synthesis during the cell cycle (Amino *et al.*, 1983). The authors claimed to obtain at least two rounds of cell dublication, each in a period of only 4 hr within the duration of a cell cycle of about 30 hr; furthermore, cell synchrony was evaluated by flow cytometry (Ando *et al.*, 1987). Changes of cell wall components in synchronized cultured cells suggested two main periods of active cell wall synthesis, which have been assigned to the G_1 phase and the G_2/M phase. The G_1 phase is the most important period of cell expansion, with the main net increase in wall components, whereas in the G_2/M phase cell plate formation occurs; in these two periods characteristic changes of the Golgi apparatus, the most important compartment for polysaccharide synthesis, were also described (Hirose and Komamine, 1989; Amino *et al.*, 1984; Delmer and Stone, 1988).

Changes in the levels of the UDPsugars and enzyme activities, involved in the formation and utilization of these cell wall precursors, gave several interesting insights. UDPGlc had its maximum in the G_2/M phase concomitant with the activity maximum of Glc1P-UTase, whereas sucrose-Sase peaked in the following G_1 phase. This may indicate that the pyrophosphorylase, the activity of which is about fivefold higher than that of sucrose-Sase, is the main UDPGlc-producing enzyme. The levels of UDPGal and UDPGlc-Ease were also high in the G_2/M phase, indicative of high activities in UDPGlc/UDPGal equilibration. Why the main period of cell wall synthesis (G_1 phase) has lower UDPGlc levels is not clear. The G_1 phase, however, showed maxima, although smaller in magnitude than those of the above-mentioned UDPsugars, of UDPGlcUA and UDPAra and also activity increases of UDPGlc-DHase (M/G_1) and GlcUA1P-UTase, indicating increased activities in the production of UDPGlcUA and its conversion into UDPXyl and UDPAra in the G_1 phase, although the levels of UDPXyl and the activities of UDPGlcUA-DCase and UDPXyl-Ease did not change in the cell cycle (Amino *et al.*, 1985a,b). Although many questions are still open, this is a very interesting approach for further studies.

2. S Phase-Dependent Activities

A second, more thoroughly studied, topic concerns enzyme activities and precursor synthesis devoted to the S phase, i.e., DNA synthesis. Studies with the enzyme dUTP-Hase, which cleaves dUTP into dUMP and PP_i, in root meristem from *Allium cepa* showed activity increases in the G_1/S boundary and high levels in the S phase, whereas in G_1 and G_2 activity was almost undetectable, as was also so in nondividing differentiated cells (Pardo and Gutiérrez, 1990). This enzyme may function in the degradation of dUTP formed to prohibit dUMP incorporation during DNA replication; however, as the pathway from dUDP or dCDP to dUMP (Fig. 2) is not known for higher plants, a role in the supply of dUMP for dTMP production is conceivable, provided that a respective dNDP-Kase would convert dUDP into dUTP. With a synchronously dividing callus culture from Jerusalem artichoke tubers, activities of dThd-Kase, dTMP-Kase, and DNA polymerase were shown to increase in the S phase. These increases were dependent on DNA replication, as omission of the auxin analog 2,4-dichlorophenoxyacetic acid from the medium prohibited the induction of the enzyme activities (Harland *et al.*, 1973).

Several investigations on S phase-dependent enzymes have been performed with algae, and the routes for deoxynucleotide synthesis were elucidated more thoroughly with these species than with higher plants, although algae may have their peculiarities. Thus the systematically unrelated alga, *E. gracilis*, is the only known eukaryote with a triphosphate reductase whereas all other species tested possess an NDP-Rase (Gleason and Hogenkamp, 1970). Furthermore, a route from dCMP to dUMP was described in *Chlorella*, although this is not established in higher plants (Fig. 2).

In synchronous cultures of *Scenedesmus* several enzymes involved in DNA precursor synthesis peak in the S phase; these are NDP-Rase, dTMP-Sase, dihydrofolate reductase, and dTMP-Kase, whereas the unspecific NDP-Kase revealed a high and constant activity during the whole cell cycle. Further evidence suggests that NDP-Rase and dTMP-Sase are subject to a common control mechanism (Klein and Follmann, 1988; Feller *et al.*,1980; Bachmann *et al.*, 1983; Bachmann and Follmann, 1987). The dNTP levels were very low in the G_1 and G_2 phases and peaked in the S phase; the maximum levels determined decreased in the order dTTP (4.5 pmol/10^6 cells) > dATP > dCTP > dGTP (1.5 pmol/10^6 cells). Inhibiting dTMP-Sase and thus DNA synthesis with 5FdUrd led to the depletion of the dTTP pool and a strong increase in the activity of NDP-Rase and in the level of dATP, whereas inhibition of DNA replication by arabinosylcytosine had no effect on NDP reduction. Control of enzyme levels involved in DNA precursor synthesis by the dTTP pool was suggested, which may be uncoupled from DNA synthesis (Feller *et al.*, 1980).

Earlier studies with synchronous cells of *Chlorella* on Asp-CTase suggested cell cycle dependent stabilization of the enzyme through its feedback inhibitor UMP (Vassef *et al.*, 1973) and showed the simultaneous induction of dCMP-Nase and dTMP-Kase in the S phase (Shen and Schmidt, 1966); the latter enzymes may function in establishing the dTTP pool as a regulatory signal for DNA replication. dThd incorporation studies during the life cycle of *Chlamydomonas* suggested continuous activity of dTMP-Kase in the chloroplast and derepression of the cytoplasmic enzyme only in the sexual reproductive cycle (Chiang *et al.*, 1975). Investigations of the circadian rythmicity of cell division with *Euglena*, following NAD-Kase and NADP-Pase activities, led the conclusion that a Ca^{2+}-transport system, Ca^{2+}/calmodulin, NAD-Kase, and NADP-Pase may represent "clock" components that constitute a self-sustained circadian oscillating loop (Laval-Martin *et al.*, 1990); this model was first derived from studies of NAD(P) pools and enzyme activities in *Lemna gibba* (Goto, 1983). It should also be mentioned that trigonelline, *N*-methylnicotinic acid (Fig. 10), is a plant hormone, shown to be present in pea seedlings, which promotes cell arrest in the G_2 phase (Tramontano *et al.*, 1982, 1983a,b).

VI. Concluding Remarks

This article is an experiment performed by those primarily interested in nucleotide metabolism and its implications for cell growth. Our approach, which tries to define "nucleotide dynamics" and to give general insight, is different from those in several excellent recent reviews, devoted to special topics of plant physiology, which included only parts of nucleotide metabolism.

Section II, entitled Nucleotide Synthesis, Utilization, and Degradation, is in fact an extended introduction; it is intended to prepare for the understanding of the following topics and contains the available data on the pathways, their regulation, the enzymes, and nucleotide pools. Although there are plant-specific parts, e.g., the utilization of the carbon source, most of the biosynthetic routes in plants are identical to those known from bacteria and animal tissue; hence we confined this section to the plant-related knowledge and only in a few cases discussed data from bacterial and animal metabolism. We further listed available K_m values, as these data are suitable for comparative discussions and for first attempts to combine the cellular pools with metabolite dynamics.

Section III deals with the cellular topography of pathways and pools. These topics urgently need future efforts to confirm known hypotheses and to answer the many questions on the localization of enzymes and pools. Although subcellular fractionation of plant cells has its peculiar difficulties, with the recently developed techniques separation of cellular compartments has been greatly improved. Progress in this field would further open the path for new investigations on the intracellular transport of nucleotides and their precursors and thus help to elucidate how the different compartments are supplied with their nucleotide pools.

Section IV is an attempt to combine nucleotide dynamics with metabolic regulation; it contains data on the plant-specific topics of energy metabolism, carbon source utilization, and the role of NDPsugars in glycoprotein and polysaccharide synthesis. Whereas these topics are currently well investigated by several groups, investigation of the regulation of general nucleotide metabolism, its dependency on the carbon, nitrogen, and phosphate source, and its relation to growth or senescence is a widely underdeveloped field. Also, the interrelation between *de novo* and salvage routes and its implication in growth, development, and senescence are only poorly elucidated.

Section V is concerned with nucleotide dynamics in *in vitro* cultured plant cells. We discussed the numerous papers devoted to this topic and tried to correlate nucleotide dynamics with the different phases of growth and their regulation. We also emphasized the role of the UDPsugars and the effect of P_i-starvation. In spite of the many investigations, this field needs new ideas of approach for further progress, as suspension cultured cells are well suited to study plant growth on a cellular basis. Section V also contains available data on nucleotide dynamics within the "cell cycle," which were enriched by respective reports on algae. This field is in its very beginning and needs further progress.

The numerous papers and data that had to be considered for this extended review were indeed the reason why only in a few cases comparative discussions with bacterial and animal topics could be included. Furthermore, there is no similar recent review on nucleotide dynamics in the field of animal tissues, which would have been very helpful. From animal tissue and also from animal cell cultures, a huge number of papers is available now and reviews are urgently required. The creation of a comprehensive picture would be less difficult in this field than with plants; such a review could stimulate discussion, especially in the field of plant cell cultures (Section V). Furthermore, application of molecular genetics in the field of nucleotide metabolism is also urgently required in the case of plants, as it was successfully performed with bacteria (Neuhard and Nygaard, 1987).

Acknowledgments

We are grateful to the editors for proposing the title of this review; we would not have had the idea of writing on nucleotide dynamics, a subject that was a great challenge. We also appreciate our many colleagues who sent us preprints and recent reprints that were of great value for us. Furthermore, we thank Mrs. C. Lippelt for preparing the graphs and Mrs. H. Starke for typing the manuscript. Last, but not least, we appreciate the help of several of our co-workers in the handling of the more than 1000 papers, i.e., in copying, managing, listing, and preparing the reference list; these are Ilona Becker, Brigitte Kornak, Karin Plank-Schumacher, Hella Schöne, Claudia Wylegalla, Perry Grabo, Thomas Gronewald, and Andreas Wagner.

References

Abel, S., and Glund, K. (1986). *Physiol. Plant.* **66,** 79–86.
Abel, S., Blume, B., and Glund, K. (1990). *Plant Physiol.* **94,** 1163–1171.
Al-Baldawi, N. F., and Brown, E. G. (1983). *Phytochemistry* **22,** 419–421.
Amino, S., Fujimura, T., and Komamine, A. (1983). *Physiol. Plant.* **59,** 393–396.
Amino, S., Takeuchi, Y., and Komamine, A. (1984). *Physiol. Plant.* **60,** 326–332.
Amino, S., Takeuchi, Y., and Komamine, A. (1985a). *Physiol. Plant.* **64,** 111–117.
Amino, S., Takeuchi, Y., and Komamine, A. (1985b). *Physiol. Plant.* **64,** 197–201.
Anderson, J. D. (1979). *Plant Physiol.* **63,** 100–104.
Ando, S., Shimizu, T., Kodama, H., Amino, S., and Komamine, A. (1987). *Agric. Biol. Chem.* **51,** 1443-1445.
ap Rees, T. (1987). *Biochem. Plants* **12,** 87–115.
ap Rees, T. (1988). *Biochem. Plants* **14,** 1–33.
ap Rees, T., Leja, M., MacDonald, F. D., and Green, J. H. (1984). *Phytochemistry* **23,** 2463–2468.
Aragón, J. J., and Lowenstein, J. M. (1980). *Eur. J. Biochem.* **110,** 371–377.
Arditti, J., and Tarr, J. B. (1979). *Am. J. Bot.* **66,** 1105–1113.
Ashihara, H. (1977a). *Z. Pflanzenphysiol.* **85,** 383–392.
Ashihara, H. (1977b). *Z. Pflanzenphysiol.* **83,** 379–392.
Ashihara, H. (1978). *Z. Pflanzenphysiol.* **87,** 225–241.
Ashihara, H. (1986). *Z. Naturforsch., C: Biosci.* **41C,** 529–531.
Ashihara, H. (1990). *Curr. Sci.* **59,** 939–941.
Ashihara, H., and Komamine, A. (1974). *Plant Sci. Lett.* **2,** 119–123.
Ashihara, H., and Kubota, H. (1986). *Physiol. Plant.* **68,** 275–281.
Ashihara, H., and Nobusawa, E. (1981). *Z. Pflanzenphysiol.* **104,** 443–458.
Ashihara, H., and Nygaard, P. (1989). *Physiol. Plant.* **75,** 31–36.
Ashihara, H., and Tokoro, T. (1985). *J. Plant Physiol.* **118,** 227–235.
Ashihara, H., and Ukaji, T. (1985). *Int. J. Biochem.* **17,** 1275–1277.
Ashihara, H., and Ukaji, T. (1986). *J. Plant Physiol.* **124,** 77–85.
Ashihara, H., Li, X.-N., and Ukaji, T. (1988). *Ann. Bot. (London)* **61,** 225–232.
Atkins, C. A., Ritchie, A., Rowe, P. B., McCairns, E., and Sauer, D. (1982). *Plant Physiol.* **70,** 55–60.
Atkins, C. A., Shelp, B. J., and Storer, P. J. (1985). *Arch. Biochem. Biophys.* **236,** 807–814.
Atkins, C. A., Storer, P. J., and Shelp, B. J. (1989). *J. Plant Physiol.* **134,** 447–452.
Avelange, M. H., and Rébeillé, F. (1991). *Planta* **183,** 158–163.

Avelange, M. H., Thiéry, J. M., Sarrey, F., Gans, P., and Rébeillé, F. (1991). *Planta* **183,** 150–157.

Bachmann, B., and Follmann, H. (1987). *Arch. Biochem. Biophys.* **256,** 244–252.

Bachmann, B., Hofmann, R., and Follmann, H. (1983). *FEBS Lett.* **152,** 247–250.

Backer, A. I. (1990). Ph.D. thesis, Technical University of Braunscheig, Braunscheig, Germany.

Baker, A., and Leaver, C. J. (1985). *Nucleic Acids Res.* **13,** 5857–5867.

Barankiewicz, J., and Paszkowski, J. (1980). *Biochem. J.* **186,** 343–350.

Barber, D. J., and Thurman, D. A. (1978). *Plant Cell Environ.* **1,** 305–306.

Becker, M. A., Raivio, K. O., and Seegmiller, J. E. (1979). *Adv. Enzymol.* **49,** 281–306.

Beevers, L. (1982). *Encycl. Plant Physiol.* **13A,** 103–123.

Billich, A., and Witzel, H. (1986). *Biol. Chem. Hoppe-Seyler* **367,** 291–300.

Billich, A., Stockhowe, U., and Witzel, H. (1986). *Biol. Chem. Hoppe-Seyler* **267,** 267–278.

Birkenhead, K., Walker, D., and Foyer, C. (1982). *Planta* **156,** 171–175.

Black, C. C., Xu, D. P., Sung, S. S., Mustardy, L., Paz, N., and Kormanik, P. P. (1987). *In* "Phosphate Metabolism and Cellular Regulation in Microorganisms" (A. Torriani-Gorini, F. G. Rothman, S. Silver, A. Wright, and E. Yagil, eds.), pp. 264–268. Am. Soc. Microbiol., Washington, D.C.

Bligny, R., Foray, M.-F., Roby, C., and Douce, R. (1989). *J. Biol. Chem.* **264,** 4888–4895.

Bligny, R., Gardestrom, P., Roby, C., and Douce, R. (1990). *J. Biol. Chem.* **265,** 1319-1326

Boland, M. J., and Schubert, K. R. (1983). *Arch. Biochem. Biophys.* **220,** 179–187.

Boland, M. J., Blevins, D. G., and Randall, D. D. (1983). *Arch. Biochem. Biophys.* **222,** 435–441.

Bolwell, G. P., and Northcote, D. H. (1983). *Biochem. J.* **210,** 497–507.

Bolwell, G. P., Dalessandro, G., and Northcote, D. H. (1985). *Phytochemistry* **24,** 699–702.

Bown, A. W., and Nicholls, F. (1985). *Plant Physiol.* **79,** 928–934.

Brawerman, G., and Chargaff, E. (1955). *Biochim. Biophys. Acta* **16,** 524–532.

Bressan, R. A., Murray, M. G., Gale, J. M., and Ross, C. W. (1978) *Plant Physiol.* **61,** 442–446.

Brown, E. G. (1963). *Biochem. J.* **88,** 498–504.

Brown E. G. (1965). *Biochem. J.* **95,** 509–514.

Brown, E. G., and Short, K. C. (1969). *Phytochemistry* **8,** 1365–1372.

Brunngraber, E. F. (1978). *In* "Methods in Enzymology" (P. A. Hoffee and M. E. Jones, eds.), Vol. 51, pp. 387–395. Academic Press, New York.

Brunngraber, E. F., and Chargaff, E. (1970). *J. Biol. Chem.* **245,** 4825–4831.

Bryan, J.K. (1990). *Plant Physiol.* **92,** 785–791.

Carter, S. G., and Tipton, C. L. (1986). *Phytochemistry* **25,** 33–37.

Champigny, M.-L. (1977). *Proc. Int. Congr. Photosynth.* **4,** 479–488.

Chen, C. M., and Petschow, B. (1978). *Plant Physiol.* **62,** 871–874.

Chen, Y.-R., Datta, N., and Roux, S. J. (1987). *J. Biol. Chem.* **262,** 10689-10694.

Chiang, K. S., Eves, E., and Swinton, D. C. (1975). *Dev. Biol.* **42,** 53–63.

Chou, I.-N., Zeiger, J., and Rapaport, E. (1984). *Proc. Natl. Acad. Sci. U.S.A.* **81,** 2401–2405.

Chrispeels, M. J. (1985). *Plant Physiol.* **78,** 835–838.

Ciccarelli, B., and Brown, A. (1988). *Can. J. Bot.* **66,** 1616–1620.

Cole, S. C. J., and Yon, R. J. (1984). *Biochem. J.* **221,** 289– 296.

Combes, A., Lafleuriel, J., and Le Floc'h, F. (1989). *Plant Physiol. Biochem.* **27,** 729–736.

Copeland, L., and Turner, J. F. (1987). *Biochem. Plants* **11,** 107–128.

Cox, B. J., Turnock, G., and Street, H. E. (1973). *J. Exp. Bot.* **24,** 159–174.

Dancer, J. E., and ap Rees, T. (1989). *Planta* **178,** 421–424.

Dancer, J., Veith, R., Feil, R., Komor, E., and Stitt, M. (1990a). *Plant Sci.* **66,** 59–63.

Dancer, J., Neuhaus, H. E., and Stitt, M. (1990b). *Plant Physiol.* **92,** 637–641.
Day, D. A., Arron, G. P., and Laties, G. G. (1979). *J. Exp. Bot.* **30,** 539–549.
De Gunst, M. C. M., Harkes, P. A. A., Val, J., Van Zwet, W. R., and Libbenga, K. R. (1990). *Enzyme Microb. Technol.* **12,** 61–71.
de Klerk-Kiebert, Y. M., and van der Plas, L. H. W. (1984) *Plant Sci. Lett.* **33,** 155–162.
Delmer, D. P. (1972). *J. Biol. Chem.* **247,** 3822–3828.
Delmer, D. P., and Stone, B. A. (1988). *Biochem. Plants* **14,** 373–420.
Deng, Q. I., and Ives, D. H. (1975). *Biochim. Biophys. Acta* **377,** 84–94.
Dennis, D. T. (1987). "The Biochemistry of Energy Utilization in Plants." Chapman & Hall, New York.
Dieter, P., and Marmé, D. (1984). *J. Biol. Chem.* **259,** 184–189.
Dieter, P., and Marmé, D. (1986). *Plant Cell Physiol.* **27,** 1327–1333.
Dietz, K.-J., and Heber, U. (1984). *Biochim. Biophys. Acta* **767,** 432–443.
Dietz, K.-J., and Heber, U. (1986). *Biochim. Biophys. Acta* **848,** 392–401.
Doehlert, D. C. (1987). *Plant Sci.* **52,** 153–157.
Doehlert, D. C. (1990). *Plant Physiol.* **93,** 353–355.
Doehlert, D. C., and Huber, S. C. (1984). *Plant Physiol.* **76,** 250–253.
Doehlert, D. C., and Huber, S. C. (1985). *Biochim. Biophys. Acta* **830,** 267–273.
Donaldson, R. P. (1982). *Arch. Biochem. Biophys.* **215,** 274–279.
Doree, M. (1973). *Phytochemistry* **12,** 2101–2108.
Doremus, H. D. (1986). *Arch. Biochem. Biophys.* **250,** 112–119.
Doremus, H. D., and Blevins, D. G. (1988a). *Plant Physiol.* **87,** 36–40.
Doremus, H. D., and Blevins, D. G. (1988b). *Plant Physiol.* **87,** 41–45.
Doremus, H. D., and Jagendorf, A. T. (1985). *Plant Physiol.* **79,** 856–861.
Doremus, H. D., and Jagendorf, A. T. (1987). *Plant Physiol.* **83,** 657–658.
Douce, R., and Neuburger, M. (1989). *Annu. Rev. Plant Physiol. Plant Mol. Biol.* **40,** 371–414.
Douce, R., Brouquisse, R., and Journet, E.-P. (1987). *Biochem. Plants* **11,** 177–211.
Drew, M. C., Saglio, P. H., and Pradet, A. (1985). *Planta* **165,** 51–58.
Dry, I. B., and Wiskich, J. T. (1982). *Arch. Biochem. Biophys.* **217,** 72–79.
Dry, I. B., Bryce, J. H., and Wiskich, J. T. (1987). *Biochem. Plants* **11,** 213–252.
Duff, S. M. G., Moorhead, G. B. G., Lefebvre, D. D., and Plaxton, W. C. (1989). *Plant Physiol.* **90,** 1275–1278.
Dutta, I., Dutta, P. K., Smith, D. W., and O'Donovan, G. A. (1991). *J. Chromatogr.* **536,** 237–243.
Eastwell, K. C., and Stumpf, P. K. (1982). *Biochem. Biophys. Res. Commun.* **108,** 1690–1694.
Echeveria, E., Boyer, C. D., Thomas, P. A., Liu, K.-C., and Shannon, J. C. (1988). *Plant Physiol.* **86,** 786–792.
Edlund, B. (1971). *FEBS Lett.* **13,** 56–58.
Edwards, J., ap Rees, T., Wilson, P. M., and Morrell, S. (1984). *Planta* **162,** 188–191.
Edwards, J., Green, J. H., and ap Rees, T. (1988). *Phytochemistry* **27,** 1615–1620.
Entwistle, G., and ap Rees, T. (1988). *Biochem. J.* **255,** 391– 396.
Erecinska, M., Stubbs, M., Miyata, Y., Ditre, C. M., and Wilson, D. F. (1977). *Biochim. Biophys. Acta* **462,** 20–35.
Feingold, D. S. (1982). *Encycl. Plant Physiol.* **13A,** 3–76.
Feingold, D. S., and Avigad, G. (1980). *Biochem. Plants* **3,** 101–170.
Feingold, D. S., and Barber, G. A. (1990). *In* "Methods in Plant Biochemistry" (P. M. Dey, ed.), Vol. 2, pp. 39–78. Academic Press, San Diego, California.
Feller, W., Schimpf-Weiland, G., and Follmann, H. (1980). *Eur. J. Biochem.* **110,** 85–92.
Fernandez-Gomez, M. E., Stockert, J. C., Gonzalez-Fernandez, A., and Lopez-Saez, J. F. (1970). *Chromosoma* **29,** 1–11.

Fleischer, B. (1989). *In* "Methods in Enzymology" (S. Fleischer and B. Fleischer, eds.), Vol. 174, pp. 173–179. Academic Press, San Diego, California.
Flügge, U. I., and Heldt, H. W. (1984). *Trends Biochem. Sci.* **9,** 530–533.
Follmann, H. (1974). *Angew. Chem., Int. Ed. Engl.* **86,** 624–634.
Foyer, C., Walker, D., and Latzko, E. (1982). *Z. Pflanzenphysiol.* **107,** 457–465.
Frehner, M., Pozueta-Romero, J., and Akazawa, T. (1990). *Plant Physiol.* **94,** 538–544.
Frost, D. J., Read, S. M., Drake, R. R., Haley, B. E., and Wasserman, B. P. (1990). *J. Biol. Chem.* **265,** 2162–2167.
Fujimori, N., and Ashihara, H. (1990). *Phytochemistry* **29,** 3513–3516.
Gans, P., and Rébeillé, F. (1988). *Arch. Biochem. Biophys.* **260,** 109–117.
Gardeström, P. (1987). *FEBS Lett.* **212,** 114–118.
Gardeström, P., and Wigge, B. (1988). *Plant Physiol.* **88,** 69–76.
Gerhardt, R., Stitt, M., and Heldt, H. W. (1987). *Plant Physiol.* **83,** 399–407.
Gleason, F. K., and Hogenkamp, H. P. C. (1970). *J. Biol. Chem.* **245,** 4894–4899.
Golaszewski, T., Rytel, M., Rogozinski, J., and Szarkowski, J. W. (1975). *FEBS Lett.* **58,** 370–373.
Goller, M., Hampp, R., and Ziegler, H. (1982). *Planta* **156,** 255–263.
Goto, K. (1983). *Z. Naturforsch., C: Biosci.* **39C,** 73–84.
Gout, E., Bligny, R., Roby, C., and Douce, R. (1990). *Proc. Natl. Acad. Sci. U.S.A.* **87,** 4280–4283.
Graham, D. (1980). *Biochem. Plants* **2,** 525–575.
Graham, D., and Cooper, J. E. (1967). *Aust. J. Biol. Sci.* **20,** 319–327.
Grivell, A. R., and Jackson, J. F. (1976). *Biochem. J.* **155,** 571–581.
Gronewald, J. W., and Hanson, J. B. (1982). *Plant Physiol.* **69,** 1252–1256.
Gross, P., and ap Rees, T. (1986). *Planta* **167,** 140–145.
Grummt, F., Paul, D., and Grummt, I. (1977). *Eur. J. Biochem.* **76,** 7–12.
Grummt, I., and Grummt, F. (1976). *Cell* **7,** 447–453.
Grzelczak, Z., and Buchowicz, J. (1975). *Phytochemistry* **14,** 329–331.
Guranowski, A. (1979a). *Arch. Biochem. Biophys.* **196,** 220–226.
Guranowski, A. (1979b). *Biochim. Biophys. Acta* **569,** 13–22.
Guranowski, A. (1982). *Plant Physiol.* **70,** 344–349.
Guranowski, A., and Barankiewicz, J. (1979). *FEBS Lett.* **104,** 95–98.
Guranowski, A., and Pawelkiewicz, J. (1977). *Eur. J. Biochem.* **80,** 517–523.
Guranowski, A., and Schneider, Z. (1977). *Biochim. Biophys. Acta* **482,** 145–158.
Guranowski, A. B., Chiang, P. K., and Cantoni, G. L. (1981). *Eur. J. Biochem.* **114,** 293–299.
Gustafson, G. L., and Gander, J. E. (1972). *J. Biol. Chem.* **247,** 1387–1397.
Hampp, R. (1985). *Plant Physiol.* **79,** 690–694.
Hampp, R., Goller, M., and Ziegler, H. (1982). *Plant Physiol.* **69,** 448–455.
Hampp, R., Goller, M., and Füllgraf, H. (1984). *Plant Physiol.* **75,** 1017–1021.
Hampp, R., Goller, M., Füllgraf, H., and Eberle, I. (1985). *Plant Cell Physiol.* **26,** 99–108.
Harland, J., Jackson, J. F., and Yeoman, M. M. (1973). *J. Cell Sci.* **13,** 121–138.
Hatch, M. D. (1966). *Biochem. J.* **98,** 198–203.
Hatch, M. D. (1982). *Aust. J. Plant Physiol.* **9,** 287–296.
Hause, B., and Wasternack, C. (1988). *Planta* **176,** 51–59.
Hayashi, T., and Matsuda, K. (1981). *Agric. Biol. Chem.* **45,** 2907–2908.
Hayashi, T., Read, S. M., Bussell, J., Thelen, M., Lin, F.-C., Brown, R. M., and Delmer, D. P. (1987). *Plant Physiol.* **83,** 1054–1062.
Hayashi, T., Koyama, T., and Matsuda, K. (1988). *Plant Physiol.* **87,** 341–345.
Heber, U. W., and Santarius, K. A. (1965). *Biochim. Biophys. Acta* **109,** 390–408.
Heber, U., Takahama, U., Neimanis, S., and Shimizu-Takahama, M. (1982). *Biochim. Biophys. Acta* **679,** 287–299.
Heber, U., Neimanis, S., Dietz, K. J., and Vill, J. (1987). *Photosynth. Res.* **3,** 293–299.

Heldt, H. W., and Flügge, U. I. (1987). *Biochem. Plants* **12,** 49–85.
Hiles, R. A., and Byerrum, R. U. (1969). *Phytochemistry* **8,** 1927–1930.
Hill, M., Dupaix, A., Nhiri, M., Guyen, L., and Arrio, B. (1988). *FEBS Lett.* **230,** 47–50.
Hillyard, D., Rechsteiner, M., Manlapaz-Ramos, P., Imperial, J. S., Cruz, L. J., and Olivera, B. M. (1981). *J. Biol. Chem.* **256,** 8491–8497.
Hirose, F., and Ashihara, H. (1983a). *Z. Pflanzenphysiol.* **110,** 183–190.
Hirose, F., and Ashihara, H. (1983b). *Z. Naturforsch., C: Biosci.* **38C,** 375–381.
Hirose, F., and Ashihara, H. (1983c). *Z. Pflanzenphysiol.* **110,** 135–145.
Hirose, F., and Ashihara, H. (1984). *Physiol. Plant.* **60,** 532–538.
Hirose, S., and Komamine, A. (1989). *New Phytol.* **111,** 599–605.
Hirschberg, C. B., and Snider, M. D. (1987). *Annu. Rev. Biochem.* **56,** 63–87.
Hondo, T., Hara, A., and Funaguma, T. (1983). *Plant Cell Physiol.* **24,** 1535–1543.
Hooks, M. A., Clark, R. A., Niemann, R. H, and Roberts, J. K. M. (1989). *Plant Physiol.* **89,** 963–969.
Hopper, J. E., and Dickinson, D. B. (1972). *Arch. Biochem. Biophys.* **148,** 523–535.
Hovemann, B., and Follmann, H. (1979). *Biochim. Biophys. Acta* **561,** 42–52.
Huber, S. C., and Akazawa, T. (1986). *Plant Physiol.* **81,** 1008–1013.
Huber, S. C., and Edwards, G. E. (1976). *Biochim. Biophys. Acta* **440,** 675–687.
Huber, S. C., and Huber, J. L. (1990). *Arch. Biochem. Biophys.* **282,** 421–426.
Ingold, E., and Seitz, H. U. (1986). *Z. Naturforsch., C: Biosci.* **41C,** 409–420.
Isherwood, F. A., and Selvendran, R. R. (1970). *Phytochemistry* **9,** 2265–2269.
Jackson, J. F., and Atkinson, M. R. (1966). *Biochem. J.* **101,** 208–213.
John, K. V., Schutzbach, J. S., and Ankel, H. (1977). *J. Biol. Chem.* **252,** 8013–8017.
Johnson, R. D., and Waller, G. R. (1974). *Phytochemistry* **13,** 1493–1500.
Kanabus, J., Bressan, R. A., and Carpita, N. C. (1986). *Plant Physiol.* **82,** 363–368.
Kanamori-Fukuda, I., and Ashihara, H. (1981). *J. Exp. Bot.* **32,** 69–78.
Kanamori, I., Ashihara, H., and Komamine, A. (1979). *Z. Pflanzenphysiol.* **93,** 437–448.
Kanamori, I., Ashihara, H., and Komamine, A. (1980). *Z. Pflanzenphysiol.* **96,** 7–16.
Kansara, M. S., Ramdas, J., and Srivastava, S. K. (1989). *J. Plant Physiol.* **134,** 603–607.
Kapoor, M., and Waygood, E. R. (1965). *Can. J. Biochem.* **43,** 143–151.
Kaushal, G. P., and Elbein, A. D. (1986). *Plant Physiol.* **82,** 748–752.
Kaushal, G. P., and Elbein, A. D. (1987). *Biochemistry* **26,** 7953–7960.
Kaushal, G. P., Szumilo, T., and Elbein, A. D. (1988). *Biochem. Plants* **14,** 421–463.
Kelly, G. J. (1983). *Trends Biochem. Sci.* **8,** 38.
Kim, W. T., Franceschi, V. R., Okita, T. W., Robinson, N. L., Morell, M., and Preiss, J. (1989). *Plant Physiol.* **91,** 217–220.
King, P. J. (1980). *Adv. Biol.* **18,** 1–38.
King, P. J., and Street, H. E. (1977). In "Plant Tissue and Cell Culture" (H.E. Street, ed.), 2nd ed., pp. 307–387. Blackwell, Oxford.
Kirsch, T., Rojahn, B., and Kindl, H. (1989). *Z. Naturforsch., C: Biosci.* **44C,** 937–945.
Kleczkowski, L. A., and Randall, D. D. (1986). *Plant Physiol.* **81,** 1110–1114.
Kleczkowski, L. A., and Randall, D. D. (1987). *J. Exp. Bot.* **38,** 1440–1445.
Kleczkowski, L. A., and Randall, D. D. (1991). *J. Exp. Bot.* **42,** 537–540.
Kleczkowski, L. A., Randall, D. D., and Zahler, W. L. (1990). *Z. Naturforsch., C: Biosci.* **45C,** 607–613.
Klein, B., and Follmann, H. (1988). *Z. Naturforsch., C: Biosci.* **43C,** 377–385.
Kobahashi, Y., Inoue, Y., Furuya, F., Shibata, K., and Heber, U. (1979). *Planta* **147,** 69–75.
Kohl, D. H., Schubert, K. R., Carter, M. B., Hagedorn, C. H., and Shearer, G. (1988). *Proc. Natl. Acad. Sci. U.S.A.* **85,** 2036–2040.
Kohl, D. H., Lin, J.-J., Shearer, G., and Schubert, K. R. (1990). *Plant Physiol.* **94,** 1258–1264.

Komossa, D., and Barz, W. (1988). *Z. Naturforsch., C: Biosci.* **43C,** 843–849.
Konstantinov, Y. M., Podsosonny, V. A., and Lutsenko, G. N. (1988). *Physiol. Plant.* **72,** 403–406.
Kosegarten, H., Judel, G. K., and Mengel, K. (1988). *J. Plant Physiol.* **133,** 126–128.
Köster, S., Upmeier, B., Komossa, D., and Barz, W. (1989). *Z. Naturforsch., C: Biosci.* **44C,** 623–628.
Krömer, S., and Heldt, H. W. (1991a). *Biochim. Biophys. Acta* **1057,** 42–50.
Krömer, S., and Heldt, H. W. (1991b). *Plant Physiol.* **95,** 1270–1276.
Krömer, S., Stitt, M., and Heldt, H. W. (1988). *FEBS Lett.* **226,** 352–356.
Kruger, N. J., and ap Rees, T. (1983). *Phytochemistry* **22,** 1891–1898.
Kubota, K., and Ashihara, H. (1990). *Biochim. Biophys. Acta* **1036,** 138–142.
Ladror, U. S., Latshaw, S. P., and Marcus, F. (1990). *Eur. J. Biochem.* **189,** 89–94.
Lang, L., Reitmann, M., Tang, J., Roberts, R. M., and Kornfeld, S. (1984). *J. Biol. Chem.* **259,** 14663–14671.
Laval-Martin, D. L., Carre, I. A., Barbera, S. J., and Edmunds , L. N., Jr. (1990). *Arch. Biochem. Biophys.* **276,** 433–441.
Lee, R. B., and Ratcliffe, R. G. (1983). *J. Exp. Bot.* **34,** 1222–1244.
Le Floc'h, F., and Lafleuriel, J. (1978). *Phytochemistry* **17,** 643–646.
Le Floc'h, F., and Lafleuriel, J. (1981a). *Phytochemistry* **20,** 2127–2129.
Le Floc'h, F., and Lafleuriel, J. (1981b). *Z. Pflanzenphysiol.* **104,** 331–340.
Le Floc'h, F., and Lafleuriel, J. (1983). *Z. Pflanzenphysiol.* **113,** 61–71.
Le Floc'h, F., and Lafleuriel, J. (1990). *Plant Physiol. Biochem.* **28,** 191–201.
Le Floc'h, F., Lafleuriel, J., and Guillot, A. (1982). *Plant Sci. Lett.* **27,** 309–316.
Leinhos, V., Krauss, G.-J., and Glund, K. (1986). *Plant Sci.* **47,** 15–20.
Leszczynska, D., Schneider, Z., Tomaszewski, M., and Mackowiak, M. (1984). *Ann. Bot. (London)* **54,** 847–849.
Li, X.-N., and Ashihara, H. (1989). *Ann. Bot. (London)* **64,** 33–36.
Li, X.-N., and Ashihara, H. (1990). *Phytochemistry* **29,** 497–500.
Liedvogel, B., and Kleinig, H. (1980). *Planta* **150,** 170–173.
Lilley, R. M., Chon, C. J., Mosbach, A., and Heldt, H. W. (1977). *Biochim. Biophys. Acta* **460,** 259–272.
Lin, F. F., and Lovatt, C. J. (1986). *In Vitro Cell. Dev. Biol.* **22,** 1–5.
Lovatt, C. J. (1983). *Plant Physiol.* **73,** 766–772.
Lovatt, C. J. (1985). *Plant Cell Environ.* **8,** 59–63.
Lovatt, C. J., and Cheng, A. H. (1984). *Plant Physiol.* **75,** 511–515.
Lovatt, C. J., Albert, L. S., and Tremblay, G. C. (1979). *Plant Physiol.* **64,** 562–569.
Lucas, K., Boland, M. J., and Schubert, K. R. (1983). *Arch. Biochem. Biophys.* **226,** 190–197.
Lucas, W. J., and Madore, M. A. (1988). *Biochem. Plants* **14,** 35–84.
Macdonald, F. D., and ap Rees, T. (1983). *Biochim. Biophys. Acta* **755,** 81–89.
Manetas, Y., Stamatakis, K., and Samaras, Y. (1986). *J. Plant Physiol.* **124,** 165–170.
Maretzki, A., and Thom, M. (1987). *Plant Physiol.* **83,** 235–237.
Maretzki, A., and Thom, M. (1988). *Plant Physiol.* **88,** 266–269.
Martin, B. A., and Tolbert, N. E. (1983). *Plant Physiol.* **73,** 464–470.
Meyer, R., and Wagner, K. G. (1985a). *Planta* **166,** 439–445.
Meyer, R., and Wagner, K. G. (1985b). *Anal. Biochem.* **148,** 269–276.
Meyer, R., and Wagner, K. G. (1985c). *Physiol. Plant.* **65,** 439–445.
Meyer, R., and Wagner, K. G. (1986). *Physiol. Plant.* **67,** 666–672.
Michalke, B., and Schnabl, H. (1987). *J. Plant Physiol.* **130,** 243–253.
Miersch, J., Krauss, G.-J., and Metzger, U. (1986). *J. Plant Physiol.* **122,** 55–66.
Miginiac-Maslow, M., and Hoarau, A. (1982). *Z. Pflanzenphysiol.* **107,** 427–436.

Miginiac-Maslow, M., Mathieu, Y., Nato, A., and Hoarau, A. (1981). *Proc. Int. Congr. Photosynth.* **5,** 977–984.
Miginiac-Maslow, M., Nguyen, J., and Hoarau, A. (1986). *J. Plant Physiol.* **123,** 69–77.
Milla, M., Capasso, J., and Hirschberg, C. B. (1989). *Biochem. Soc. Trans.* **17,** 447–448.
Mitsui, K., and Ashihara, H. (1988). *Plant Cell Physiol.* **29,** 1177–1183.
Mitsui, T., Sado, T., Terada, A., and Igaue, I. (1989). *Agric. Biol. Chem.* **53,** 1991–1993.
Mittal, R., Das, J., and Sharma, C. B. (1988). *Plant Sci.* **55,** 93–101.
Miyamoto, J., Ishigami, T., Hayashi, T., Nakajima, T., Ichishima, E., and Matsuda, K. (1989). *Agric. Biol. Chem.* **53,** 1987–1988.
Moffatt, B., and Somerville, C. (1988). *Plant Physiol.* **86,** 1150–1154.
Moore, B. d., Ku, M. S. B., and Edwards, G. E. (1984). *Plant Sci. Lett.* **35,** 127–138.
Morell, M., and Copeland, L. (1985). *Plant Physiol.* **78,** 149–154.
Morrow, D. L., and Lucas, W. J. (1987). *Plant Physiol.* **84,** 565–567.
Munch-Petersen, A. (1983). "Metabolism of Nucleotides, Nucleosides and Nucleobases in Microorganisms." Academic Press, New York.
Murata, T. (1977). *Agric. Biol. Chem.* **41,** 1995–2002.
Muto, S. (1983). *Z. Pflanzenphysiol.* **109,** 385–393.
Muto, S., Miyachi, S., Usuda, H., Edwards, G. E., and Bassham, J. A. (1981). *Plant Physiol.* **68,** 324–328.
Negishi, O., Ozawa, T., and Imagawa, H. (1985a). *Agric. Biol. Chem.* **49,** 251–253.
Negishi, O., Ozawa, T., and Imagawa, H. (1985b). *Agric. Biol. Chem.* **49,** 887–890.
Negishi, O., Ozawa, T., and Imagawa, H. (1985c). *Agric. Biol. Chem.* **49,** 2221–2222.
Negishi, O., Ozawa, T., and Imagawa, H. (1988). *Agric. Biol. Chem.* **52,** 169–175.
Neuburger, M., and Douce, R. (1983). *Biochem. J.* **216,** 443–450.
Neuburger, M., Day, D. A., and Douce, R. (1985). *Plant Physiol.* **78,** 405–410.
Neuhard, J., and Nygaard, P. (1987). *In* "*Escherichia coli* and *Salmonella typhimurium:* Cellular and Molecular Biology" (F. C. Neidhard, J. L. Ingraham, K. B. Low, B. Magasanik, M. Schaechter, and H. E. Umbarger, eds.), Vol. 1, pp. 445–473. Am. Soc. Microbiol., Washington, D.C.
Neuhaus, H. E., Krause, K.-P., and Stitt, M. (1990a). *Phytochemistry* **29,** 3411–3415.
Neuhaus, H. E., Quick, W. P., Siegl, G., and Stitt, M. (1990b). *Planta* **181,** 583–592.
Ngernprasirtsiri, J., Takabe, T., and Akazawa, T. (1989). *Plant Physiol.* **89,** 1024–1027.
Nguyen, J., and Feierabend, J. (1978). *Plant Sci. Lett.* **13,** 125–132.
Nielsen, E., and Cella, R. (1988). *Plant Cell Physiol.* **29,** 503–508.
Nobusawa, E., and Ashihara, H. (1983). *Int. J. Biochem.* **15,** 1059–1065.
Northcote, D. H. (1982). *Encycl. Plant Physiol.* **14A,** 637–655.
Nougarède, A., Landré, P., Rembur, J., and Hernandez, M. N. (1985). *Can. J. Bot.* **63,** 309–323.
Nürnberger, T., Abel, S., Jost, W., and Glund, K. (1990). *Plant Physiol.* **92,** 970–976.
Nygaard, P. (1972). *Physiol. Plant.* **26,** 29–33.
O'Neal, T. D., and Naylor, A. W. (1976). *Plant Physiol.* **57,** 23–28.
Okamura, S., Sueki, K., and Nishi, A. (1975). *Physiol. Plant.* **33,** 251–255.
Okita, T. W., Greenberg, E., Kuhn, D. N., and Preiss, J. (1979). *Plant Physiol.* **64,** 187–192.
Oliver, D. J., Neuburger, M., Bourguignon, J., and Douce, R. (1990). *Physiol. Plant.* **80,** 487–491.
Ong, B. L., and Jackson, J. F. (1972a). *Biochem. J.* **129,** 571–581.
Ong, B. L., and Jackson, J. F. (1972b). *Biochem. J.* **129,** 583–593.
Orgen, W. L., and Krogmann, D. W. (1965). *J. Biol. Chem.* **240,** 4603–4608.
Otozai, K., Taniguchi, H., and Nakamura, M. (1973). *Agric. Biol. Chem.* **37,** 531–537.
Palmer, J. M., and Moller, I. M. (1982). *Trends Biochem. Sci.* **7,** 258–261.

Pardo, E. G., and Gutiérrez, C. (1990). *Exp. Cell Res.* **186,** 90–98.
Parish, R. W. (1972). *Planta* **104,** 247–251.
Parker, N. F., and Jackson, J. F. (1981). *Plant Physiol.* **67,** 363–366.
Pérez-Vicente, R., Pineda, M., and Cárdenas, J. (1988). *Physiol. Plant.* **72,** 101–107.
Polya, G. M. (1974). *Proc. Natl. Acad. Sci. U.S.A.* **71,** 1299–1303.
Pradet, A., and Raymond, P. (1983). *Annu. Rev. Plant Physiol.* **34,** 199–224.
Prakash, S., Singh, P., and Naik, M. S. (1986). *J. Plant Physiol.* **122,** 303–314.
Prasher, D. C., Carr, M. C., Ives, D. H., Tsai, T. C., and Frey, P. A. (1982). *J. Biol. Chem.* **257,** 4931–4939.
Preiss, J. (1982). *Annu. Rev. Plant Physiol.* **33,** 431–454.
Preiss, J. (1988). *Biochem. Plants* **14,** 181–254.
Preisser, J., and Komor, E. (1988). *Plant Physiol.* **88,** 259–265.
Price-Jones, M. J., and Harwood, J. L. (1983). *Biochem. J.* **216,** 627–631.
Quick, P., Neuhaus, E., Feil, R., and Stitt, M. (1989). *Biochim. Biophys. Acta* **973,** 263–271.
Quigley, F., Rickauer, M., and Tanner, W. (1984). *Plant Sci. Lett.* **35,** 43–49.
Rainbird, R. M., and Atkins, C. A. (1981). *Biochim. Biophys. Acta* **659,** 132–140.
Rapaport, E., Garcia-Blanco, M. A., and Zamecnik, P. C. (1979). *Proc. Natl. Acad. Sci. U.S.A.* **76,** 1643–1647.
Rasmusson, A. G., and Moller, I. M. (1990). *Plant Physiol.* **94,** 1012–1018.
Raymond, P., Gidrol, X., Salon, C., and Pradet, A. (1987). *Biochem. Plants* **11,** 129–176.
Rébeillé, F., and Hatch, M. D. (1986). *Arch. Biochem. Biophys.* **249,** 171–179.
Rébeillé, F., Bligny, R., and Douce, R. (1982). *Arch. Biochem. Biophys.* **219,** 371–378.
Rébeillé, F., Bligny, R., Martin, J.-B., and Douce, R. (1983). *Arch. Biochem. Biophys.* **225,** 143–148.
Reinbothe, H., Miersch, J., and Mothes, K. (1981). *Compr. Biochem.* **19A,** 51–163.
Reynolds, P. H. S., Boland, M. J., Blevins, D. G., Randall, D. D., and Schubert, K. R. (1982). *Trends Biochem. Sci.* 366–368.
Reynolds, P. H. S., Blevins, D. G., and Randall, D. D. (1984). *Arch. Biochem. Biophys.* **229,** 623–631.
Richard, J. P., Prasher, D. C., Ives, D. H., and Frey, P. A. (1979). *J. Biol. Chem.* **254,** 4339–4341.
Riedell, W. E., and Miernyk, J. A. (1988). *Plant Physiol.* **87,** 420–426.
Roberts, J. K. M., Lane, A. N., Clark, R. A., and Nieman, R. H. (1985). *Arch. Biochem. Biophys.* **240,** 712–722.
Robinson, S. P., and Wiskich, J. T. (1977). *Plant Physiol.* **59,** 422–427.
Roby, C., Martin, J.-B., Bligny, R., and Douce, R. (1987). *J. Biol. Chem.* **262,** 5000–5007.
Rodionova, M. A., Kholodenko, N. Y., and Makarov, A. D. (1978). *Fiziol. Rast.* (*Moscow*) **25,** 731–734.
Roeske, C. A., and Chollet, R. (1989). *Plant Physiol.* **90,** 330–337.
Ross, C., and Murray, M. G. (1971). *Plant Physiol.* **48,** 626–630.
Ross, C. W. (1981). *Biochem. Plants* **6,** 169–205.
Rustin, P., Neuburger, M., Douce, R., and Lance, C. (1987). *In* "Plant Mitochondria" (A. L. Moore and R. B. Beechey, eds.), pp. 89–92. Plenum, New York.
Sabina, R. L., Dalke, P., Hanks, A. R., Magill, J. M., and Magill, C. W. (1981). *Can. J. Biochem.* **59,** 899–905.
Salerno, G. L. (1986). *Plant Sci.* **44,** 111–117.
Sanchez, J., and Heldt, H. W. (1989). *Bot. Acta* **102,** 186–188
Sanchez, J., and Heldt, H. W. (1990). *Plant Physiol.* **92,** 684–689.
Santarius, K. A., and Heber, U. (1965). *Biochim. Biophys. Acta* **102,** 39–54.
Sasamoto, H., and Ashihara, H. (1988). *Int. J. Biochem.* **20,** 87–92.

Sasamoto, H., Saito, K., and Ashihara, H. (1987). *Ann. Bot.* (*London*) **60,** 417–420.

Sawert, A. (1988). Ph.D. thesis, Technical University of Braunschweig, Braunschweig, Germany.

Sawert, A., Backer, A. I., Plank-Schumacher, K.-H., and Wagner, K. G. (1987). *J. Plant Physiol.* **127,** 183–186.

Sawert, A., Backer, A. I., and Wagner, K. G. (1988). *Plant Cell Physiol.* **29,** 61–65.

Sawhney, S. K., Naik, M. S., and Nicholas, D. J. D. (1978). *Nature* (*London*) **272,** 647–648.

Scheibe, R. (1987). *Physiol. Plant.* **71,** 393–400.

Schmidt, C. L., Danneel, H. J., Schultz, G., and Buchanan, B. B. (1990). *Plant Physiol.* **93,** 758–766.

Schubert, K. R. (1986). *Annu. Rev. Plant Physiol.* **37,** 539–574.

Schvartzman, J. B., Kriner, D. B., and van't Hof, J. (1984). *Exp. Cell Res.* **150,** 379–389.

Sharma, C. B., Mittal, R., and Tanner, W. (1986). *Biochim. Biophys. Acta* **884,** 567–577.

Shelp, B. J., and Atkins, C. A. (1983). *Plant Physiol.* **72,** 1029–1034.

Shelp, B. J., Atkins, C. A., Storer, P. J., and Canvin, D. T. (1983). *Arch. Biochem. Biophys.* **224,** 429–441.

Shen, S. R. C., and Schmidt, R. R. (1966). *Arch. Biochem. Biophys.* **115,** 13–20.

Shibata, H., Ochiai, H., Sawa, Y., and Miyoshi, S. (1986). *Plant Physiol.* **80,** 126–129.

Shimazaki, A., and Ashihara, H. (1982). *Ann. Bot.* (*London*) **50,** 531–534.

Shimazaki, A., Hirose, F., and Ashihara, H. (1982). *Z. Pflanzenphysiol.* **106,** 191–198.

Shimizu, T., Clifton, A., Komamine, A., and Fowler, M. W. (1977). *Physiol. Plant.* **40,** 125–129.

Shimosaka, M., Fukuda, Y., Murata, K., and Kimura, A. (1984). *Agric. Biol. Chem.* **48,** 1303–1310.

Simon, P., Dieter, P., Bonzon, M., Greppin, H., and Marmé, D. (1982). *Plant Cell Rep.* **1,** 119–122.

Simon, P., Bonzon, M., Greppin, H., and Marmé, D. (1984). *FEBS Lett.* **167,** 332–338.

Singh, P., and Naik, M. S. (1984). *FEBS Lett.* **165,** 145–150.

Smith, A. M., Neuhaus, H. E., and Stitt, M. (1990). *Planta* **181,** 310–315.

Stitt, M. (1990). *Annu. Rev. Plant Physiol. Plant Mol. Biol.* **41,** 153–185.

Stitt, M., and Grosse, H. (1988). *J. Plant Physiol.* **133,** 392–400.

Stitt, M., Lilley, R. M., and Heldt, H. W. (1982). *Plant Physiol.* **70,** 971–977.

Stitt, M., Wirtz, W., and Heldt, H. W. (1983a). *Plant Physiol.* **72,** 767–774.

Stitt, M., Gerhardt, R., Kürzel, B., and Heldt, H. W. (1983b). *Plant Physiol.* **72,** 1139–1141.

Stitt, M., Wirtz, W., Gerhardt, R., Heldt, H. W., Spencer, C., Walker, D., and Foyer, C. (1985). *Planta* **166,** 354–364.

Stitt, M., Huber, S., and Kerr, P. (1987). *Biochem. Plants* **10,** 327–409.

Stitt, M., Wilke, I., Feil, R., and Heldt, H. W. (1988). *Planta* **174,** 217–230.

Suzuki, T., and Takahashi, E. (1975). *Biochem. J.* **146,** 79–85.

Suzuki, T., and Takahashi, E. (1976). *Phytochemistry* **15,** 1235–1239.

Suzuki, T., and Takahashi, E. (1977). *Drug Metab. Rev.* **6,** 213–242.

Suzuki, T., and Waller, G. R. (1988). *In* "Analysis of Nonalcoholic Beverages" (H. F. Linskens and J. F. Jackson, eds.), pp. 184–220. Springer-Verlag, Berlin.

Swinton, D. C., and Chiang, K. S. (1979). *Mol. Gen. Genet.* **176,** 399–409.

Taguchi, H., Nishitani, H., Okumura, K., Shimabayashi, Y., and Iwai, K. (1989). *Agric. Biol. Chem.* **53,** 1543–1549.

Takahama, U., Shimizu-Takahama, M., and Heber, U. (1981). *Biochim. Biophys. Acta* **637,** 530–539.

Tarr, J. B., and Arditti, J. (1982). *Plant Physiol.* **69,** 553–556.

Theimer, R. R., and Beevers, H. (1971). *Plant Physiol.* **47,** 246–251.

Thom, M., and Maretzki, A. (1985). *Proc. Natl. Acad. Sci. U.S.A.* **82,** 4697–4701.

Thom, M., and Maretzki, A. (1987). *J. Plant Physiol.* **130,** 413–421.
Thom, M., and Maretzki, A. (1989). *Plant Physiol. Biochem.* **27,** 87–92.
Thom, M., Leigh, R. A., and Maretzki, A. (1986). *Planta* **167,** 410–413.
Thomas, R. J., and Schrader, L. E. (1981). *Phytochemistry* **20,** 361–371.
Thomas, R. J., Meyers, S. P., and Schrader, L. E. (1983). *Phytochemistry* **22,** 1117–1120.
Tobin, A., Djerdjour, B., Journet, E., Neuburger, M., and Douce, R. (1980). *Plant Physiol.* **66,** 225–229.
Torossian, K., and Maclachlan, G. (1987). *Biochim. Biophys. Acta* **925,** 305–313.
Tramontano, W. A., Hartnett, C. M., Lynn, D. G., and Evans, L. S. (1982). *Phytochemistry* **21,** 1201–1206.
Tramontano, W. A., Lynn, D. G., and Evans, L. S. (1983a). *Phytochemistry* **22,** 343–346.
Tramontano, W. A., Lynn, D. G., and Evans, L. S. (1983b). *Phytochemistry* **22,** 673–678.
Tritz, G. J. (1987). *In* "*Escherichia coli* and *Salmonella typhimurium:* Cellular and Molecular Biology" (F. C. Neidhard, J. L. Ingraham, K. B. Low, B. Magasanik, M. Schaechter, and H. E. Umbarger, eds.), Vol. 1, pp. 557–563. Am. Soc. Microbiol., Washington, D.C.
Turner, J. F., and Turner, D. H. (1980). *Biochem. Plants* **2,** 279–316.
Ukaji, T., and Ashihara, H. (1986). *Z. Naturforsch., C: Biosci.* **41C,** 1045–1051.
Ukaji, T., and Ashihara, H. (1987a). *Int. J. Biochem.* **19,** 1127–1131.
Ukaji, T., and Ashihara, H. (1987b). *Ann. Bot. (London)* **60,** 109–114.
Ukaji, T., and Ashihara, H. (1987c). *Plant Physiol., Suppl.* **83,** 975.
Ukaji, T., Hirose, F., and Ashihara, H. (1986). *J. Plant Physiol.* **125,** 191–197.
Upmeier, B., Thomzik, J. E., and Barz, W. (1988a). *Phytochemistry* **27,** 3489–3493.
Upmeier, B., Gross, W., Köster, S., and Barz, W. (1988b). *Arch. Biochem. Biophys.* **262,** 445–454.
Usuda, H. (1988). *Plant Physiol.* **88,** 1461–1468.
Usuda, H. (1990). *Curr. Res. Photosynth.* **4,** 255–258.
Vassef, A. A., Flora, J. B., Weeks, J. G., Bibbs, B. S., and Schmidt, R. R. (1973). *J. Biol. Chem.* **248,** 1976–1985.
Vella, J., and Copeland, L. (1990). *Physiol. Plant.* **78,** 140–146.
Voss, M., and Weidner, M. (1988). *Planta* **173,** 96–103.
Wagner, R. (1987). *Pharm. Unserer Zeit* **16,** 53–59.
Wagner, R., and Wagner, K. G. (1984). *Phytochemistry* **23,** 1881–1883.
Wagner, R., and Wagner, K. G. (1985). *Planta* **165,** 532–537.
Wagner, R., Feth, F., and Wagner, K. G. (1986a). *Physiol. Plant.* **68,** 667–672.
Wagner, R., Feth, F., and Wagner, K. G. (1986b). *Planta* **167,** 226–232.
Wagner, R., Feth, F., and Wagner, K. G. (1986c). *Planta* **168,** 408–413.
Walker, D. A., and Sivak, M. N. (1986). *Trends Biochem. Sci.* **11,** 176–179.
Walther, R., Krauss, G. J., and Reinbothe, H. (1981). *Biochem. Physiol. Pflanz.* **176,** 116–128.
Walther, R., Wald, K., Glund, K., and Tewes, A. (1984). *J. Plant Physiol.* **116,** 301–311.
Walther, R., Koch, G., Wasternack, C., and Neumann, D. (1989). *Biochem. Physiol. Pflanz.* **185,** 201–209.
Wasternack, C. (1975). *Plant Sci. Lett.* **4,** 353–360.
Wasternack, C. (1980). *Pharmacol. Ther.* **8,** 629–651.
Wasternack, C. (1982). *Encycl. Plant Physiol.* **14B,** 264–301.
Wasternack, C., and Benndorf, R. (1983). *Biol. Zentralbl.* **102,** 1–16.
Watt, M. P., Gray, V. M., and Cresswell, C. F. (1987). *Planta* **172,** 548–554.
Webb, M. A., and Newcomb, E. H. (1987). *Planta* **172,** 162–175.
Weber, G., Burt, M. E., Jackson, R. C., Prajda, N., Lui, M. S., and Takeda, E. (1983). *Cancer Res.* **43,** 1019–1023.
Weil, M., and Rausch, T. (1990). *Plant Physiol.* **94,** 1575–1581.

Weiner, H., Stitt, M., and Heldt, H. W. (1987). *Biochim. Biophys. Acta* **893,** 13–21.
Wendler, R., Veith, R., Dancer, J., Stitt, M., and Komor, E. (1990). *Planta* **183,** 31–39.
Wilder, J. P., Sae-Lee, J. A., Mitchell, E. D., and Gholson, R. K. (1984). *Biochem. Biophys. Res. Commun.* **123,** 836–841.
Willmitzer, L., and Wagner, K. G. (1982). *In* "ADP-Ribosylation Reactions: Biology and Medicine" (O. Hayaishi and K. Ueda, eds.), pp. 241–252. Academic Press, New York.
Winkler, R. G., Blevins, D. G., Polacco, J. C., and Randall, D. D. (1987). *Plant Physiol.* **83,** 585–591.
Winkler, R. G., Blevins, D. G., Polacco, J. C., and Randall, D. D. (1988). *Trends Biochem. Sci.* **13,** 97–100.
Wirtz, W., Stitt, M., and Heldt, H. W. (1980). *Plant Physiol.* **66,** 187–193.
Wolcott, J. H., and Ross, C. (1967). *Plant Physiol.* **42,** 275–279.
Woldegiorgis, G., Voss, S., Shrago, E., Werner-Washburne, M., and Keegstra, K. (1983). *Biochem. Biophys. Res. Commun.* **116,** 945–951.
Woo, K. C., Atkins, C. A., and Pate, J. S. (1980). *Plant Physiol.* **66,** 735–739.
Wylegalla, C., Meyer, R., and Wagner, K. G. (1985). *Planta* **166,** 446–451.
Wyngaarden, J. B. (1976). *Adv. Enzyme Regul.* **14,** 25–42.
Xu, D.-P., Sung, S.-J. S., Loboda, T., Kormanik, P. P., and Black, C. C. (1989). *Plant Physiol.* **90,** 635–642.
Yamashita, Y., and Ashihara, H. (1988). *Z. Naturforsch., C: Biosci.* **43C,** 827–834.
Yazaki, Y., Asukagawa, N., Ishikawa, Y., Ohta, E., and Sakata, M. (1988). *Plant Cell Physiol.* **29,** 919–924.
Yeh, G. C., and Phang, J. M. (1988). *J. Biol. Chem.* **263,** 13083–13089.
Yeh, G. C., Roth, E. F., Phang, J. M., Harris, S. C., Nagel, R. L., and Rinaldi, A. (1984). *J. Biol. Chem.* **259,** 5454–5458.
Yin, Z.-H-, Dietz, K.-J., and Heber, U. (1990). *Planta* **182,** 262–269.
Yokozawa, T., and Oura, H. (1986). *Agric. Biol. Chem.* **50,** 1317–1319.
Yon, R. J. (1984). *Biochem. J.* **221,** 281–287.
Yoshino, M., and Murakami, K. (1980). *Z. Pflanzenphysiol.* **99,** 331–338.
Yoshino, M., and Murakami, K. (1981). *Biochim. Biophys. Acta* **672,** 16–20.

An addendum to this chapter appears on page 291.

Compartmentation of Nitrogen Assimilation in Higher Plants[1]

K. A. Sechley,* T. Yamaya,† and A. Oaks,*

* Department of Botany, University of Guelph, Guelph, Ontario, Canada N1G 2W1

† Department of Agricultural Chemistry, Tohoku University, Sendai 981, Japan

I. Introduction

Nitrogen, a major constituent in plants, is found principally in nucleic acids, proteins, and chlorophyll. It is generally considered to be the main rate-limiting element in plant growth and because of this much research has been directed to (1) understanding the nitrogenase reaction that converts atmospheric dinitrogen to NH_4^+, (2) understanding the implications of adding nitrogen fertilizer to agricultural lands, and (3) understanding the metabolic processes involved in nitrogen utilization within the plant.

NO_3^- and NH_4^+ are the common forms of nitrogen added to the soil. However, because of the ubiquitous distribution of *Nitrosomonas* and *Nitrobacter* species, NO_3^- is, in fact, the dominant form of nitrogen in most soils (Haynes, 1986; Marschner, 1986). NO_3^- taken up from the soil is either converted to NH_4^+ in the roots by the action of nitrate (NR) and nitrite (NiR) reductases or is transported to the shoot before assimilation. Typically, temperate legume plants reduce NO_3^- in the roots whereas other plant species, cereals, for example, transfer most of their accumulated NO_3^- to the shoots (Andrews, 1986; Atkins and Beevers, 1990; Pate, 1980, 1989), where it is reduced. It may even be that NO_3^- is assimilated more efficiently in the leaves of these "other plants," or that plants that do assimilate NO_3^- in their roots have also developed the capacity in roots to compete effectively for fixed carbon derived from photosynthate (Layzell *et al.*, 1990; Oaks, 1992).

[1] We would like to dedicate this review to Harry Beevers in honor of his retirement and in recognition of his many contributions to the understanding of compartmentation in plant metabolism.

In contrast to NO_3^- additions, when plants are given NH_4^+, that NH_4^+ is reduced in the roots. Experimentally this is demonstrated by the lack of NH_4^+ in the xylem sap and by the accumulation of glutamine and asparagine (Lewis, 1986). If NH_4^+ is the dominant form of nitrogen in the soil, as it is in acid soils, or as it is, at least temporarily, when ammonium-based fertilizers or urea are added to the agricultural soils, the root system must be able to overcome a potential ammonium toxicity and it must also have the capacity to direct more of the carbon assimilated in the leaf system to root metabolism.

Carbon and nitrogen partitioning has been considered in detail by Pate (1980, 1989) and by Atkins and Beevers (1990). It is well established that reduced nitrogen compounds in addition to photosynthate circulate in the plant and that the final destination of both nitrogen and carbon is determined by the competition between various metabolic sinks. The root system, younger growing leaves, or the developing fruit represent major sinks. Because the metabolic intensity of these various sinks alters with age, so too the direction of carbon and nitrogen mobilization is altered as well. Nodules on soybean roots represent a metabolic sink, and are strong enough to attract sufficient photosynthate to permit a rapid conversion of dinitrogen first into NH_4^+ and then into organically combined nitrogen (e.g., the amides or ureides). When pod development starts, however, they outcompete the nodules for available nutrients and the nodules begin to senesce (Thibodeau and Jaworski, 1975). This type of interaction is apparent in all plants. However, it has been most effectively studied in an experimental setting in soybean plants (Staswick, 1990; Thibodeau and Jaworski, 1975).

In addition to supplying carbon for nitrogen assimilation, carbohydrates derived from photosynthesis supply the energy (in the form of ATP) and reductant (NADH, NADPH, or reduced ferredoxin) required for the assimilation of nitrogen. These requirements are illustrated superficially in Figs. 1 and 2.

NO_3^- uptake, itself, is an energy-requiring process (Glass, 1988). Its reduction to ammonia in leaves requires both NADH and reduced ferredoxin (Beevers and Hageman, 1980). In roots, NAD(P)H is required, both in the reductase reaction (Long and Oaks, 1990; D. M. Long and A. Oaks, unpublished observations), and in the reduction of ferredoxin (Emes and Bowsher, 1991). The assimilation of the resultant NH_4^+ into glutamine requires ATP (Fig. 1a). The amide nitrogen of glutamine is used in many reactions (see Joy, 1988, for a detailed summary) (Fig. 1b). In plants, asparagine (Rognes, 1975), carbamoyl phosphate (Davis, 1986; Shargool *et al.*, 1988), ureide biosynthesis (Vance and Griffith, 1990), and the formation of tryptophan (Gilchrist and Kosuge, 1980) and histidine (Miflin, 1980) have been shown to require the amide nitrogen of glutamine. In these

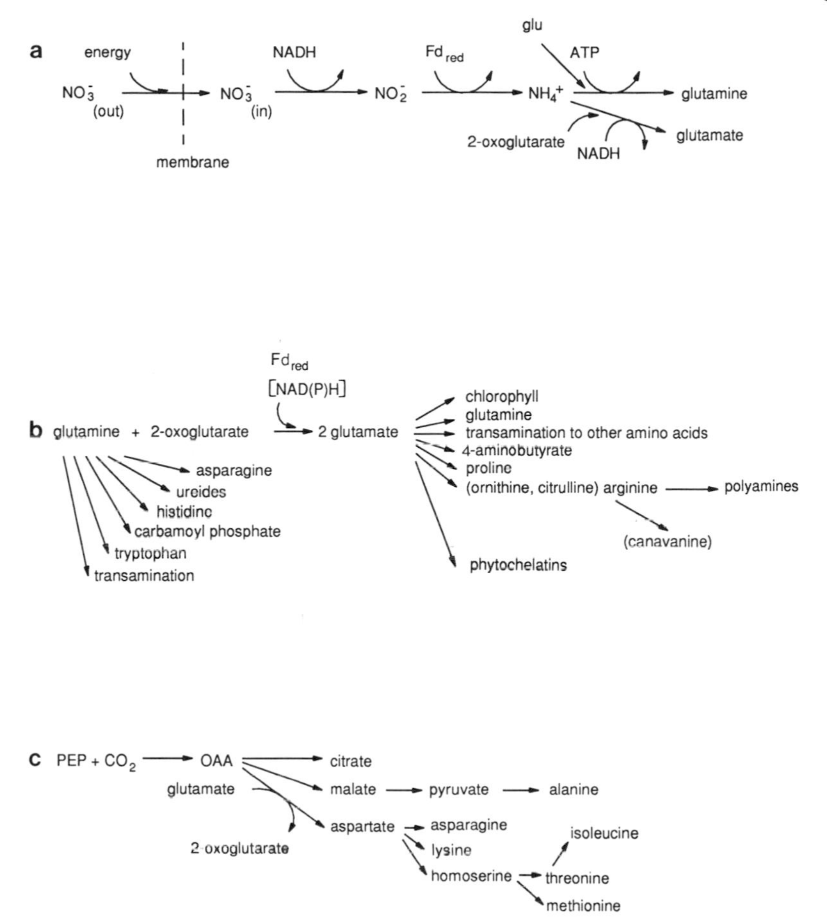

FIG. 1 Outline of metabolic reactions involved in the synthesis of amino acids. (a) Processes leading to the synthesis of NH_4^+ and glutamine or glutamate. The reactions, in order from left to right, are (NADH) nitrate reductase, (ferredoxin) nitrite reductase (Fd_{red}), glutamine synthetase (ATP requiring), and (NADH) glutamate dehydrogenase. (b) Possible metabolic fates of glutamine and glutamate within the cell. The ferredoxin [NAD(P)H]-dependent reaction is glutamate synthase. (c) Synthesis of alanine and the aspartate family of amino acids.

processes glutamate is regenerated. Transamination of glutamine (Ireland, 1986) results in the formation of 2-oxoglutaramate, which can be deamidated to produce NH_4^+ and 2-oxoglutarate (Meister, 1953; Kretovich, 1958). NH_4^+ could also be incorporated directly into glutamate via the glutamate dehydrogenase reaction (Fig. 1a); however this mode of action is at the moment controversial (Lea *et al.*, 1990; Oaks and Hirel, 1985; Miflin and Lea, 1977; Rhodes *et al.*, 1989).

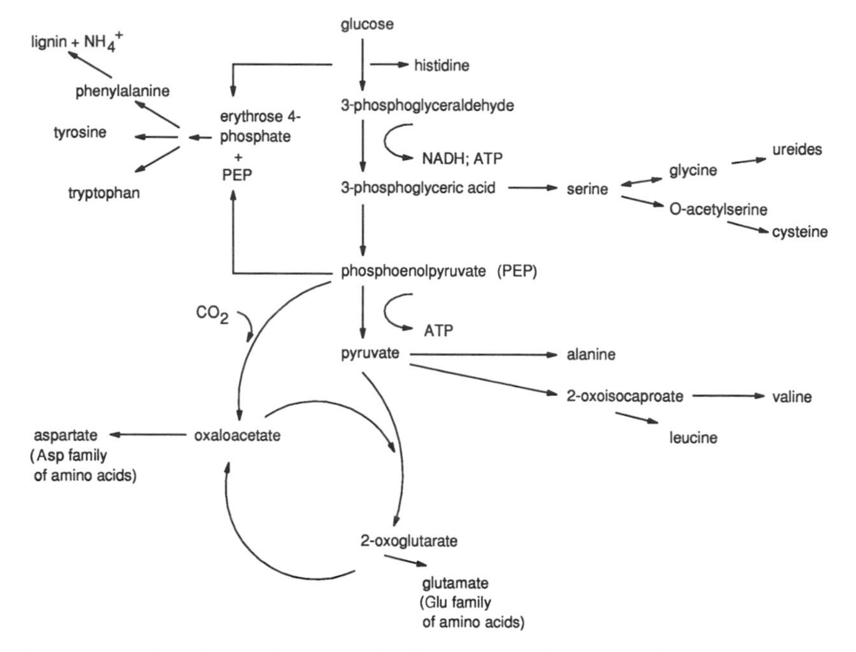

FIG. 2 Origins of carbon skeletons required for amino acid synthesis. These compounds are derived from glycolysis, the oxidative pentose phosphate pathway, and the TCA cycle.

Glutamine can also react with 2-oxoglutarate to yield two glutamate molecules and NAD^+, $NADP^+$, or oxidized ferredoxin (Lea and Miflin, 1974). This glutamate (Fig. 1b) then serves as a carbon and nitrogen source in the synthesis of glutamine, proline and arginine, and phytochelatins (Rauser, 1990), and by way of arginine biosynthesis the polyamines as well (Galston and Sawhney, 1990). It is also a nitrogen donor in many transaminase reactions. When biosynthetic reactions such as these are favored, the acid pools of the tricarboxylic acid cycle (TCA) are replenished by the carboxylation of phosphoenolpyruvate (Kornberg, 1966). The product oxaloacetate can then be used in the synthesis of citrate, malate, or aspartate (Fig. 1c). Aspartate nitrogen comes principally from glutamate in a transamination reaction. Aspartate carbon and nitrogen are converted further to asparagine and the aspartate-derived amino acids (lysine, homoserine, threonine, methionine, and isoleucine) (Bryan, 1980, 1990). In plants the synthesis of shikimate-derived amino acids (phenylalanine, tyrosine, and tryptophan) drains carbon from both the glycolytic pathway and the oxidative pentose phosphate pathway (Fig. 2). This is a significant drain, principally because of the association of phenylalanine with the synthesis of lignin and many secondary metabolites (Gilchrist and Kosuge, 1980; Hahlbrock and Scheel, 1989; Poulsen and Verpoorte, 1991).

Where NADPH is required to drive reduction reactions, for example, the reduction of ferredoxin in root tissues (Emes and Bowsher, 1991), it is generated by the oxidative pentose phosphate pathway (Butt and Beevers, 1961; Emes and Bowsher, 1991; Oji *et al.,* 1985). In leaves NADPH generated by the photosynthetic electron transport chain can be used to reduce ferredoxin. It can also be shuttled out of the chloroplast supplying reductant to the cytosol (e.g., Scheibe, 1990). In the dark NADPH can also be used to reduce ferredoxin (Davis and San Pietro, 1977). In the reactions leading from glucose to pyruvate (Fig. 2), NADH and ATP are generated. When the biosynthesis of amino acids is a dominant reaction, the utilization of glucose proceeds at a faster rate because of the modulation of fructose-2,6-bisphosphate, which regulates the partitioning of carbon, either toward the synthesis or oxidation of sucrose (Stitt, 1990).

Serine, glycine, and cysteine can also be derived from the carbon of glycolysis in the cytosol or, in cases where the glycolytic enzymes have been found in the chloroplast (Klein, 1986), serine and glycine are synthesized in that organelle as well. Larsson and Albertsson (1979) established that 3-phosphoglycerate could be a precursor initially of 3-phosphoserine and ultimately of serine and glycine (Fig. 2; also see Fig. 5). Thus the chloroplast or cytosol could be an important site of synthesis of these two amino acids. In green leaves the peroxisome is a major source of glycine when photorespiration is active (Fig. 7; Section VI,A), and that glycine is a precursor of serine in a reaction that is localized in the mitochondria (Oliver *et al.,* 1990a,b). It is, however, not the only source because mutants in specific reactions of the photorespiratory cycle that limit glycine or serine biosynthesis are viable under conditions where photorespiration is repressed (Kleczkowski and Givan, 1988). Cysteine is derived from serine via *0*-acetylserine and the occurrence of the required enzymes for its synthesis is documented in plants (Giovanelli *et al.*, 1980). Alanine carbon is derived directly from pyruvate or indirectly from malate. Its synthesis probably involves a transaminase reaction in which either glutamine, glutamate (Joy, 1988), or 4-aminobutyrate (Streeter and Thompson, 1972a,b) is the nitrogen donor. The synthesis and turnover of alanine can be very rapid, an observation that suggests that its biosynthesis is closely related to the carbon flow in photosynthesis (Joy, 1988). Pyruvate also supplies carbon destined for valine and leucine biosynthesis (Bryan, 1990). The carbon of the ureides is derived from glycine and the one-carbon pool, and the nitrogen from glutamine in transamidation reactions as well as from aspartate (Atkins and Beevers, 1990).

The assimilation of carbon and nitrogen are tightly coupled processes.

For example, the synthesis of many amino acids takes place in the chloroplasts, where there is a supply of carbon and nitrogen, in addition to energy and reductant. In addition, there are membrane transporters that permit the entrance and exit of specific compounds that are essential for nitrogen assimilation. In this article we will consider (1) important reactions of nitrogen assimilation and current initiatives leading to new insights in the role of several reactions involved in nitrogen metabolism, (2) the localization of these reactions, and, finally, (3) the advantage to the organism of such compartmentation. The major emphasis will be on leaf metabolism and in particular the role of chloroplasts, peroxisomes, and mitochondria in utilizing carbon and nitrogen intermediates involved in photorespiration.

II. The Production of NH_4^+

A. The NO_3^- Uptake–NO_3^- Reductase System

The conversion of NO_3^- to NH_4^+ involves at least three proteins that are induced by NO_3^-, a permease that permits the selective uptake of NO_3^- from the medium (soil), nitrate reductase, and nitrite reductase (Fig. 1a). Nitrate reductase is the best characterized of these proteins and its properties have been recently described by Caboche and Rouzé (1990), Kleinhofs and Warner (1990), and Solomonson and Barber (1990). The permease has not yet been characterized, however, there have been several reports of membrane proteins that are induced by NO_3^- and which may have a permease function (McClure *et al.*, 1987; Ward *et al.*, 1988). There is also some evidence suggesting that nitrate reductase itself could be in part a membrane protein involved in the uptake of NO_3^- (Butz and Jackson, 1977; Ward *et al.*, 1988). However, Warner and Huffaker (1989) have shown that barley mutants that lack all nitrate reductase activity still have a normal capacity to take up NO_3^- from the medium. This suggests that either the permease and NR are either two distinct proteins with nonoverlapping functions or that the permease and NR activities are combined in one protein, yet do not share the same active site.

It has been assumed that NO_3^--specific permeases are active in NO_3^- uptake by cells and in the transfer of that NO_3^- from the cytosol to the vacuole (Clarkson, 1986; Glass, 1988; Larsson and Ingemarsson, 1989) or to cells in the vascular bundle. Mutants have been characterized in *Arabidopsis* that lack the capacity to take up NO_3^- from the medium but the actual protein coded for by the mutant gene has not yet been characterized (Doddema *et al.*, 1978; Doddema and Telkamp, 1979).

Recently, Omata *et al.* (1989) identified a 45-kDa protein that is present in NO_3^--grown cells of *Synechococcus* PCC 7942 (*Anacystis nodulans* R2) but which is absent in NH_4^+-grown cells. They introduced the gene for kanamycin resistance into the gene coding for the 45-kDa protein, and in this constructed mutant (M45) the 45-kDa protein was not made under any growth conditions. In addition the cells were pale green in color when grown in NO_3^- levels of less than 20 m*M* KNO_3. At higher levels of NO_3^- (40–70 m*M*) the color was normal. Growth in medium containing 1 m*M* $(NH_4)_2SO_4$ was almost normal. Thus it is clear that there are two processes involved in the uptake of NO_3^-, a high-affinity process active when NO_3^- levels in the medium are low and a low-affinity system active when high levels of NO_3^- are included in the medium. Recently, Unkles *et al.* (1991) also identified a protein in *Aspergillus nidulans* that has properties similar to Omata's NO_3^- transport protein.

Evidence from traditional physiological experiments also suggests at least two factors involved in the uptake of NO_3^-. One is a high-affinity process (active with NO_3^- concentrations of less than 1 m*M*) that is energy dependent (Glass *et al.*, 1990) and the second is a mechanism that is constitutive rather than inductive, and which is active at higher levels of NO_3^- (Siddiqi *et al.*, 1990). It will be interesting to see if Omata's gene can restore the NO_3^- uptake properties of Doddema's *Arabidopsis* mutants, which are putatively permease minus (Doddema and Telkamp, 1979; Doddema *et al.*, 1978). Omata (1991) also sequenced the gene for the 45-kDa protein. Homologies between this gene and the genes for the NO_3^- and NO_2^- reductase proteins will help to elucidate overlapping functions and regulatory properties of these three proteins.

Depending on the level of NO_3^- administered to the system, NO_3^- can be stored in the root, transferred to the leaf to be stored in the vacuole (Martinoia *et al.*, 1981, 1986; Granstedt and Huffaker, 1982), or be reduced in the roots or leaves. It has long been known that when ions (metabolites) are taken up from the medium that they are not equally distributed in all cells of a particular tissue (see Oaks and Bidwell, 1970, for appropriate references and discussion). This phenomenon has also been observed with the uptake of NO_3^- (Rufty *et al.*, 1986). With low levels of NO_3^- (200 μ*M*), NO_3^- accumulation and NR protein is detected only in epidermal cells. With higher levels of NO_3^- (20 m*M*), NR protein is found in the cortical cells as well. This observation has recently been confirmed at the mRNA level (D. M. Long, A. Oaks, and S. Rothstein, unpublished observations). Thus, it could be that NO_3^- and NR are confined to the epidermal layer of the root as proposed by Rufty *et al.* (1986), or that NO_3^- is efficiently transferred to the stele such that the level of NO_3^- in the intervening cells is below both the detection level of NO_3^- and the inductive threshold for NR. Both experiments (Rufty's and Long's) were performed with maize seedling roots, which are NO_3^- exporters (Pate, 1973). It would be useful to repeat this experiment with roots that export asparagine, that is, in a situation where the plant root has a major role in NO_3^- reduction.

Levels of NO_3^- in the tissue do not necessarily correlate with levels of NR activity as measured in *in vitro* assays (Chantarotwong *et al.*, 1976). Ferrari *et al.* (1973) demonstrated with tobacco suspension cultures that NR activity disappears long before the internal NO_3^- is lost when cells are transferred to a medium lacking NO_3^- . Shaner and Boyer (1976a,b) also showed that leaf NR activity was dependent on NO_3^- transported to the leaf rather than on NO_3^- levels in the leaf. Such experiments indicate that there is more than one pool of NO_3^- in plant cells. These may be thought of as active pools, which respond directly to the external NO_3^- supply, and storage pools, which may be located in the vacuole. When plants are transferred to a NO_3^--free medium, it is assumed that the active pool disappears very quickly and a concomitant loss of NR activity is seen (Ferrari *et al.*, 1973). Subsequently the NO_3^- in the storage pool responds to a NO_3^- deficit cue and there is a loss of that NO_3^- as well (Oaks, 1974; Somers *et al.*, 1983).

The potential regulation of NO_3^- in the storage pools has been examined in detail in leaves by Aslam *et al.* (1973, 1976, 1984). It is known, for example, that NO_3^-, accumulated in leaves during a previous light treatment, is not reduced very efficiently by NR in the dark and consequently remains at high levels (Abrol *et al.*, 1983; Aslam and Huffaker, 1984; Aslam *et al.*, 1973; Canvin and Atkins, 1974; Steingrover *et al.*, 1986). Aslam *et al.* (1973) designed an experiment to study the effect of light on the utilization of this stored NO_3^-. After preloading barley leaves with NO_3^- in the dark, they excised the leaves and placed them in vials in the light or in the dark. The NO_3^- levels in the leaves exposed to the light treatment declined and there was a modest induction of NR activity. In the dark there was no change in internal NO_3^- and only when external NO_3^- was added was there an induction of NR activity. This light-dependent release of NO_3^- probably does not reflect an energy requirement because glucose did not replace the light requirement when NO_3^- was not supplied in the external medium (Aslam *et al.*, 1976). The solution to this observation awaits the preparation of physiologically sound vacuoles so that NO_3^- exchange can be examined *in vitro*.

Nitrate reductase, the first enzyme in the sequence of reactions involved in the assimilation of NO_3^-, was shown by Evans and Nason (1953) to require reduced pyridine nucleotides (NADH or NADPH) and by Tang and Wu (1957) to be substrate inducible. Early results from Hageman's laboratory showed that light was also required for the induction (Beevers and Hageman, 1980; Hageman and Flesher, 1960). More recently, light has been shown to act at the level of transcription (Melzer *et al.*, 1989; Somers *et al.*, 1983).

End products of NO_3^- assimilation, NH_4^+ and amino acids, have been shown to inhibit the induction of NR in *Neurospora crassa* (Marzluf, 1981). In higher plants similar inhibitions with low levels of NH_4^+ or amino acids have not been observed (Oaks *et al.*, 1977, 1979). Even though in tissue culture systems amino acids clearly inhibit the induction of NR (Filner, 1966), on closer investigation the uptake of NO_3^-, and not its conversion to NO_2^-, was the phase of metabolism most sensitive to amino acid additions (Heimer and Filner, 1970). Of the amino acids tested, glutamine appears to hold the key in the *Neurospora* system (Marzluf, 1981; Hurlburt and Garrett, 1988), and perhaps in higher plants as well (Oaks, 1974; Deng *et al.*, 1991; H. Nagawa, personal communication).

Physical separation of cellular constituents (Dalling *et al.*, 1972a,b; Miflin, 1974; Oaks and Gadal, 1979) suggests that nitrate reductase is located in the cytosol. Experiments at the histochemical level have been less definitive (see Solomonson and Barber, 1990, for an extensive discussion of this point). Vaughn and Campbell (1988) using a polyclonal antibody specific for NR, demonstrated that NR was in the cytosol of mesophyll cells in the maize leaf, a location in agreement with results obtained with organelle and tissue separations (Moore and Black, 1976; Rathnam and Edwards, 1976). However, this same group also showed that their monospecific antibody recognized the chloroplast glyceraldehyde-3-phosphate dehydrogenase (NADPH) (Gowri and Campbell, 1989). It is possible that their antibody also recognized phosphoenolpyruvate carboxylase because this is a major protein in maize leaves and it is known to be a contaminant of the purification process at least to the Blue Sepharose stage (A. Oaks and V. J. Goodfellow, unpublished observations). The distribution of colloidal gold, if this were true, would be similar to the distribution reported by Vaughn and Campbell (1988) for NR. Caboche's group has also examined this point. With both poly- and monoclonal antibodies prepared against NR they observed a membrane-bound protein epitope that was recognized by their antibody in both wild-type *Nicotiana plumbaginifolia* and mutant plants that lacked the NR protein. There was no evidence of a localization in the cytosol where one would expect to find NR (M. Caboche, personal communication).

The localization of NR in the cytosol poses a logistic problem because the enzyme is most active in green leaves in the light, and the source of reductant (NADPH) in the light is produced within the chloroplast, yet leaf NR typically mediates an NADH-dependent reaction. Klepper *et al.* (1971) were the first to demonstrate a possible shuttle mechanism in which 3-phosphoglyceraldehyde, made in the chloroplast with the available NADPH and ATP, is released from the chloroplast. In the cytosol, it is converted to phosphoglyceric acid with the production of NADH and ATP [Fig. 5, translocator H (see Table I) and associated processes]. The phosphoglyceric acid is then shuttled back into the chloroplast, and the NADH in the cytosol is used for the reduction of NO_3^- to NO_2^- (or for other reductions that occur in the cytosol). Later, House and Anderson (1980) demonstrated that malate and oxaloacetate shuttle between the chloroplast and cytosol (Fig. 5; translocator E). Again NADPH is used to produce malate in the chloroplast, and NADH is produced by the oxidation of malate in the cytosol. Woo and Canvin (1979) have also suggested that a malate/oxaloacetate shuttle between the mitochondria and the cytosol could produce the NADH in the cytosol required for the reduction of NO_3^-. However, the relative contribution of this shuttle in supplying cytosolic malate in the light has been questioned (Dry *et al.*, 1987; Gardström and Wigge, 1988; see Section V, A and B).

In roots, NR can use both NADH or NADPH as reductants (Long and Oaks, 1990; Warner *et al.*, 1987). Either of these reductants can be supplied by the glucose-6-phosphate or 6-phosphogluconate dehydrogenases that are present in the cytosol (Emes and Fowler, 1979; Oji *et al.*, 1985) or by a malate shuttle in which malate produced in the mitochondria is oxidized to oxaloacetate in the cytosol (Naik and Nicholas, 1986). The *in vivo* assay for NR involves using endogenously generated reductant, and differences in the properties of this assay in the root or leaf reflect the availability of reductant within either organ. NR can be measured under aerobic conditions *in vivo* in roots (Aslam and Huffaker, 1982), whereas anaerobic conditions, or inhibitors of mitochondrial electron transport, must be used to measure the *in vivo* NR in leaves (Woo and Canvin, 1979). The reason for this difference could be explained by the availability of

TABLE I
List of Plastid and Mitochondrial Translocators

Plastid translocators	Letter[a]	Mitochondrial translocators	Letter[a]
Dicarboxylate	A	Amino acid	I
Glutamate:malate	B	Dicarboxylate	J
Glutamine:glutamate	C	Malate:oxaloacetate	K
Glycolate:glycerate	D	Ornithine:citrulline	L
Malate:oxaloacetate	E	2-Oxoglutarate:dicarboxylate	M
Ornithine:arginine	F	Pyruvate	N
2-Oxoglutarate:malate	G	Tricarboxylate	O
Phosphate	H		

[a] Letters refer to Figs. 4, 5, and 7.

cytosolic NADPH in the root, or a more favorable competitive advantage of root NR for the mitochondrial NADH.

Nitrite reductase [Fig. 1a; Fig. 5, reaction 32 (see Table II)], which mediates the third reaction in this trilogy, converts NO_2^- to NH_4^+. It uses ferredoxin as reductant (Joy and Hageman, 1966), is located in the chloroplast or root plastid (Dalling *et al.*, 1972a,b; Borchert *et al.*, 1989), and is induced by NO_3^- (Back *et al.*, 1988; Beevers and Hageman, 1980). Anderson and Done (1978) demonstrated that NO_2^- enters the chloroplast, where it is first reduced to NH_4^+ followed by its incorporation into glutamine. This uptake proceeds at a greater rate in the light than in the dark (Brunswick and Cresswell, 1988). It was initially thought that the reduction of NO_2^- could proceed only in the light because of a requirement for reduced ferredoxin and that this step was really the block in the overall reduction of NO_3^- in the light. However, Davis and San Pietro (1977) showed that NADPH is capable of reducing ferredoxin, and Aslam and Huffaker (1984) demonstrated that if sufficient carbohydrate was available either NO_3^- or NO_2^- could be reduced in the dark. Thus four possibilities for the limitation of NO_3^- reduction in the dark have been demonstrated: (1) transfer of NO_3^- into the cell or from the vacuole to the cytosol, (2) supply of reductant for NR, (3) supply of reductant for NiR, or (4) the uptake of NO_2^- by the chloroplast.

B. Nitrogenase

Nitrogenase mediates the reaction converting dinitrogen (N_2) to NH_4^+. It is found in a few free-living bacteria (*Klebsiella, Clostridium, Azotobacter, Azospirillum, Bacillus,* and some of the cyanobacteria) (Russell, 1977). The gene for this protein is also found in *Rhizobium* and *Bradyrhizobium* species but, except under very abnormal circumstances, is expressed only after these bacterial species have infected specific host plants. During the infection a nodule is formed on the root. The boundaries of the nodule and the peribacteroid membrane have special properties of controlling not only the flow of metabolites between the bacteroid and the host tissue, but also the level of oxygen (Vance and Griffith, 1990; Layzell *et al.*, 1990).

The enzyme, nitrogenase, is sensitive to oxygen, and is housed in the bacteroid portion of the nodule where biochemical and morphological features have evolved to protect it from oxygen (Layzell *et al.*, 1990). The enzyme produces NH_4^+, which diffuses to the host portion of the nodule where it is converted to glutamine via (a nodule-specific) glutamine synthetase. The amide nitrogen of glutamine is then transferred to asparagine or to the ureides, depending on the nitrogen product that is exported by the root system of the host plant. Vance and Griffith (1990) described this

TABLE II
List of Enzymes

Reaction number[a]	Enzyme	EC number	Ref.
1	Aconitase (cytosolic)	4.2.1.3	Brouquisse *et al.* (1987)
2	Aldehyde dehydrogenase (proposed)	1.2.1.3	Davies and Asker (1985)
3	ω-Amidase	3.5.1.3	Meister (1953)
4	Arginine iminohydrolase	3.5.3.6	Ludwig (1991)
5	Asparaginase	3.5.1.1	Sieciechowicz *et al.* (1985)
6	Aspartate aminotransferase	2.6.1.1	Givan (1980)
7	Carbamate kinase	2.7.2.2	Ludwig (1991)
8	Carbamoyl-phosphate synthase	2.7.2.5	Shargool *et al.* (1978)
9	Catalase	1.11.1.6	Havir and McHale (1989)
10	Citrate synthase	4.1.3.7	Wiskich (1980)
11	Glutamate dehydrogenase	1.4.1.2	Rhodes *et al.* (1989)
12	Glutamate decarboxylase	4.1.1.15	Streeter and Thompson (1972b)
13	Glutamate-glyoxylate aminotransferase	2.6.1.4	Nakamura and Tolbert (1983)
14	Glutamate/4-aminobutyrate:pyruvate aminotransferase	2.6.1.2/14	Wallace *et al.* (1984)
15	Glutamate synthase [ferredoxin/NAD(P)H]	1.4.1.13	Lea and Miflin (1974)
16	Glutaminase	3.5.1.2	Ireland (1986)
17	Glutamine synthetase	6.3.1.2	O'Neal and Joy (1973)
18	Glyceraldehyde-3-phosphate dehydrogenase (NADH)	1.2.1.12	Scheibe (1990)
19	Glyceraldehyde-3-phosphate dehydrogenase (NADPH)	1.2.1.9	Scheibe (1987)
20	Glycerate kinase	2.7.1.31	Kleczkowski *et al.* (1985)
21	Glycine decarboxylase complex	2.1.2.10	Oliver *et al.* (1990b)
22	Glycolate oxidase	1.1.3.15	Kerr and Groves (1975)

23	Glyoxylate reductase (NADPH)	1.1.1.79	Givan *et al.* (1988b)
24	Hydroxypyruvate reductase (NADH)	1.1.1.81	Kleczkowski *et al.* (1990)
25	Hydroxypyruvate reductase (NADPH)	1.1.1.81	Kleczkowski *et al.* (1988)
26	Isocitrate dehydrogenase (NADP; cytosolic)	1.1.1.42	Chen and Gadal (1990b)
27	Isocitrate lyase (cytosolic)	4.1.3.1	Zelitch (1988)
28	Malate dehydrogenase (NADH)	1.1.1.37	Ebbighausen *et al.* (1987)
29	Malate dehydrogenase (NADPH)	1.1.1.82	Hatch *et al.* (1984)
30	Malic enzyme	1.1.1.39	Day *et al.* (1984)
31	Nitrate reductase [NADH/NAD(P)H]	1.6.6.1/2	Evans and Nason (1953)
32	Nitrite reductase	1.7.7.1	Joy and Hageman (1966)
33	Ornithine carbamoyltransferase	2.1.3.3	Ludwig (1991)
34	Phosphoglycerate dehydrogenase	1.1.1.95	Gaudillere *et al.* (1983)
35	Phosphoglycerate phosphatase	3.1.3.33	Larsson and Albertsson (1979)
36	Phosphoglycolate phosphatase	3.1.3.13	Richardson and Tolbert (1961)
37	Phosphohydroxypyruvate phosphatase	—	—
38	Phosphohydroxypyruvate:glutamate transaminase	2.6.1.52	Larsson and Albertsson (1979)
39	Phosphoserine phosphatase	3.1.3.3	Larsson and Albertsson (1979)
40	Phosphoenolpyruvate carboxylase	4.1.1.31	Schuller *et al.* (1990a,b)
41	Pyruvate decarboxylase	4.1.1.1	Davies and Asker (1985)
42	Pyruvate dehydrogenase	1.2.4.1	Schuller and Randall (1989)
43	Pyruvate kinase	2.7.1.40	Plaxton (1988)
44	Rubisco	4.1.1.39	Andrews *et al.* (1973)
45	Serine hydroxymethyltransferase	2.1.2.1	Oliver *et al.* (1990a)
46	Serine-glyoxylate transaminase	2.6.1.45	Nakamura and Tolbert (1983)
47	Transhydrogenase	1.6.1.1	Schmitt and Edwards (1983)
48	Triose-phosphate isomerase	5.3.1.1	Schiebe (1990)
49	Ureidoglycolate amidohydrolase	4.3.2.3	Wells and Lees (1991)

[a] Numbers refer to reactions in Figs. 3–7 and 9.

reaction as a "second major process" by which plants acquire nitrogen, and although it is specific to a few plants, traditional agricultural practices of crop rotation ensure that this method of nitrogen acquisition is important for a broad spectrum of plants.

C. Phenylalanine Ammonia-Lyase

Phenylalanine is deaminated to yield cinnamic acid and ammonium in a reaction that initiates phenylpropanoid metabolism. This pathway leads to the synthesis of lignin, suberin, and other wall-associated phenolic compounds as well as flavonoids, isoflavonoids, coumarins, and soluble esters (Hahlbrock and Scheel, 1989). As a result phenylalanine ammonia-lyase (PAL) activity is important during cotyledon and leaf development (Jahnen and Hahlbrock, 1988; Hahlbrock *et al.*, 1971) and secondary product synthesis, for example, during pathogen infection (Chappell and Hahlbrock, 1984). During these events the flux of carbon, and hence NH_4^+ release, mediated by PAL is potentially very large. As summarized by Hanson and Havir (1981), the flux of carbon can be as great as 22 μmol/g fresh weight per hour. This compares with estimated fluxes of carbon in photosynthesis of 100–150 μmol/g fresh weight per hour, and of NO_3^- reduction of 10–15 μmol/g fresh weight per hour (Keys *et al.*, 1978). Ammonia release by the action of PAL is probably localized in active centers of secondary product metabolism; nevertheless it can only be emphasized that the NH_4^+ derived from the action of PAL can be a truly major source of secondary cellular nitrogen.

PAL activity can be specifically inhibited by α-aminooxy-β-phenylpropionic acid (Amrhein and Gödeke, 1977). Judicious use of this inhibitor could help in modifying the flux of carbon through the phenylpropanoid pathway and this, in turn, could help in determining how this pathway impinges on basic nitrogen metabolism.

PAL is found in plastids, mitochondria, microbodies, and microsomes. However, there is some question as to whether this represents true localization or artifacts incurred during the physical separation of the organelles (Hanson and Havir, 1981). Immunochemical localization of PAL in developing parsley seedlings indicates that PAL protein is detected in all cells, especially epidermal and gland-forming cells, and that it decreases with seedling age (Jahnen and Hahlbrock, 1988). To our knowledge colloidal gold immunochemistry has not been used to determine localizations within the cell. If there are massive releases of NH_4^+ in localized regions of high PAL activity in the cells of green leaves, then there should also be major levels of glutamine synthetase (or some alternative enzyme) for trapping the NH_4^+ released by the action of that PAL.

D. Asparaginase, ω-Amidase, and Glutaminase

The amides, asparagine and to a lesser degree glutamine, are important nitrogen transport compounds in higher plants (Atkins and Beevers, 1990; Pate 1980, 1989; Schubert, 1986). The degradation of asparagine in growing organs supplies nitrogen in the form of NH_4^+ and amino groups as well as carbon for synthetic processes. Asparagine is degraded through two metabolic routes in leaves: either transamination, catalyzed by serine-glyoxylate aminotransferase (Ireland and Joy, 1983) (Fig. 6, reaction 46), or hydrolysis via asparaginase (Sieciechowicz *et al.*, 1985)(Fig. 6, reaction 5). The transamination of asparagine in leaves introduces nitrogen into the photorespiratory pathway (Ta *et al.*, 1985), and results in the transfer of amino nitrogen of asparagine to glyoxylate for the synthesis of glycine and serine. This enzyme also catalyzes the amination of pyruvate from asparagine to produce alanine (Ireland and Joy, 1983; Murray *et al.*, 1987) (Fig. 3) or 4-hydroxy-2-oxobutyrate to synthesize homoserine (Joy and Prabha, 1986; Sieciechowicz *et al.*, 1988). The transamination of asparagine is likely to be important during leaf development when the supply of asparagine from the root system is high (Sieciechowicz and Joy, 1988), and when the requirement for amino acids for synthetic processes is great. The significance of the contribution of asparagine nitrogen in photorespiration will be discussed (Section VI,A). This route of asparagine utilization may also be important in mature tissues (which lack asparaginase activity) because asparagine accumulates in leaves in the dark, or in response to elevated pools of NH_4^+ or cyanide (see Section IV,C). The transamination of asparagine does not exhibit a diurnal variation in activity (K. A. Sechley and K. W. Joy, unpublished observations), a property that

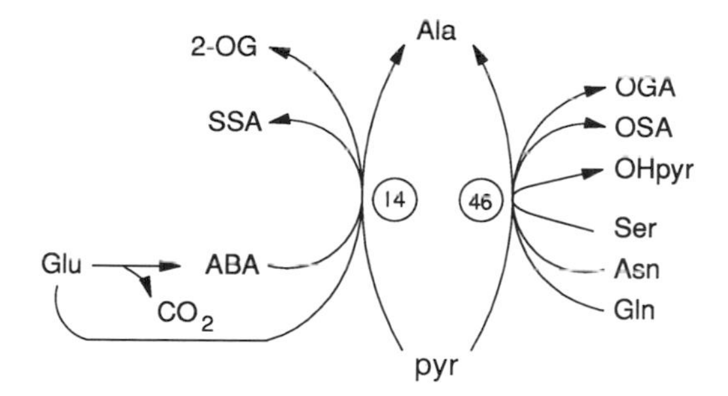

FIG. 3 Aminotransferase reactions leading to the synthesis of alanine. Reaction 14, glutamate (4-aminobutyrate):pyruvate aminotransferase; reaction 46, serine (asparagine, glutamine)-glyoxylate (pyruvate) aminotransferase (see text for details). ABA, 4-Aminobutyrate; Ala, alanine; Asn, asparagine; Gln, glutamine; Glu, glutamate; 2-OG, 2-oxoglutarate; OGA, oxoglutaramic acid; OSA, oxosuccinamate; Ser, serine; SSA, succinic semialdehyde; pyr, pyruvate; OHpyr, hydroxypyruvate.

is associated with asparaginase. The transamination of asparagine produces 2-oxosuccinamate, which is not detectable in plant extracts (Lloyd and Joy, 1978; Streeter, 1977). Therefore it may either deamidate rapidly, producing oxaloacetate and NH_4^+ in a reaction catalyzed by ω-amidase (Meister, 1953) (Fig. 6, reaction 3), or be reduced to 2-hydroxysuccinamate, which does accumulate in leaves (Lloyd and Joy, 1978). This latter compound may be deamidated to yield malate and NH_4^+, or be oxidized back to 2-oxosuccinamate (see Fig. 6). The metabolic interconversions of these compounds are not adequately understood in plants (Ta *et al.,* 1984a; Sieciechowicz *et al.,* 1988). Transamination of 2-oxosuccinamate to produce asparagine in the presence of a suitable amino donor is unlikely to occur *in vivo* because the equilibrium of this reaction favors the deamination reaction. This is an unusual property for aminotransferase reactions (Ireland and Joy, 1983), but it imparts directionality during photorespiration (see Section VI,A).

Asparagine degradation, catalyzed by asparaginase, produces aspartate and NH_4^+. This reaction is observed only in growing organs, which receive asparagine as a major source of nitrogen. For example, developing cotyledons (Atkins *et al.,* 1975), leaves (Ireland and Joy, 1981), or root tips (Chang and Farnden, 1981; Ireland and Joy, 1981) have high levels of asparaginase activity. The aspartate could then support the synthesis of the aspartate family of amino acids (lysine, methionine, threonine, and isoleucine; see Fig. 2). In developing leaves asparaginase activity displays a diurnal variation, increasing during the light and decreasing in the dark period (Sieciechowicz *et al.,* 1985). The resultant NH_4^+ would probably be utilized by glutamine synthetase (Section III,A). The brief window of asparaginase activity in young leaves, roots, and cotyledons suggests that the hydrolysis of asparagine is required to supply the nitrogen required by these organs during their development.

Much less information is available on the deamidation of glutamine in plant tissues. Two isozymes of glutaminase activity are detected in pea leaves. These activities decrease with leaf age (R. J. Ireland, personal communication) and display a diurnal variation in activity, decreasing during the light period (K. A. Sechley and K. W. Joy, unpublished observations). The possibility remains that glutaminase activity is a partial reaction of amide-utilizing reactions such as glutamine synthetase (in the presence of ADP or P_i) (Tate and Meister, 1973) or asparagine synthetase (in the presence of Cl^-) (Rognes, 1980). Glutamine is also transaminated by either of two isozymes, one preferentially utilizing glyoxylate or pyruvate as amino acceptors, and the other phenylpyruvate (Ireland, 1986). Deamidation of 2-oxoglutaramate, the transamination product of glutamine, by ω-amidase (Kretovich, 1958) has been detected in developing pea leaves (Ireland, 1986).

III. The Assimilation of NH_4^+

NH_4^+ is incorporated into organic form principally by the action of glutamine synthetase (Kanamori and Matsumoto, 1972; O'Neal and Joy, 1973; Miflin and Lea, 1977). The glutamine thus formed can be used as a substrate in a number of transamidation reactions [e.g., glutamate synthase (glutamine–2-oxoglutarate aminotransferase, GOGAT), asparagine synthetase (AS), carbamoyl-phosphate synthetase (CPS), or 5-phosphoribosyl pyrophosphate amidotransferase, which is involved in ureide biosynthesis]. In these reactions the glutamate required for the synthesis of glutamine is regenerated. Glutamine is also transaminated or used in the synthesis of protein and in the export of carbon and nitrogen to other parts of the plant. In these latter examples, glutamine is withdrawn from the system and the glutamate molecule required for its synthesis must be generated from alternative sources. GOGAT, which generates two glutamates, is the reaction most often cited in this regeneration (Lea and Miflin, 1974; Stewart *et al.*, 1980), but transaminase and other transamidase reactions (Joy, 1988) and glutamate dehydrogenase (Oaks and Hirel, 1985) could also be important in the supply of glutamate.

Other reactions, which have not attracted so much attention, may be involved in the assimilation of NH_4^+: carbamoyl-phosphate synthetase (CPS), the initial step in arginine and pyrimidine biosynthesis (Davis, 1986), and asparagine synthetase, which is a potentially important reaction where asparagine is made for export (see Section IV,C). In the case of CPS in fungi, two isoenzymes are involved, one for the synthesis of arginine and the other for the synthesis of the pyrimidines (Davis, 1986). In either case the enzyme can use glutamine or NH_4^+ as a nitrogen source. However, for each enzyme the K_m for glutamine is much lower than the K_m for NH_4^+, an observation that suggests that glutamine should be the favored nitrogen donor. Carbamoyl-phosphate synthase, involved in the synthesis of carbamoyl phosphate destined for arginine, may be located in the mitochondria (*N. crassa*) or in the cytosol (yeast) (Davis, 1986). Evidence in the literature, although much less rigorous, indicates that CPS in higher plants is predominantly located in the plastid (Shargool *et al.*, 1978; Taylor and Stewart, 1981). However, recent immunological evidence also indicates that CPS is localized in the mitochondria (Ludwig, 1991). If CPS were active in the synthesis of carbamoyl phosphate destined to be a precursor for citrulline it should use glutamine as a substrate. However, in the mitochondria, where there is no glutamine (D. Rhodes, T. Yamaya, and A. Oaks, unpublished observations) and no glutamine synthetase (Miflin, 1974), and where there is ample NH_4^+ (Yamaya and Matsumoto, 1985b), CPS may play an important role in the assimilation of NH_4^+ as proposed by Ludwig (1991). The CO_2, the other substrate required for the synthesis of carbamoyl phosphate, is also produced at relatively high rates in the mitochondria either via CO_2 released as a result of an active TCA cycle or via the decarboxylation of glycine. If this were a major reaction in the cycling of CO_2 and NH_4^+ from the mitochondria to chloroplast as proposed by Ludwig (1991), carbamoyl-phosphate synthetase would be a major reaction in the reassimilation of NH_4^+, in magnitude as important as glutamine synthetase.

Two asparagine synthetases have been identified by mutant methodology in *Escherichia coli* (Cedar and Schwarz, 1969; Humbert and Simoni, 1980) and in yeast (Ramos and Wiame, 1980). The NH_4^+ form is typically found in the wild-type *E. coli* and the glutamine form in mutants that lack the NH_4^+ form of the enzyme (Humbert and Simoni, 1980). Such stringent characterization has not been performed in higher plants, however, it is clear that the asparagine synthetase in lupine cotyledons has very different kinetic properties from the asparagine synthetase found in maize roots (Oaks and Ross, 1984). Where there is a potential massive synthesis of asparagine, in legume cotyledons (Rognes, 1975; Tsai and Coruzzi, 1990) and in nodules on soybean roots (Streeter, 1973), glutamine appears to be the major nitrogen donor. Recently, Tsai and Coruzzi (1990) have isolated cDNA clones that code for two asparagine synthetase genes in *Pisum sativum*. Both genes encode a glutamine-binding domain, suggesting that glutamine is the major source of nitrogen for asparagine. To isolate this clone they used a cDNA probe encoding human asparagine synthetase, which also has a glutamine-binding domain. It is possible, therefore, that a gene for an NH_4^+-dependent asparagine synthetase enzyme in plant material has not been detected, or that in plants asparagine synthetase is a bispecific enzyme.

Glutamate itself is required for protein synthesis and under photorespiratory conditions it may also be a major export form of nitrogen (Madore and Grodzinski, 1984). It is a principal nitrogen donor in many transaminase reactions. Certain transaminases are enhanced under special environmental conditions. The glutamic-pyruvic transaminase, for example, is enhanced under anaerobic conditions and alanine itself may accumulate (Good and Crosby, 1989; Vanlerberghe *et al.*, 1991). Under anoxic conditions 4-aminobutyrate, the decarboxylation product of glutamate, also accumulates (Streeter and Thompson, 1972a). This would lead to an extra drain on the glutamate pool. Glutamate is also the carbon and nitrogen precursor for arginine, proline, and chlorophyll synthesis (Bryan, 1990; Kannangara *et al.*, 1978). The synthesis of any of these compounds can reach major proportions under environmental conditions typically encountered in natural systems or during development. The synthesis of proline, for example, is enhanced where there is severe drought (Bogness *et al.*, 1976), and the synthesis of arginine that takes place during cotyledon development in many legume species (Micallef and Shelp, 1989b; Thompson, 1980). Thus, under conditions that promote the synthesis of proline, arginine, aspartate, alanine, 4-aminobutyrate, or chlorophyll, the flux of carbon and nitrogen through glutamate could be relatively high. The question that arises and which will be considered later is whether the accepted GS/GOGAT pathway is sufficient to supply the glutamate carbon in amounts required to meet such emergencies.

A. Glutamine Synthetase

The purification and properties of glutamine synthetase have been reviewed by Stewart *et al.* (1980), McNally and Hirel (1983), and Lea *et al.* (1990). Molecular aspects were also reviewed by Forde and Cullimore (1989). The native glutamine synthetase protein has a molecular weight of 350 kDa and is composed of eight identical or almost identical subunits (Table III). In leaves it is found in chloroplasts (GS_2; Fig. 5, reaction 17) and in the cytosol (GS_1). The proportion of GS_1 and GS_2 polypeptides varies with species (McNally *et al.*, 1983) for reasons we do not understand. The proportion also varies with the developmental stage. In etiolated plants, for example, the GS_1 isozyme is the dominant form in leaf tissues (Guiz *et al.*, 1979; Mann *et al.*, 1979, 1980; Tobin *et al.*, 1985). GS_2 appears when the plants are transferred to the light, and in time this form becomes the dominant form. The localization of GS_2 in pea (O'Neal and Joy, 1973) and spinach (Ericson, 1985) chloroplasts was established by both cell fractionation and by immunocytochemical localizations (Hirel *et al.*, 1982). Observations localizing GS_2 in the chloroplast in tomato (Botella *et al.*, 1988) and soybean (Branjeon *et al.*, 1989) leaves have recently been confirmed at the electron microscope level using immunogold techniques. By gently rupturing chloroplasts, Coruzzi's group has also established that the GS_2 in tobacco (*N. plumbaginifolia*) (Tingey and Coruzzi, 1987) and in pea leaves (Tingey *et al.*, 1987) is located in the stroma of the chloroplast. They are coded for by nuclear genes, and the initial product found with *in vitro* translation is somewhat larger (49kDa) than the GS_2 form isolated from chloroplasts (44kDa; Table III). Thus a signal peptide is cleaved from the GS_2 precursor on transfer across the chloroplast membrane. Edwards *et al.* (1990) and Forde *et al.* (1989) have created chimeric genes by fusing the promoter regions of either GS_1 or GS_2 with the reporter gene β-galactosidase (*GUS*). These *GUS* fusion constructs were then introduced into separate tobacco or *Lotus corniculatus* plants and the activity of GUS protein assayed during plant development. The detection of GUS activity is therefore dependent on the regulation of

TABLE III

Glutamine Synthetase Subunits Obtained from Leaves, Roots, and Nodules[a]

Species	Leaf		Root		Nodule	Ref.[b]
	Chloroplast (GS_2)	Cytosol (GS_1)	Plastid (GS_2)	Cytosol (GS_1)		
Non legumes						
Spinacea oleraceae	44	—	ND[c]	ND	—	1
Nicotiana plumbagnifolia	44 (49)[d]	—	—	38 (38)[d]	—	2
Oryza sativa "Sasanishiki"	44 (49)[d]	41	—	41 (41)[d]	—	3
Oryza sativa	45	45	—	45	—	4
Zea mays (W64A × W18E)	40	38	40	38	—	5
Legumes						
Pisum sativum						
"Sparkle"	44 (49)[d]	38	—	38	38, 37[e]	6
"Little Marvel"	—	—	44	38	ND	7
"Improved Laxton"	48	43	48	43	ND	5
Phaseolus vulgaris	45	43	—	43	43[e]	8
Glycine max	44	40, 38	—	40, 38, 42, 41	44, 40, 38, 42, 41	9
Storage roots						
Cichorium intybus	ND	ND	42	39	—	10

[a] Relative molecular mass: M_r (in kDa).

[b] 1, Erickson (1985); 2, Tingey and Coruzzi (1987); 3, Hayakawa *et al.* (1990); 4, Hirel and Gadal (1980); 5, Oaks and Sechley (unpublished data); 6, Tingey *et al.* (1987); 7, Vézina *et al.* (1987); 8, Forde and Cullimore (1989); 9, Sengupta-Gopalan and Pitas (1986); 10, Sechley *et al.* (1991).

[c] ND, Not determined.

[d] Value in parentheses is relative molecular mass of polypeptides made *in vitro*.

[e] There are other polypeptides of the same size, but with charge differences.

either GS_1 or GS_2 expression. They observed that GS_2 is expressed in cells with chloroplasts whereas GS_1 is found in phloem elements in the leaf of these transgenic plants. These observations support the results of Wallsgrove *et al.* (1987) and Blackwell *et al.* (1987), which suggest that GS_2 is localized where it could play a major role in the assimilation of photorespiratory NH_4^+. Using a different approach, Kamachi *et al.* (1991) found in senescing rice leaves that GS_2 (activity and protein) disappears in parallel with the loss of other chloroplast proteins. GS_1, on the other hand, remains high during the senescence period. This observation and that of Edwards *et al.* (1990) suggest that GS_1 is involved in the synthesis of glutamine that is destined for transport. Combined, these studies support the contention that chloroplastic and cytosolic glutamine synthetases have nonoverlapping functions. If the two enzymes do, in fact, have completely nonoverlapping functions, then the survival of those plants with only GS_2 (e.g., spinach or tobacco) or GS_1 (e.g., *Latheracea* species) in their leaves (McNally *et al.*, 1983) needs a more current explanation. Experiments by

Wallsgrove *et al.* (1987) do, however, suggest some overlap in function. In their experiments with glutamine synthetase and GOGAT mutants, they found on transfer from high CO_2 to air that the deficiency symptoms appeared much earlier in the GOGAT than in the glutamine synthetase mutants. One interpretation of this observation is that the phloem-localized (cytosolic) GS_1 can be active to some extent in the assimilation of photorespiratory NH_4^+. On the other hand, within the population of glutamine synthetase-deficient mutants identified there exists an array of extractable glutamine synthetase activities (from 8 to 160% of the wild-type activity) (Blackwell *et al.*, 1987). Therefore there is always some background GS_2 activity that may contribute to NH_4^+ assimilation under conditions of high CO_2.

Early reports (Hirel and Gadal, 1980; Suzuki *et al.*, 1981) suggest that root GS is localized within the cytosol. There was also one early study that suggests that a small portion of the root GS is in the plastid fraction (Miflin, 1974). It was not until recently that Vézina *et al.* (1987; Vézina and Langlois, 1989) established that about 50% of the glutamine synthetase is in the plastid fraction in the root system of *P. sativum* cv. "Little Marvel" (Table III) when the plants are grown in the presence of NO_3^-. We have also shown a seasonal variation in the abundance of GS_2 in chicory roots (Sechley *et al.*, 1991) and have observed that whereas most of the pea root ("Improved Laxton") glutamine synthetase is plastidic, most of the maize root glutamine synthetase is cytosolic (A. Oaks and K. A. Sechley, unpublished observations). There seems to be variability in the occurrence of the different glutamine synthetase isozymes even between different cultivars of *P. sativum*. The cultivar "Sparkle," for example, has no plastidic glutamine synthetase (Tingey *et al.*, 1987), whereas "Improved Laxton" has about 80% of its glutamine synthetase in the plastid (A. Oaks and K. A. Sechley, unpublished observations). These observations suggest that plants that export asparagine (e.g., *P. sativum*) have a glutamine synthetase localized in the plastid whereas plants that are essentially NO_3^- exporters (e.g., *Zea mays, Oryza sativa*), but which do export glutamine as well, have most of their root glutamine synthetase in the cytosol. This point needs to be investigated in more detail both by comparing root systems with different capacities for reducing nitrogen and by altering the capacity to reduce nitrogen by making transgenic plants.

The regulation of glutamine synthetase is not well understood in higher plants. Additions of NO_3^- and NH_4^+ have been shown to influence the level of glutamine synthetase activity in tissue culture systems (Loyola-Vargas and Sanchez de Jiménez, 1986; Hayakawa *et al.*, 1990), and in *Lemna minor* (Rhodes *et al.*, 1975). More recently, Hirel *et al.* (1987) have established that NH_4^+ induces glutamine synthetase in soybean nodules. This observation has recently been confirmed in transgenic *Lotus cornicu-*

latus and *Nicotiana tabaccum* plants (Miao *et al.*, 1991). In these experiments chimeric genes (containing the promoter of cytosolic glutamine synthetase from either soybean roots or nodules, fused to the reporter gene *GUS*) were expressed in a root-specific fashion in both species of plant. It was expressed in the root apices and in the vascular bundle of the mature root of *Lotus* and *Nicotiana* (Miao *et al.*, 1991). In the legume background, the gene was turned on by the addition of $(NH_4)_2SO_4$, but in the tobacco plant, *GUS* was expressed in the absence of added NH_4^+. In *Phaseolus vulgaris* NH_4^+ (Cock *et al.*, 1990) or NO_3^- (Swarup *et al.*, 1991) has no effect on the induction of glutamine synthetase. It appears, therefore, that differences in the promoter regions of the various genes encoding glutamine synthetase polypeptides lead to important differences in their expression.

Vézina's *et al.* (1987; Vézina and Langlois, 1989) have shown that NO_3^- additions lead to an increase in root plastids and to increases in the abundance and activity of GS_2 in uninfected roots of *P. sativum* (cv. "Little Marvel"). These changes were observed in mature regions of the roots and not in the root tip region. We have also found a GS_2 in roots of *P. sativum* ("Improved Laxton") that is only mildly enhanced by NO_3^-. A GS_1 form of the enzyme, however, is enhanced by NO_3^- additions. In *Z. mays* the cytosolic form of glutamine synthetase is the dominant form of the enzyme and it is induced by NO_3^- additions (A. Oaks and K. A. Sechley, unpublished observations). The glutamine synthetase protein in the shoot is not affected by NO_3^- additions to either *Pisum* or *Zea*. Light has been shown to affect levels of GS_2 activity in rice (Hirel and Gadal, 1980), barley leaves (Mann *et al.*, 1979, 1980), and *P. vulgaris* cotyledons (Swarup *et al.*, 1991). Recently Edwards and Coruzzi (1989) have established that light and possibly NH_4^+ are required for the full induction of GS_2 in pea leaves. At least part of the light effect is mediated by phytochrome, a characteristic of several other nuclear-encoded proteins destined for the chloroplast (Taylor, 1989). The cytosolic glutamine synthetase (GS_1) is not influenced by light (Tingey *et al.*, 1987), but may be induced by NO_3^- or NH_4^+ (Kozaki *et al.*, 1991).

From these few experiments and from those with transgenic plants cited above, it is clear that there are different environmental cues for the induction of GS_1 and GS_2. Thus, although characteristics of the monomers for the different glutamine synthetase's seem to be largely conserved, this is not true for the regulatory functions.

B. Glutamate Synthase

The purification and properties of glutamate synthase (glutamine–2-oxoglutarate amino transferase; GOGAT) has been reviewed by Stewart *et al.* (1980), Suzuki and Gadal (1984), and Lea *et al.* (1990). It has not been studied as intensively at the molecular level as GS even though it has a crucial role in maintaining appropriate levels of glutamate. In higher plants there are two major isoforms of GOGAT, one requiring NADH or NADPH as reductant and the other ferredoxin (Fig. 1b; Fig. 5, reaction 15). The basic reaction was first discovered in bacteria by Tempest *et al.* (1970), and subsequently in carrot tissue cultures by Dougall (1974) and in chloroplasts by Lea and Miflin (1974). Although an NADH or an NAD(P)H-bispecific enzyme is present in most tissues, a GOGAT that uses reduced ferredoxin as reductant is the major form of the enzyme in both green (Lea and Miflin, 1974) and nongreen tissues (Suzuki and Gadal, 1982; Suzuki *et al.,* 1983, 1984, 1988; Matoh *et al.,* 1980a; Sechley *et al.,* 1991). The ferredoxin GOGAT has been purified from leaves of a number of plant species (see Suzuki and Gadal, 1984, for early references; Matoh *et al.,* 1979; Hirasawa and Tamura, 1984; Marquez *et al.,* 1988). The molecular size of the native protein ranges from 140 to 244 kDa. The spinach and barley enzymes appear to contain a single polypeptide, whereas the rice enzyme was reported to be a homodimer with subunits of 115 kDa (Suzuki and Gadal, 1982). However, recent studies with rice tissue cultures indicate that it has a single polypeptide with a molecular weight of 160 kDa (Hayakawa *et al.,* 1990; T. Yamaya, unpublished observations). Based on absorbance spectra of the purified enzyme, it is an iron–sulfur flavoprotein. Recently, Sakakibara *et al.* (1991) cloned and characterized cDNA responsible for the coding of the ferredoxin-dependent GOGAT and have demonstrated with this clone that Fd-GOGTAT message is enhanced by light treatment.

The NAD(P)H-GOGAT is much less active than the ferredoxin-GOGAT and, except in legume nodules, has received much less attention. It has been partially purified from several tissues (Matoh *et al.,* 1980a,b; Chiu and Shargool, 1979; Avila *et al.,* 1987). It, too, is an iron–sulfur flavoprotein, and it has a single subunit of 230 kDa. The NADH-GOGAT from root nodules of *P. vulgaris* comprises two distinct isoenzymes that can be separated by ion-exchange chromatography (Chen and Cullimore, 1988). In this case, and in the case of NADH-GOGAT isolated from alfalfa nodules (Anderson *et al.,* 1989), the protein is a monomer with a molecular weight of abut 200 kDa. There have also been reports of a ferredoxin-requiring GOGAT in alfalfa (Suzuki *et al.,* 1988) and soybean (Suzuki *et al.,* 1984) nodules. As with NADH GOGAT, the activities of ferredoxin-GOGAT increase dramatically with nodule development.

Results derived from cell fractionations indicate that ferredoxin-GOGAT is localized in the chloroplasts in green leaves (Matoh and Takahashi, 1981; Wallsgrove *et al.,* 1982). The NADH-GOGAT is also largely located in chloroplasts in leaves (Matoh and Takahashi, 1981), in plastids in roots (Suzuki *et al.,* 1981), and in plastids in nodules (Awonaike *et al.*, 1981; Boland *et al.*, 1982; Shelp *et al.*, 1983; Shelp and Atkins, 1984). There is, however, one report based on chemical data rather than cell fractionation that suggests that an NADH-GOGAT is located in the cytosol (Hecht *et al.*, 1988). In 1988, Botella *et al.* presented direct evidence on the localization of ferredoxin-GOGAT in the chloroplasts of tomato leaves, using monospecific antibodies raised against ferredoxin-GOGAT and immunogold techniques. The enzyme is confined to the chloroplast stroma in mesophyll cells of tomato leaves.

Compared to glutamine synthetase, there is much less information at the molecular level on the isoenzymes of GOGAT. Antibodies have been produced to ferredoxin-GOGAT prepared from leaves of a number of plant species (Suzuki *et al.*, 1982, 1985; Botella *et al.*, 1988; Hayakawa *et al.*, 1990) and an antibody against NADH-GOGAT has recently been obtained from alfalfa root nodules (Anderson *et al.*, 1989). T. Hayakawa and T. Yamaya (unpublished observations) recently obtained monospecific antibodies to the NADH-GOGAT in rice leaves. Antibodies to ferredoxin-GOGAT do not recognize NADH-GOGAT, indicating that the two proteins are quite distinct (Suzuki *et al.*, 1982), and conversely antibodies to the NADH-GOGAT do not recognize the ferredoxin protein (T. Hayakawa and T. Yamaya, unpublished observations). The ferredoxin-GOGAT obtained from maize root bears enough similarity to the leaf protein to be inactivated by antibody prepared against the leaf enzyme (Suzuki *et al.*, 1985). However, results with the Ouchterlony double-immunodiffusion technique also indicate that there are important differences between the root and leaf enzyme. The antibody to the alfalfa nodule NADH-GOGAT cross-reacts with the NADH-GOGAT in nodules from a number of legume species, but lacks reactivity with the NADH-GOGAT of alfalfa roots and leaves (Anderson *et al.*, 1989). Thus distinct isoforms appear to be induced in root tissue in response to infection with *Rhizobium*.

As with the plastid glutamine synthetase, the ferredoxin-GOGAT activity is enhanced in leaves during greening (Suzuki *et al.*, 1982; Matoh and Takahashi, 1981; Wallsgrove *et al.*, 1982; Suzuki *et al.*, 1987). The induction is more pronounced in C_3 than in C_4 plants (Suzuki and Gadal, 1984). In maize the increase in ferredoxin-GOGAT activity is accompanied by an increase in the amount of protein and the level of NADH-GOGAT activity declined during greening (Suzuki *et al.*, 1987).

As with GS_2, the major function of the ferredoxin-GOGAT appears to be in the reassimilation of photorespiratory NH_4^+ (Kendall *et al.*, 1986; Keys *et al.*, 1978; Somerville and Ogren, 1980b). As with other photorespiratory mutants, mutants of ferredoxin-GOGAT are lethal under conditions that support photorespiration, but the mutants survive under conditions that repress photorespiration, an observation that indicates that ferredoxin-

GOGAT is not the only mechanism for the synthesis of glutamate in green leaves. These mutants contain wild-type levels of NADH-GOGAT activity (Kendall *et al.*, 1986) and it has been suggested that levels of this enzyme in *Arabidopsis* (Morris *et al.*, 1989), or in barley (Kendall *et al.*, 1986) mutants that lack the ferredoxin-enzyme, are sufficient to handle the flux of NH_4^+ (or glutamine) derived from other sources. Using mutant methodology, similar arguments have been put forth for *Neurospora* mutants that lack glutamate dehydrogenase (Hummelt and Mora, 1980). In *E. coli* glutamate dehydrogenase or glutamate synthase can compensate for each other (Rosenfeld and Brenchley, 1983). *Escherichia coli* mutants that lack GOGAT can grow on NH_4^+, but only when that NH_4^+ is not rate limiting. This observation suggests that glutamate dehydrogenase can provide the glutamate under specific environmental conditions.

The function of the ferredoxin-GOGAT in roots or other nongreen tissues was a puzzle until a ferredoxin-like protein was identified in nongreen tissues (Ninomiya and Sato, 1984; Suzuki *et al.*, 1985; Wada *et al.*, 1986). This protein, although not identical to leaf ferredoxin, does bear some antigenically similar epitopes (Suzuki *et al.*, 1985). Wada *et al.* (1986) purified the ferredoxin from the storage root of radish and this protein differs from the leaf ferredoxin in amino acid content and in the N-terminal sequence. However, it does have an absorption spectrum typical of leaf ferredoxin and it mediates NADP reduction in broken chloroplasts from spinach.

Oji *et al.* (1985, 1989) and Emes and Bowsher (1991) established that root ferredoxin as well as the root GOGAT are localized in the root plastid. Furthermore, a ferredoxin-NADP reductase in maize roots bears some antigenic similarity to the ferredoxin-NADP reductase purified from spinach leaves (Suzuki *et al.*, 1985). The root protein could therefore transfer electrons from NADH or NADPH to the root ferredoxin. Oji *et al.* (1985, 1989) suggested that the NADPH required for the GOGAT reaction was supplied by glucose 6-phosphate and 6-phosphogluconate dehydrogenases. Emes's group later demonstrated that these two enzymes and the root ferredoxin are located in pea root plastids and are induced by NO_3^- additions to growth medium (Emes and Bowsher, 1991). The phosphate translocator in root plastids mediates the entry of glucose 6-phosphate. This is a unique property not associated with phosphate translocators in chloroplasts (Borchert *et al.*, 1989). Recent studies with rice cell cultures also demonstrate that levels of ferredoxin-GOGAT as well as GS_2 are enhanced in cells grown with NO_3^- as the sole source of nitrogen (Hayakawa *et al.*, 1990). Localization of this enzyme as well as NiR and GS_2 in plastids suggests that ferredoxin-GOGAT is involved in the primary assimilation of NO_3^-.

C. Glutamate Dehydrogenase

Glutamate dehydrogenase (GDH) catalyzes the reductive amination of 2-oxoglutarate in the presence of NADH or NADPH, or the oxidative deamination of glutamate (Fig. 1a). Its activity, occurrence, and possible function in higher plants were summarized by Stewart *et al.* (1980), Srivastava and Singh (1987), and Lea *et al.* (1990). There are two major types of GDH, an NADH-requiring form that is found in the mitochondria, and an NADPH form in the chloroplast. Molecular and genetic studies on plant GDH are still minimal (Bascomb *et al.,* 1987; Magalhaes *et al.,* 1990), although recently there have been several reports of antibody production to pure GDHs (Kamachi *et al.,* 1991; Laurière *et al.,* 1981; Loulakakis *et al.,* 1990).

The plant GDH is a metalloprotein having a molecular size in the range of 208 to 270 kDa which is composed of identical subunits of 46 to 58.5 kDa in size (Miflin and Lea, 1982). There is, however, a discrepancy in the literature regarding the number of subunits. GDH in *Pisum* seeds (Kindt *et al.,* 1980) and *Arabidopsis thaliana* (Cammaerts and Jacobs, 1983) is thought to be a hexamer, whereas Scheid *et al.* (1980) reported that the enzyme from *Lemna* and *Pisum* seeds has a tetrameric configuration. GDH is frequently separated into 7 (Fawole, 1977; Scheid *et al.,* 1980) to 14 forms (Hartmann *et al.,* 1973; Mazurowa *et al.,* 1980) with native polyacrylamide gel electrophoresis (PAGE). According to Cammaerts and Jacobs (1983), de Vienne (1983), and Magalhaes *et al.* (1990), this is the result of random association of two or more subunits into a hexamer complex.

The amination reaction of GDH is activated by Ca^{2+} over a concentration range of 1 μM to 10 mM *in vitro* (Chou and Splittstoesser, 1972; Joy, 1973, Furuhashi and Takahashi, 1982; Yamaya *et al.,* 1984). Other divalent cations, such as Zn^{2+} and Mn^{2+}, are also effective on the activation of NADH-GDH (Srivastava and Singh, 1987). The apparent K_m of the enzyme for NH_4^+ (which ranges from 5 to 70 mM) is much higher than the K_m with glutamine synthetase, and this fact is often used to question the aminating role of GDH in plant tissues. However, this is not a good argument because the two enzymes are localized in different compartments within the cell. In addition, the K_m value for NH_4^+ is influenced by many factors, i.e., it decreases manyfold at low concentrations of NH_4^+ and NADH (Nauen and Hartmann, 1980; Palich and Gerlitz, 1980; Yamaya *et al.,* 1984). Saturation curves for NADH-GDH shows normal Michaelis–Menten kinetics for NH_4^+ in the presence of Ca^{2+}, but, in the absence of Ca^{2+} a marked substrate inhibition is observed (Furuhashi and Takahashi, 1982; Yamaya *et al.,* 1984). Other mechanisms for the regulation of GDH may also exist. In yeast, for example, GDH activity is covalently modified through a phosphorylation–dephosphorylation mechanism (Hemmings, 1982; Uno *et al.*, 1984). We have noted that seasonal variation in GDH activity in roots is not correlated with the seasonal distribution in the abundance of GDH protein as detected by Western analysis (Sechley *et al.*, 1991). This indicates that the regulation of GDH activity is not well understood in higher plants.

NADH-GDH is found in the mitochondria of roots (Suzuki *et al.*, 1981), shoots (Bowman *et al.*, 1976; Miflin, 1974; Nauen and Hartmann, 1980; Yamaya *et al.*, 1984), and seeds (Priestly and Bruisma, 1982). In C_4 leaves, NADH-GDH is found in bundle sheath cells where photorespiration occurs but not in the mesophyll cells (Rathnam and Edwards, 1976). Because the NADH-GDH is relatively easily dissolved it is thought to be localized in the soluble matrix of the mitochondria (Bowman *et al.*, 1976; Nauen and Hartmann, 1980; Priestly and Bruisma, 1982). However, the method of disrupting mitochondria usually involves treatments such as freezing and thawing followed by sonication. Yamaya *et al.* (1984), using a gentle disruption method, osmotic shock, demonstrated that the NADH-GDH in corn shoots is loosely associated with the mitochondrial membrane. Duke *et al.* (1978) also deduced that the GDH is probably associated with membrane lipids.

NADPH-GDH is found in chloroplasts of many higher plants (Bascomb and Schmidt, 1987; Lees and Dennis, 1981; McKenzie *et al.*, 1981; Rathnam and Edwards, 1976). The NADPH-GDH is known to be tightly bound to the thylakoid membranes and it can be dissolved only by a detergent treatment (Leech and Kirk, 1968). In nongreen tissues, NADPH-GDH appears to be localized in plastids (Washitani and Sato, 1977). *In vitro* translation products encoded by NADPH-GDH mRNA from *Chlorella sorokiniana* result in the synthesis of a precursor polypeptide having a molecular size 58.5 kDa. It is processed *in vitro* to the mature size of NADPH-GDH (55.8 kDa) in the presence of a soluble fraction prepared from broken cells (Bascomb *et al.*, 1987; Prunkard *et al.*, 1986).

In 1974, Pryor described a GDH null mutant in maize and these mutants are more sensitive to the cold than are related wild-type plants (Pryor, 1990). Recently Magalhaes *et al.* (1990) characterized the mutant and wild-type GDHs in related strains of maize. In the wild type there are seven GDH isozymes on native PAGE. The array in the wild type could be accounted for by the random aggregation of two distinct polypeptides. In the mutant, one of the associating monomers is missing, and this results in only one protein band in native PAGE. The total GDH activity is 10 to 15-fold lower than the wild type. These mutants also exhibited a 40–50% lower rate of $^{15}NH_4^+$ assimilation into total reduced nitrogen and reduced shoot development. Thus it appears that glutamate dehydrogenase is functioning in the assimilation of NH_4^+. However, it is not known whether reduced shoot growth is a result, or a cause, of the decreased rates of NH_4^+ assimilation in this plant (Magalhaes *et al.,* 1990).

In lupine root nodules, it has been shown that GDH is present in eight isoforms, only one of which is identical with the root form (Ratajczak *et al.,* 1986). Immunodiffusion tests revealed that seven forms of the nodule GDH arose as the result of random association of two subunits in a hexameric complex, while the eighth form consisted of totally distinct subunits. Thus, as with GS and GOGAT, there is the appearance of a new plant-derived protein as a result of infection with *Rhizobium*.

With the discovery of GOGAT (Tempest *et al.,* 1970; Lea and Miflin, 1974), the role of GDH in the assimilation of NH_4^+ was thought to be unimportant in higher plants and in other organisms as well (see Rosenfeld and Brenchley, 1983). Similarly, time course studies with $^{15}NO_3^-$, inhibitor studies with methionine sulfoximine (an inhibitor of glutamine synthetase) or azaserine (an inhibitor of GOGAT), and the relatively high K_m for NH_4^+ indicate that GDH is not involved in the primary assimilation of NH_4^+, except perhaps under certain environmental or developmental conditions (Miflin and Lea, 1982; Oaks and Hirel, 1985; Rhodes *et al.,* 1989; Srivastava and Singh, 1987). Unlike GS_2 and GOGAT, GDH is located primarily in the mitochondria. Glutamate dehydrogenase is also localized in mitochondria in animal cells (Frieden, 1963). Although glutamate may be a substrate for respiration in animal cells it should be noted that GDH is a highly regulated protein and as a result may have other functions in the cell. Experiments by Davies and Teixeira (1975), using mitochondria from etiolated pea shoots, and by Journet *et al.* (1982), using potato tubers, suggest that glutamate is also a substrate for mitochondrial oxidation in plants. This idea is supported by recent work from Stewart's laboratory, where NMR was used to detect the fate of glutamate in carrot suspension cultures (Robinson *et al.,* 1991). Their results show that glutamate is oxidized in storage tissue cells. However, when glutamate metabolism is examined in cells that are reducing, rather than oxidizing, reserves, evidence for the amination of 2-oxoglutarate is obtained. Aminooxyacetate (an inhibitor of transaminase reactions) inhibits the oxidation of glutamate, but not of 2-oxoglutarate or other potential substrates of mitochondrial oxidations (Yamaya amd Matsumoto, 1985b; Yamaya *et al.,* 1984, 1986). This suggests that 2-oxoglutarate derived from transamination reactions, and not via GDH, is the true substrate for mitochondrial oxidations. Glutamate is also decarboxylated to 4-aminobutyrate, which is ultimately converted to succinate (Streeter and Thompson, 1972 a,b). This succinate is also a potential substrate for mitochondrial oxidation. In addition, when [^{14}C]citrate (Hartmann and Ehmke, 1980), $^{15}NH_4^+$, or [^{15}N]glycine (Yamaya *et al.,* 1986) are supplied as substrates to intact (state 4) mitochondria, [^{14}C]- or [^{15}N]glutamate is a product. In these experiments, however, the rate of synthesis of glutamate was low, certainly much lower than would be required to account for the assimilation of the NH_4^+ generated by photorespiration. It could be that some reaction other than GDH, the supply of carbon, for example, could be limiting the production of glutamate in the mitochondria. Evidence that would support this contention are (1) that pyruvate dehydrogenase activity is reduced in the light (Budde and Randall, 1990), (2) that citrate may be exported from the mitochondria in the light to support glutamate synthesis in the chloroplast (Chen and Gadal, 1990a), and (3) that the addition of 2-oxoglutarate to mitochondrial preparations enhances the production of [^{15}N]glutamate (T. Yamaya and A. Oaks, unpublished observations).

D. Assessment of GS, GOGAT, and GDH Reactions

From the experiments described here it is clear that GS_2 and GOGAT are active in maintaining the levels of glutamate required to permit the cycling of carbon in the photorespiratory cycle. It is also clear that the functions of GS_1 and GS_2 do not overlap. Their cellular and tissue localization differ, and GS_1 does not appear replace the function of GS_2 in mutants that lack GS_2. It is also clear that GDH does not replace glutamine synthetase when inhibitors of glutamine synthetase are added to the system. However, it should be noted that inhibitors of glutamine synthetase activity have extensive secondary effects within the cell (see Section VI,A; Sauer *et al.*, 1987; Sieciechowicz *et al.*, 1989; Walker *et al.*, 1984a,b). It is our contention that because GS and GDH are in different parts of the cell they probably have quite different functions. At the present time we do not adequately understand the regulation, or know the real function of GDH in higher plants.

IV. Amino Acid Biosynthesis

In several recent reviews (Bryan, 1980, 1990; Coruzzi, 1991) the biochemical and molecular aspects of amino acid biosynthesis have been well covered. Most of the amino acids are synthesized in chloroplasts in leaves (Bryan, 1990; Wallsgrove *et al.*, 1983), an observation that is not surprising because major proteins such as the large subunit of ribulosebisphosphate carboxylase (Rubisco) and the 32-kDa protein are made there as well (Ellis, 1981). There is some, but less substantial, evidence that suggests that the plastid fraction in roots is also a major site of amino acid biosynthesis (Miflin, 1974; Burdge *et al.*, 1979; K. G. Wilson, J. M.

Widholm, and A. Oaks, unpublished observations). In this article we will concentrate on the biosynthesis of alanine, arginine, asparagine, glycine, and serine (also see Section VI,A), which are directly or indirectly involved with processes that shuttle nitrogen and carbon between the chloroplast, mitochondrion, and peroxisome.

A. Alanine

The flux of nitrogen involved in photorespiration is an order of magnitude above that involved in primary nitrogen assimilation (Wallsgrove *et al.*, 1983). This, coupled with the realization that alanine is a major nitrogen donor during photorespiration (see Section VI,A; Betsche, 1983; Ta and Joy, 1986), and that up to 28% of the amino nitrogen of maize phloem sap is alanine (Valle and Heldt, 1991), has generated interest as to the origin of alanine within the cell. However, the reactions involved in achieving these rates of alanine synthesis are not fully understood. Alanine becomes rapidly labeled in the light when $^{14}CO_2$ is provided to isolated cells (Larsen *et al.*, 1981; Lawyer *et al.*, 1981). The formation of alanine through the direct incorporation of ammonium in the presence of NAD(P)H and pyruvate is mediated by a pure alanine dehydrogenase preparation from cyanobacteria (Rowell and Stewart, 1976). However, this reaction does not appear to be significant in higher plants (Joy, 1988). The major route through which alanine is synthesized appears to involve the transamination of pyruvate (Fig. 3) in the presence of glutamate (reaction 14) (Valle and Heldt, 1991; Wallace *et al.*, 1984), 4-aminobutyrate (reaction 14) (Streeter and Thompson, 1972a,b; Tsushida and Murai, 1987), serine, asparagine (reaction 46) (Ireland and Joy, 1983; Murray *et al.*, 1987), or glutamine (reaction 46; Ireland, 1986). The *in vitro* rates of alanine synthesis in the presence of serine or asparagine are equivalent to the rates of glycine synthesis (in the presence of the same amino donors), yet the rates of alanine synthesis are three-fold greater than glycine when glutamate is the amino donor (Murray *et al.*, 1987). This suggests that *in vivo* alanine formation may be preferred over glycine in the presence of glutamate. Indeed, a two-fold increase in the flux of ^{15}N from glutamine or glutamate into alanine is observed under photorespiratory conditions (Joy, 1988). As alanine does not appear to be synthesized within isolated mitochondria (K. W. Joy, personal communication), its synthesis in the chloroplast may be important during photorespiration.

Because the amount of alanine increases in response to several stresses (Streeter and Thompson, 1972a,b; Turpin *et al.*, 1990; Wallace *et al.*, 1984), as does alanine aminotransferase itself (Good and Crosby, 1989), the increase of alanine pools observed in several photorespiratory mutants in air (e.g., Lea and Blackwell, 1990) may arise from the preferential transamination of pyruvate with either glutamate, or its decarboxylation product 4-aminobutyrate. The flow of carbon leading toward alanine synthesis results in the production of ATP and NADH, which could be used for ammonium assimilation (Vanlerberghe *et al.*, 1990, 1991) or other processes under anaerobic conditions.

Transamination of pyruvate would be favored under conditions that limit its degradation by pyruvate dehydrogenase. This selective metabolism of pyruvate may occur *in vivo* because ammonium, ATP (Schuller and Randall, 1989), or acetyl-CoA (Wiskich, 1980) inhibit *in vitro* pyruvate dehydrogenase activity, even though they favor the synthesis of amino donors required for transamination. However, ATP inhibits pyruvate kinase activity and, unless there are other mechanisms for the synthesis of pyruvate, this would limit the availability of pyruvate for transamination. As there is an inverse relationship between the pool sizes of malate and alanine in air in several plants that are deficient in processes associated with photorespiration (e.g., Blackwell *et al.*, 1988; Murray *et al.*, 1987; Walker *et al.*, 1984a; Wallsgrove *et al.*, 1986), it is possible that pyruvate, and therefore alanine, pools are derived from the decarboxylation of malate by malic enzyme (Valle and Heldt, 1991). In this case, acetyl-CoA would limit pyruvate oxidation by inhibiting pyruvate dehydrogenase activity and at the same time ensure pyruvate synthesis through the activation of malic enzyme activity (Day *et al.*, 1984).

Malate can be synthesized from phosphoenolpyruvate via the reactions of phosphoenolpyruvate carboxylase and malate dehydrogenase (e.g., Dahlbender and Strack, 1986). Hence it could be considered a normal end product of glycolysis (Lance and Rustin, 1984). Malate may also be a product of glutamate breakdown, synthesized from the decarboxylation of glutamate, transamination of 4-aminobutyrate, oxidation of succinic semialdehyde and succinate, and hydrolysis of fumarate, with the latter three reactions taking place in the mitochondria via the tricarboxylic acid (TCA) cycle (Streeter and Thompson 1972a,b; Walker *et al.*, 1984b). Alanine is involved in the photorespiratory pathway, and its role in this capacity (Section VI,A) as well as its contribution to the nitrogen balance within the peroxisome will be examined (Section V,C).

B. Arginine Biosynthesis

The pathway for arginine biosynthesis in higher plants has been described by Thompson (1980), Shargool *et al.* (1988), and Micallef and Shelp (1989a–c), and in *Neurospora* by Davis and co-workers (Davis, 1986; Davis and Weiss, 1988). In both higher plants (Vanetten *et al.*, 1963) and fungi (Davis and Weiss, 1988) arginine can be a major nitrogen storage compound as well as an intermediate in polyamine biosynthesis (via agmatine to putrescine) (Galston and Sawheny, 1990) or a precursor for protein synthesis. The synthesis of arginine requires a three-step process and involves the addition of carbamoyl phosphate to ornithine, producing citrulline, followed by the incorporation of aspartate to yield arginosuccinate, and finally the release of fumarate from arginosuccinate to produce arginine. In fungi and perhaps in higher plants as well there is a dynamic balance between arginine and ornithine. In *Neurospora* (Davis and Weiss, 1988) and in animal cells (Gamble and Lehninger, 1973) the enzymes involved in carbamoyl phosphate, ornithine, and citrulline biosynthesis are located in the mitochondria. Citrulline is released into the cytosol, where arginine is made. This arginine is a precursor for both protein synthesis and for the production of urea and ornithine via arginase activity. Ornithine can enter the vacuole, mitochondria, or via the action of ornithine aminotransferase can contribute its carbon to proline (Davis and Weiss, 1988). Ornithine can also serve as a precursor to proline biosynthesis in higher plants (Sans *et al.*, 1988; Mestichelli *et al.*, 1979) as well as serve as an intermediate in alkaloid or polyamine formation.

Unlike *Neurospora,* where carbamoyl-phosphate synthetase and ornithine transcarbamylase activities are localized in the mitochondria, these enzymes have been found in plastids obtained from either tissue culture or cotyledons of higher plants (Shargool *et al.,* 1978, 1988; Taylor and Stewart, 1981). The enzymes converting citrulline to arginine are localized in the cytosol and the enzymes involved in the degradation of arginine (arginase and ornithine aminotransferase) are in the mitochondria. In animal cells (Gamble and Lehninger, 1973) as well as in *Neurospora,* ornithine is taken up by the mitochondria and citrulline is released. From the work done to date it would appear that plants have a very different organization in the synthesis and degradation of arginine.

Recently, Ludwig (1991) proposed a scheme involving the shuttling of ornithine, citrulline, and arginine between the chloroplast and mitochondria. Ornithine enters the mitochondria where it combines with carbamoyl phosphate to yield citrulline (Fig. 4, reactions 8 and 33). The citrulline is exported in exchange for ornithine and it is converted in the cytosol to arginine (a two-step process involving arginosuccinate synthetase, which produces arginosuccinate from citrulline, aspartate, and ATP, followed by arginosuccinate lyase, which forms fumarate and arginine; Fig. 5), which would then enter the chloroplast. This proposal requires three unique reactions in the chloroplast (Fig. 5, reactions 4, 7, and 33) and, for higher plants, a unique reaction in the mitochondria (Fig. 4, reaction 8): within the chloroplast, (1) arginine iminohydrolase releases citrulline and NH_4^+; (2) the hydrolysis of citrulline by ornithine carbamoyltransferase produces

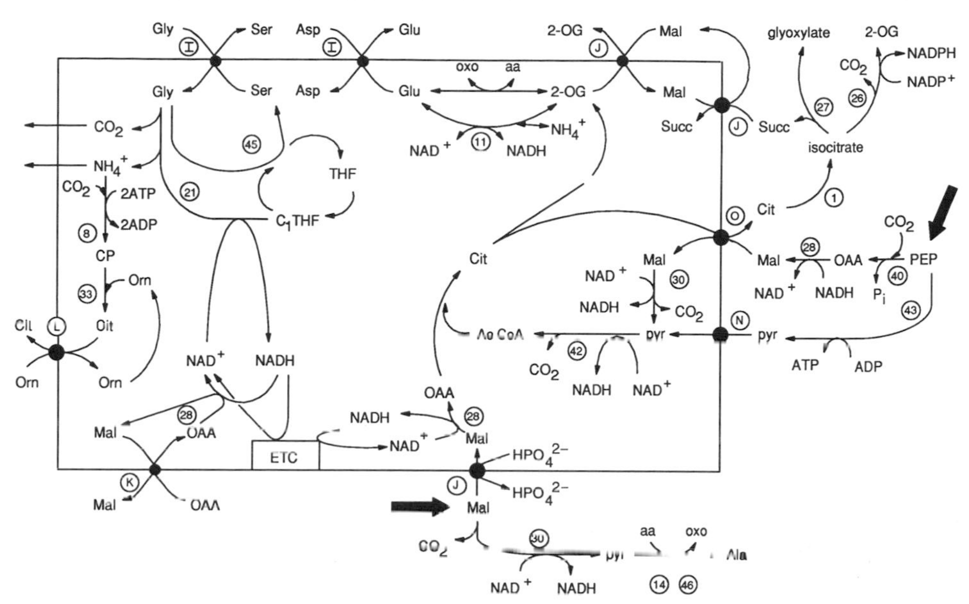

FIG. 4 Metabolic reactions related to nitrogen processing that take place within the mitochondria. Large arrows indicate inputs derived from glycolysis. See Tables II and III for a list of enzymes and translocators involved. CP, Carbamoyl phosphate; ETC, electron transport chain; Succ, succinate; THF, tetrahydrofolate; OAA, oxaloacetic acid, aa, amino acid; oxo, oxo acid.

ornithine and carbamoyl phosphate; (3) carbamoyl phosphate is catabolized by carbamate kinase in the presence of ADP to yield NH_4^+, HCO_3^-, and ATP; and (4) within the mitochondria carbamoyl-phosphate synthase assimilates NH_4^+ and CO_2 into carbamoyl phosphate. In support of this model, Ludwig has observed that chloroplasts fed arginine release NH_4^+, that carbamate kinase is active in chloroplast extracts, and that ^{14}C-labeled guanidinoarginine taken up by intact chloroplasts is metabolized to yield $^{14}CO_2$. This is all supporting evidence for the existence of a pathway for arginine degradation in the chloroplast but the uptake of arginine and the release of ornithine by chloroplasts has yet to be demonstrated. Previous workers have not detected carbamoyl-phosphate synthetase activity in isolated mitochondria (Shargool *et al.*, 1978, 1988; Taylor and Stewart, 1981), yet this protein is present in this organelle as detected by Western blotting (Ludwig, 1991). It remains to be determined whether or not mitochondria are capable of synthesizing citrulline from supplied ornithine. Furthermore, by using ^{14}C-labeled TCA cycle intermediates or glycine, the source of CO_2 for carbamoyl phosphate synthesis could be defined. Similarly, by using ^{15}N-labeled glycine it could be determined whether or not glycine nitrogen is destined for citrulline as well as serine synthesis. At

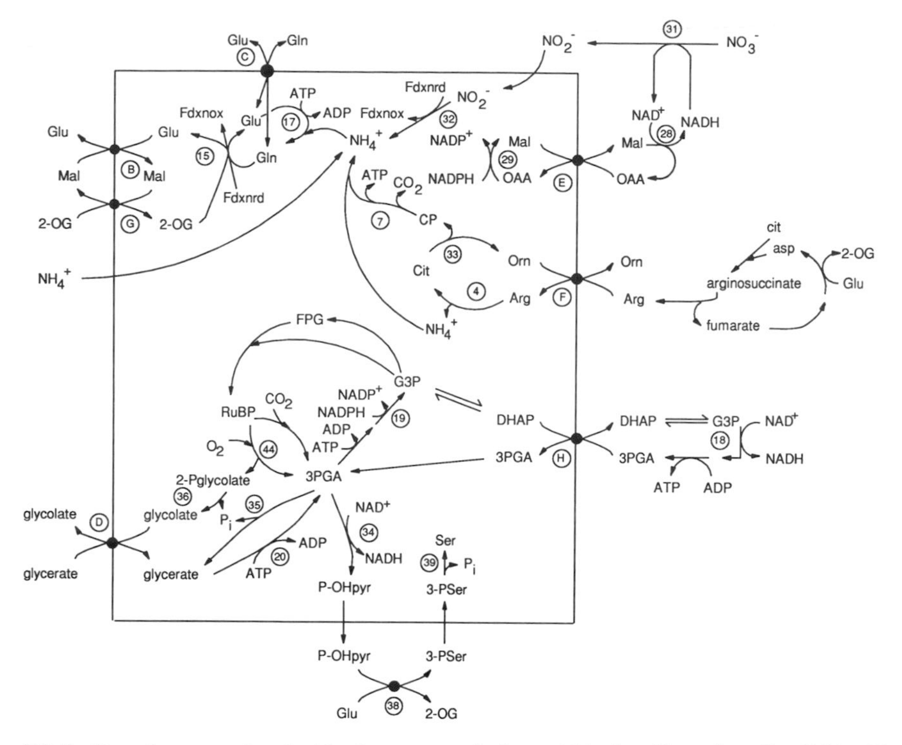

FIG. 5 Reactions associated with nitrogen metabolism within the chloroplast. See Tables II and III for a list of enzymes and translocators. Fdxnox, Oxidized form of ferredoxin; Fdxnrd, reduced form of ferredoxin; DHAP, dihydroxyacetone phosphate; 3-PGA, 3-phosphoglyceraldehyde; FPG, fructose-1,6-bisphosphate; RuBP, ribulose-1,5-bisphosphate.

this stage it is difficult to rationalize these new results with the fact that in previous studies the biosynthetic reactions in higher plants (ornithine and carbamoyl phosphate to citrulline) were compartmentalized in the plastids whereas the catabolic reactions (arginase) were localized in the mitochondria (Shargool *et al.*, 1988). What makes Ludwig's hypothesis attractive is that it presents an efficient mechanism for the transfer of both NH_4^+ and CO_2, released in the mitochondria, to the chloroplast. This aspect of Ludwig's scheme will be dealt with in photorespiration (Section VI,A).

C. Asparagine

The synthesis and metabolism of asparagine (reviewed by Sieciechowicz *et al.*, 1988) is of importance in plants, where this is a dominant form of nitrogen that is either transported from roots or leaves (Pate, 1973;

Schubert, 1986) or stored in the vacuole (Genix *et al.*, 1990; Givan, 1979; Goodchild and Sims, 1990). Asparagine has a low carbon : nitrogen ratio of 2 : 1 and this minimizes the carbon requirement for its synthesis within the root. Furthermore, as it is selectively catabolized in developing leaves, roots, or seeds it is ideally suited as a transport compound supplying nitrogen (and carbon) required for organ growth (Sieciechowicz *et al.*, 1988). Asparagine also accumulates in tissues in the presence of NH_4^+ (Givan, 1979) or as an end product of cyanide metabolism (Conn, 1979; Lea and Miflin, 1980; Sellmar *et al.*, 1988).

Two reactions catalyze the synthesis of asparagine. The first involves its synthesis from aspartate in the presence of ATP and either glutamine or NH_4^+ and is catalyzed by asparagine synthetase. This reaction is a potential enzymatic route leading to asparagine synthesis in maize roots (Stulen *et al.*, 1979), cotyledons (Rognes, 1975; Streeter, 1973), or the plant fraction of legume nodules (Scott *et al.*, 1976). Transcripts encoding glutamine-dependent activity are present in cotyledons and nodules in pea (Tsai and Coruzzi, 1990). In leaves or roots of pea, however, this activity is not discernable even though both the *in vivo* synthesis of asparagine from [^{14}C]aspartate (Joy *et al.*, 1983), and an increase in transcripts encoding asparagine synthetase (Tsai and Coruzzi, 1990) are observed in dark-adapted plants. Leaves of *Pisum* and *Asparagus* contain inhibitory components of asparagine synthetase activity and this results in the underestimation of activity from these tissues (Joy *et al.*, 1983).

A second reaction leading to asparagine synthesis involves the hydrolysis of β-cyanoalanine catalyzed by β-cyanoalanine hydrolase (Castric *et al.*, 1972). Asparagine synthesized in this manner can be considered an end product in the detoxification of cyanide (Conn, 1979; Lea and Miflin, 1980). β-Cyanoalanine is produced from cyanide and cysteine by β-cyanoalanine synthase, and the distribution of this enzyme appears to be ubiquitous in higher plants (Miller and Conn, 1980). Cyanide is formed either during the species-specific degradation of cyanogenic glycosides (Conn, 1979), or during the conversion of 1-aminocyclopropane-1-carboxylic acid to ethylene (Peiser *et al.*, 1984). The ethylene-forming enzyme responsible for ethylene biosynthesis is localized on the plasma membrane (Bouzayen *et al.*, 1990). β-Cyanoalanine synthase activity, on the other hand, appears to be localized within the mitochondrial fraction (Tittle *et al.*, 1990). Unless the activities of the ethylene-forming enzyme, β-cyanoalanine synthase, and β-cyanoalanine hydrolase are tightly coordinated, toxic effects arising from cyanide accumulation could occur. Interestingly, β-cyanoalanine synthase activity does increase in response to ethylene in roots, shoots, and cotyledons of *Pisum, Phaseolus, Glycine, Lactuca, Sinapsis, Hordeum,* and *Triticum* (Goudey *et al.*, 1989), and in maturing carnation flowers (Manning, 1986). A corresponding increase in β-cyanoalanine hydrolase activity and an associated accumulation of asparagine might also be expected, but this has not yet been determined. In maize roots, asparagine synthetase activity is highest in mature regions, whereas β-cyanoalanine synthase activity is relatively higher in the root tip (Stulen *et al.*, 1979). It would be of interest to determine the distribution of β-cyanoalanine hydrolase in this tissue.

V. Organelles and Nitrogen Metabolism

A. Mitochondrion

The oxidation of glycine is one of the major metabolic processes that takes place within the mitochondria in the light. The enzyme complexes responsible for its catabolism are present in great abundance, approaching one-third the soluble mitochondrial protein (Oliver *et al.*, 1990a). The glycine decarboxylase multienzyme complex (Fig. 4, reaction 21) is composed of four associated proteins (P, H, T, and L) that catalyze the catabolism of glycine in the presence of NAD^+ and tetrahydrofolate to produce CO_2, NH_4^+, NADH, and N^5,N^{10}-methylene tetrahydrofolate (Oliver *et al.*, 1990b). Serine is formed by serine hydroxymethyltransferase (Fig. 4, reaction 45), which condenses the methylene moiety of N^5,N^{10}-methylenetetrahydrofolate with glycine (Neuburger *et al.*, 1986). This latter enzyme is also localized within the chloroplast (Shah and Cossins, 1970).

The NADH and serine formed through the cleavage of glycine need to be consumed efficiently within the mitochondria because they are competitive inhibitors of the glycine decarboxylase multienzyme complex (Oliver *et al.*, 1990a). At least two possible routes exist for the oxidation of glycine-derived NADH. The first involves oxidation of NADH via the mitochondrial respiratory chain, which is operative in the light (Krömer *et al.*, 1988; Krömer and Heldt, 1991; Gardeström and Wigge, 1988; Weger *et al.*, 1988). The glycine-derived NADH is used in preference to TCA- or externally derived NADH (Dry *et al.*, 1983, 1987). Under photorespiratory conditions the contribution of the chloroplast in supplying ATP to the cytosol is reduced (Gardeström and Wigge, 1988; Usuda and Edwards, 1982). Therefore, the preferential oxidation of glycine ensures the export of ample ATP from the mitochondria under these conditions (Gardeström and Wigge, 1988; Krömer and Heldt, 1991). Second, NADH may be shuttled out of the mitochondria as malate by way of a malate-oxaloacetate transporter (Ebbighausen *et al.*, 1987) (Fig. 4, translocator K). However, the estimated rates of malate efflux from the mitochondria are below those required to support hydroxypyruvate reduction in the peroxisome (Dry *et al.*, 1987). The export of NADH from the mitochondria would occur when the intramitochondrial concentration of ADP decreases limiting NADH oxidation by the electron transport chain. To account for the fact that glycine is oxidized simultaneously with malate or oxaloacetate, Wiskich *et al.* (1990) proposed that enzyme complexes within the mitochondria are compartmentalized into domains. These metabolons permit the concurrent oxidation of glycine-derived NADH in the presence of oxaloacetate via malate dehydrogenase to produce malate for export, while elsewhere malate is converted to oxaloacetate, during tricarboxylic acid cycle activity (Fig. 4, reaction 28).

Ammonium is also formed during the decarboxylation of glycine. In this regard it is interesting to note that mitochondrial oxidative phosphorylation is resistant to high concentrations of NH_4^+. The rate of oxygen uptake in isolated mitochondria in the presence of malate is unaffected by up to 45 m*M* NH_4^+; the oxidation of citrate, succinate, or glycine is resistant up to 20 m*M* NH_4^+ (Yamaya and Matsumoto, 1985a). This suggests that *in vivo,* high concentrations of NH_4^+ may be present within the mitochondria, permitting its assimilation by either glutamate dehydrogenase (Rhodes *et al.*, 1989) (Fig. 4, reaction 11) or carbamoyl-phosphate synthetase (Fig 4, reaction 8; Ludwig, 1991). On the other hand, NH_4^+ does inhibit the *in vitro* activity of pyruvate dehydrogenase (Schuller and Randall, 1989) and alters the extractable activities of several TCA cycle enzymes (Wakiuchi *et al.*, 1971). Hence its interorganellar concentration is likely to be very tightly regulated. The metabolic fate of NH_4^+ within the mitochondria (or cell) is not fully understood. The recent suggestion that carbamoyl-phosphate synthetase might play a role in its assimilation (Ludwig, 1991) satisfies many of the problems that potentially arise if NH_4^+ were freely diffusible (see Section VI,B).

B. Chloroplast

Chloroplasts are the major site for amino acid biosynthesis within the cell. They are capable of synthesizing homoserine, lysine, threonine, isoleucine, and alanine from [^{14}C]aspartate or malate (Mills *et al.*, 1980), or phenylalanine, tyrosine, and tryptophan from $^{14}CO_2$ (Buchholz and Schultz, 1980). However, these processes are not fully characterized. The conversion of sulfate to sulfide, and its assimilation into cysteine by cysteine synthase also takes place in chloroplasts (Anderson, 1980). Alanine and glycine are early products of $^{14}CO_2$ fixation in isolated chloroplasts (Buchholz *et al.*, 1979) or mesophyll cells (Lawyer *et al.*, 1981) and they might be formed via transamination of glutamate or aspartate. However, the synthesis of alanine precedes the synthesis of both glutamate or aspartate (Lawyer *et al.*, 1981). The solution to this enigma may be found in a recent report by Andrews and Kane (1991). They found that pyruvate is a direct product of the action of Rubisco. This pyruvate could then give rise to alanine via a transamination reaction and would explain a synthesis of alanine which precedes the synthesis of both glutamate or aspartate in whole cells (Lawyer *et al.* 1981). The processes leading to alanine or glycine synthesis within the chloroplast are not adequately understood. Chloroplasts are also capable of synthesizing serine either from glycine (and methylenetetrahydrofolate) in a reversible reaction catalyzed by serine hydroxymethyltransferase (Shah and Cossins, 1970), or from 3-phosphoserine (Larsson and Albertsson, 1979). The synthesis of arginine, leucine, valine, and proline within the chloroplast is less well documented (Wallsgrove *et al.*, 1983). The formation of methionine and asparagine within the chloroplast is not documented at all and they are probably synthesized elsewhere.

Several energy-intensive reactions are involved in reducing inorganic nitrogen and incorporating this nitrogen into an organic form and these reactions are located within the chloroplast. The assimilation of NH_4^+ within the chloroplast is light stimulated and closely associated with O_2 evolution (Anderson and Done, 1977). This is further supported by the observations that the glutamine synthetase (Fig. 5, reaction 17) located in the chloroplast utilizes photosynthetically derived ATP (Jordan and Givan, 1979) and that the subsequent deamidation of glutamine in the presence of 2-oxoglutarate by glutamate synthase (Fig. 5, reaction 15) is dependent on ferredoxin (or NADPH; Lea and Miflin, 1974). The reduction of nitrite to NH_4^+ via nitrite reductase (Fig. 5, reaction 32) also requires reduced ferredoxin, which again is derived from photosynthetic electron transport.

The influx of NH_4^+ (Kleiner, 1981) into the chloroplast does not appear to be regulated to any degree. However, transport of other substrates across the inner chloroplast membrane is mediated via protein translocation and the flux of metabolites is determined by the relative concentrations of compounds on either side of this membrane. The uptake of NO_2^- is influenced by stromal pH, several cations, and is inhibited by sulfite or *N*-ethylmaleimide. This suggests that the uptake process is protein mediated (Brunswick and Cresswell, 1988). Glutamine may leave (or enter) the chloroplast in exchange for glutamate and this process is mediated by the glutamine:glutamate translocator (Fig. 5, translocator C; Yu and Woo, 1988). The transport of glutamate as interceded by this translocator can be biochemically differentiated from that mediated by the glutamate:malate translocator (Fig. 5, translocator B). In the latter case glutamate counterexchanges with malate, which can then exchange with 2-

oxoglutarate via the 2-oxoglutarate : malate translocator (Fig. 5, translocator G). The net result is that glutamate and 2-oxoglutarate exit and enter the chloroplast, respectively, with malate cycling to facilitate their exchange (Woo *et al.*, 1987).

A supply of reductant within the cytosol is required for a range of cytosolic reactions and, notably within the context of this article, for the reduction of nitrate reductase (Fig. 5, reaction 31) and hydroxypyruvate (Fig. 6, reaction 24). This is mediated through two translocators, one that counterexchanges malate for oxaloacetate (Fig. 5, translocator E; Hatch *et al.*, 1984) and the other, dihydroxyacetone-phosphate for glyceraldehyde-3-phosphate via the triosephosphate translocator (Fig. 5, translocator H; Flügge and Heldt, 1984). This latter translocator also permits the furnishing of ATP within the cytosol. The provision of reductant of chloroplast origin is required for hydroxypyruvate reduction because under photorespiratory conditions the contribution of mitochondria toward the export of malate is limited (Dry *et al.*, 1987; see Section V,A). Redox equivalents translocated in the form of malate could be generated by chloroplastic NADPH-dependent malate dehydrogenase (Fig. 5, reaction 29; Scheibe, 1987). Under photorespiratory conditions, the relative contribution of triosephosphates in providing reductant for hydroxypyruvate reduction is not significant because their production and export is reduced (Gardeström and Wigge, 1988; Usuda and Edwards, 1982). Under these conditions carbon is still leaving the chloroplast in the form of glycolate, which is exchanged for glycerate (Fig. 5, translocator D), and this indirectly results in a supply of cytosolic ATP through the preferential oxidation of glycine in mitochondria (Dry *et al.*, 1983; Gardeström and Wigge, 1988; see Section V,A).

Recently, evidence for a model that minimizes the extent of ammonium diffusion and concentrates both ammonium and CO_2 within the chloroplast was proposed (Ludwig, 1991). This model necessitates the entry of arginine to, and the exit of ornithine from, the chloroplast via an uncharacterized process (Fig. 5, translocator F). However, the catabolism of arginine within the chloroplast by arginine iminohydrolase, ornithine carbamoyltransferase, and carbamate kinase (Fig. 5, reactions 4, 7, and 33), which results in the liberation of ammonium, CO_2, and ATP has been documented (Ludwig, 1991).

C. Peroxisome

Peroxisomes contain at least one H_2O_2 generating oxidase (Fig. 6, reaction 22) along with catalase (Fig. 6, reaction 9; de Duve, 1969). Even though the generation and oxidation of H_2O_2 within this organelle ensures that metabolic perturbations resulting from this potentially toxic compound are minimized (see Section VI,B), H_2O_2 may still be important in the nonenzymatic peroxidation of glyoxylate and hydroxypyruvate within the peroxisome itself (Grodzinski, 1978, 1979; Walton and Butt, 1981; Zelitch and Ochoa, 1953) (see Fig. 6). This point is contentious because it has been postulated that peroxidation may occur only under extreme growth conditions or within mutants that lack catalase activity (Kendall *et al.*, 1983; Lea *et al.*, 1989). On the other hand, a new class of mutants that contain 40% more extractable catalase activity than wild-type plants (Zelitch, 1989, 1990), and which are more efficient at photosynthesis in air, have been characterized (Zelitch, 1987, 1989). This demonstrates that peroxidation of organic acids does occur in wild-type plants under typical growth conditions.

Peroxisomal membranes are permeable to most intermediates of the photorespiratory pathway, hydroxypyruvate, glycerate, glycolate, glycine, serine, and malate, but not to glyoxylate or NADH (Anderson and Butt, 1986). As peroxisomal reduction of hydroxypyruvate requires redox equivalents of cytosolic origin, the provision of the high flux of pyridine nucleotide required within the peroxisome may arise through the oxidation of malate via the abundant peroxisomal malate dehydrogenase (Fig. 6,

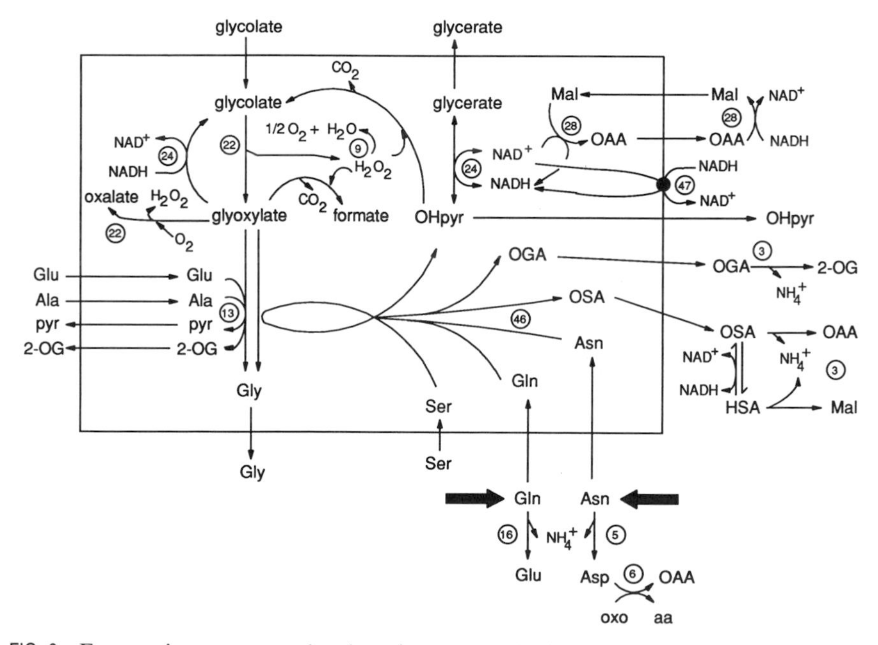

FIG. 6 Enzymatic processes related to nitrogen metabolism within the peroxisome. Large arrows indicate compounds arising from transport processes. See Table II for a list of enzymes. Refer to Fig. 8 for processes involved in the synthesis of metabolites that are related to these reactions. HSA, Hydroxysuccinamate.

reaction 28; Tolbert, 1981), or via a peroxisomal membrane transhydrogenase (Fig. 6, reaction 47; Schmitt and Edwards, 1983).

Glyoxylate is known to inhibit and activate the activities of several chloroplastic, mitochondrial, and (in the presence of NH_4^+) peroxisomal enzymes *in vitro* (see Section VI,B). This may be of little physiological significance because the synthesis of glyoxylate is localized within the peroxisome and it is metabolized efficiently by the peroxisomal transaminases (Fig. 6, reactions 13 and 46). However, conditions that restrict glyoxylate transamination (lack of amino donors, inhibition of either glyoxylate, glutamate, or serine transaminases, or mutational deficiency of either enzyme) would result in its accumulation and leakage into the cytosol, and from there it may enter without hindrance into the chloroplast or mitochondrion (Halliwell, 1984).

Glutamate and serine are classically portrayed as the preferred amino donors for glyoxylate transamination (Artus *et al.*, 1986; Husic *et al.*, 1987; Keys, 1980; Tolbert, 1980); however (see Section VI,A), other amino acids play an active role in the amination of glyoxylate and these include alanine (Betsche, 1983; Betsche and Eising, 1986; Ta and Joy, 1986), asparagine (Ireland and Joy, 1983; Ta *et al.*, 1985), and glutamine (Ireland, 1986). Deamidation products of asparagine and glutamine transamination are 2-oxosuccinamic acid and 2-oxoglutaramic acid, respectively (Fig. 6, reaction 3). These deamidation reactions and the subsequent metabolism of their products are not well characterized (Kretovitch, 1958; Meister, 1953; Ta *et al.*, 1984a,b) and require further study (see Section II,D).

VI. Compartmentation and the Control of Carbon and Nitrogen Flow in Photorespiration

An apparently unavoidable property of the Rubisco active site is that in addition to catalyzing the formation of the carboxylation product, 3-phosphoglycerate, it is capable in the presence of O_2 and ribulose-1,5-bisphosphate of producing 3-phosphoglycerate and 2-phosphoglycolate (Andrews *et al.*, 1973). The light-dependent formation of 2-phosphoglycolate results in the shunting of two carbons out of the Calvin cycle and eventually to the evolution of CO_2, leading to a net loss of fixed carbon to the plant. Classic portrayals of this process (Artus *et al.*, 1986; Husic *et al.*, 1987; Ogren, 1984; Tolbert, 1980) suggest that in C_3 plants three-quarters of the carbon diverted from the Calvin cycle may be returned back via the operation of the oxidative photosynthetic carbon, or C_2 cycle (Fig. 7), with the remaining 25% of the carbon evolved as CO_2. Although this represents a loss of carbon, this is less than the observed loss in unicellular algae,

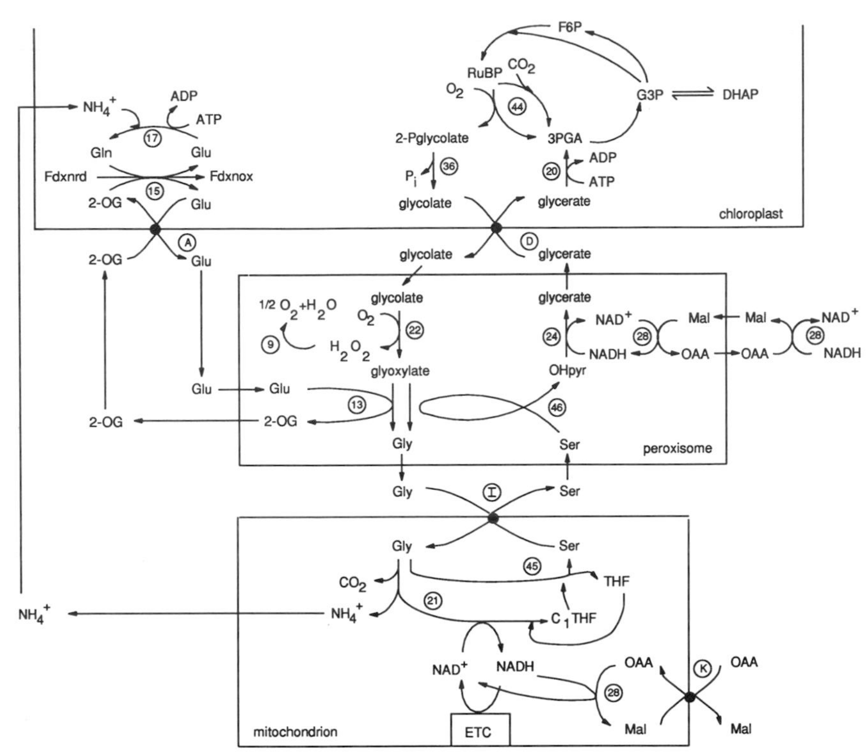

FIG. 7 Classical portrayal of the photorespiratory pathway. Compare this scheme with Figs. 3–6 and 9 (alanine synthesis, mitochondrion, chloroplast, peroxisome, and glycolate/glyoxylate synthesis, respectively—outlining reactions that are currently known to contribute or remove carbon and nitrogen from this pathway).

where glycolate is excreted into the medium when algal cells are transferred from high levels of CO_2 (5%) to atmospheric levels (Tolbert *et al.*, 1985). However, in algal cells grown continuously in low ambient Co_2, mechanisms for concentrating CO_2 develop. This, then, reduces the oxygenase component of Rubisco in these cells (Tolbert *et al.*, 1985).

The operation of the oxidative photosynthetic carbon pathway results in the release of substantial amounts of NH_4^+. Estimates suggest that the flux of nitrogen through this pathway may exceed the rate of primary nitrate assimilation by up to 10-fold (Lea *et al.*, 1989; Wallsgrove *et al.*, 1983). Because NH_4^+ is volatile (Harper *et al.*, 1987; Parton *et al.*, 1988) and toxic (Givan, 1979; Lea and Ridley, 1989) it must be efficiently reassimilated. The photorespiratory nitrogen cycle is one method by which this goal could be achieved (Keys *et al.*, 1978).

These two interrelated pathways are thought to operate in a closed, cyclic manner ensuring that carbon and nitrogen losses are kept to a minimum (Andrews and Lorimer, 1987; Husic *et al.*, 1987; Keys *et al.*, 1978; Keys, 1986; Lea and Blackwell, 1990; Ogren, 1984; Tolbert, 1980). However, the enzymatic constituents of both the C_2 and the associated nitrogen cycles are distributed in the chloroplast, cytosol, peroxisome, and mitochondrion. This arrangement requires extensive and efficient shuttling of metabolites between these compartments. However, the diffusion of these metabolites may be minimized by close oppression of the organelles (Frederick and Newcomb, 1969) (Fig. 8).

The classical closed system shows glycolate channeled from chloroplast to peroxisome, glycine from peroxisome, to mitochondria, serine from mitochondria to peroxisome, and ammonium from mitochondria to chloroplast (Fig. 7). Recently several of the reactions that at one time were considered specific to these organelles have been localized within the cytosol as well. This allows the replenishment of metabolites that partici-

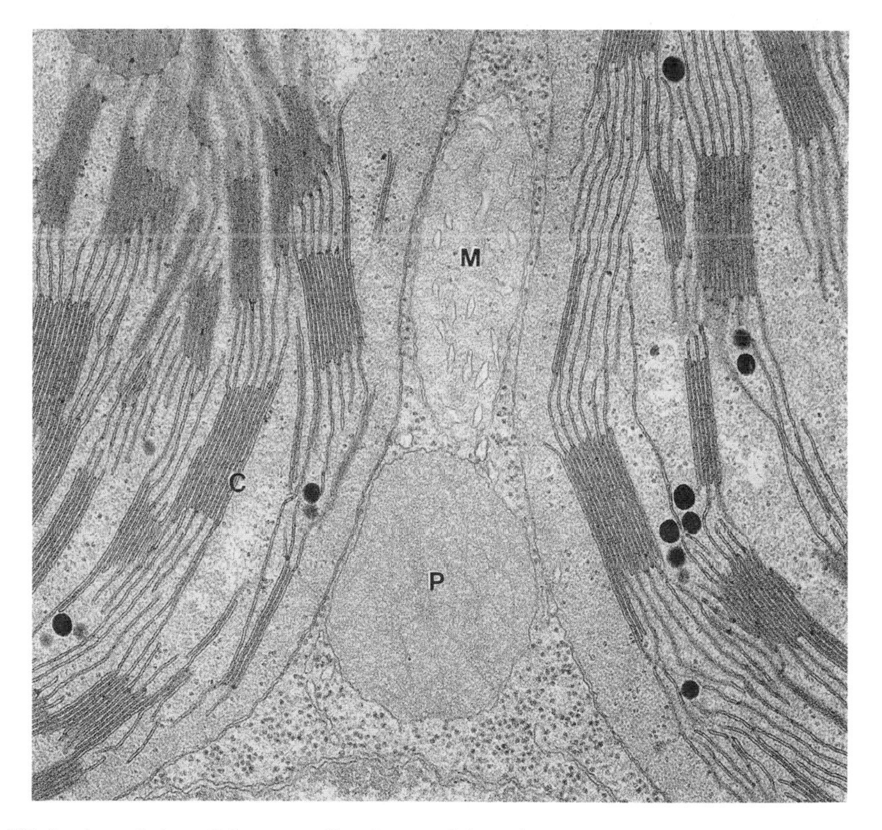

FIG. 8 Association of the organelles that participate in photorespiration. Chloroplast (C), mitochondria (M), and peroxisome (P) (see Frederick and Newcomb, 1969). (Electron micrograph courtesy of E. H. Newcomb.)

pate within the photorespiratory pathway to be bled off to other reactions. Furthermore, the contribution of external sources of carbon (Kleczkowski *et al.*, 1986; Winkler *et al.*, 1987; Zelitch, 1973, 1988) or nitrogen (Betsche, 1983; Betsche and Eising, 1986; Ireland, 1986; Ta *et al.*, 1985; Ta and Joy, 1986) into this pathway, or the removal of carbon (Davies and Asker, 1985; Grodzinski, 1979; Walton and Butt, 1981; Zelitch and Ochoa, 1953) and nitrogen (Givan *et al.*, 1988a; Lea *et al.*, 1989; Madore and Grodzinski, 1984) from this pathway, demonstrates that the "stoichiometric" view of photorespiration is limited and needs to be "opened" to truly reflect physiological processes.

A. A Reconsideration of the C_2 and Associated Photorespiratory Nitrogen Cycle

The individual steps of the C_2 cycle have been identified by the use of inhibitors (Zelitch, 1964; Tolbert, 1971) and by the selection of conditional lethal mutants, principally in *Arabidopsis* (Somerville, 1986) and barley (Lea *et al.*, 1989). Deficiency mutants lacking phosphoglycolate phosphatase (Fig. 7, reaction 36), catalase (reaction 9), glycine decarboxylase (reaction 21), serine hydroxymethyltransferase (reaction 45), serine-glyoxylate aminotransferase (reaction 46), NADH-dependent hydroxypyruvate reductase (reaction 24), ferredoxin-dependent glutamate synthase (reaction 15), and chloroplastic glutamine synthetase (reaction 17) have so far been characterized. An interesting observation is that most of the photorespiratory mutants selected so far exhibit deficiencies related to photorespiratory nitrogen, and not carbon, metabolism (Somerville, 1986; Somerville and Ogren, 1981, 1982b). This is puzzling because the method of screening for photorespiratory mutants is based on their ability to grow under conditions of high CO_2 but not air (Blackwell *et al.*, 1987, 1988; Kendall *et al.*, 1983, 1986; Murray *et al.*, 1987; Somerville and Ogren, 1979, 1980a,b, 1981; Wallsgrove *et al.*, 1986). The occurrence of a photorespiratory mutant identifies the associated deficient process as one that is related solely to photorespiration. This process is, therefore, not required for regular plant growth (Kendall *et al.*, 1986; Sommerville and Ogren, 1980a,b, 1982a, 1983). The corollary of this argument should also be true, namely, that the reactions involved in the oxidative photosynthetic carbon pathway for which mutants have not been obtained participate in essential metabolic processes.

The oxygenase reaction of Rubisco results in the synthesis of 2-phosphoglycolate (Fig. 7, reaction 44), a compound that inhibits triose-phosphate isomerase (Wolfenden, 1970) and phosphofructokinase (Kelly and Latzko, 1976) activities. 2-Phosphoglycolate is catabolized by phosphoglycolate phosphatase (Fig. 7, reaction 36), within the chloroplast to produce glycolate (Richardson and Tolbert, 1961). Glycolate is transported across the plastid envelope in a carrier-dependent manner (Fig. 7, translocator D), as indicated by inhibition of the transport caused by *N*-ethylmaleimide, glycerate, glyoxylate, or lactate (Howitz and McCarty, 1983, 1985, 1987).

Selection of mutants lacking phosphoglycolate phosphatase activity (or other activities of the photorespiratory cycle) in *Arabidopsis* (Somerville and Ogren, 1979) or barley (Hall *et al.*, 1987) is facilitated by the fact that although they grow under conditions of low O_2 (or high CO_2) they are lethal when the plants are grown in air. In the case of the phosphoglycolate phosphatase deficiencies the plants accumulate 2-phosphoglycolate when they are grown in air and this results in an inhibition of CO_2 fixation. The occurrence of these mutants has been used to demonstrate that wild-type phosphatase activity is essential for the disposal of photorespiratory-derived 2-phosphoglycolate, and, because the level of glycolate is low in these mutants it has been proposed that 2-phosphoglycolate is the physiologically significant precursor for glycolate synthesis (Hall *et al.*, 1987; Husic *et al.*, 1987; Ogren, 1984; Somerville and Ogren, 1979, 1982b; Somerville, 1986). However, these conclusions may be of limited physiological significance because secondary effects are associated with the accumulation of 2-phosphoglycolate and its inhibition of Calvin cycle activity. For example, glycine and serine are derived both from glycolate and from other, less well-defined precursors (Keys, 1980; Kleczkowski and Givan, 1988), yet their pool sizes are greatly reduced when the phosphoglycolate phosphatase is missing. Other metabolic routes leading to glycolate formation have been reported (see Fig. 9); [2-^{14}C]glycolate is synthesized from [3-^{14}C]pyruvate at substantial rates in the light and in air through a nonphotorespiratory series of reactions involving pyruvate decarboxylase, malic enzyme, malate dehydrogenase, citrate synthase, aconitase, isocitrate lyase, and glyoxylate reductase (Zelitch, 1988). Isozymes of malate dehydrogenase (Fig. 7, reaction 28), aconitase (Fig. 9, reaction 1; Brouquisse *et al.*, 1987), isocitrate lyase (Fig. 9, reaction 27; Zelitch, 1988), and NADPH-dependent glyoxylate reductase (Fig. 9, reaction 23; Givan *et al.*, 1988b) are all localized within the cytosol as well as in mitochondria or peroxisomes. Glycolate may also be synthesized from the decarboxylation of hydroxypyruvate (Davies and Asker, 1985) catalyzed by pyruvate decarboxylase (Fig. 9, reaction 41) and aldehyde dehydrogenase (Fig. 9, reaction 2), or from the reduction of glyoxylate arising from ureide breakdown (Wells and Lees, 1991; Winkler *et al.*, 1987, see below) by a cytosolic NADPH-dependent glyoxylate reductase (Fig. 9, reaction 23; Givan *et al.*, 1988b; Kleczkowski *et al.*, 1986, 1988). The synthesis of glycolate through these reactions results in the input of carbon into the photosynthetic oxidative carbon cycle in addition to that arising from the oxygenase reaction of Rubisco.

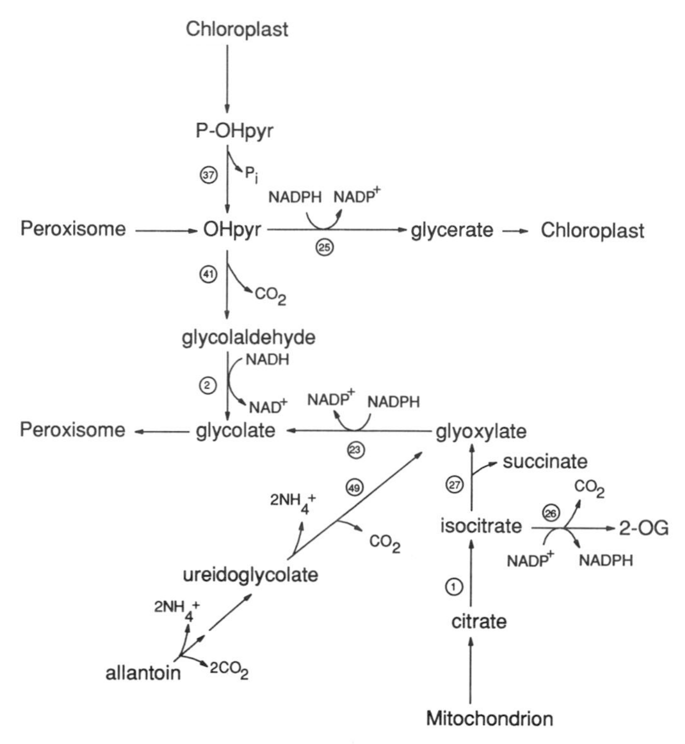

FIG. 9 Alternate sources for the synthesis of glycolate and glyoxylate within the cytosol of the cell.

Glycolate enters the peroxisome freely because the metabolism of [1-^{14}C]glycolate supplied to intact peroxisomes does not exhibit any latency, that is, the extent of its metabolism is equivalent whether or not the organelle is intact (Anderson and Butt, 1986). It is oxidized by glycolate oxidase (Fig. 7, reaction 22) to yield glyoxylate and H_2O_2 (Kerr and Groves, 1975), and the H_2O_2 is decomposed by catalase (Fig. 7, reaction 9). Catalase activity itself is regulated by CO_2 concentration. As the ambient CO_2 concentration rises the catalase activity declines (Havir and McHale, 1987, 1989). This observation suggests that catalase is required by the plant under conditions of low ambient CO_2. In fact, plants selected for enhanced photosynthetic activity in air and at elevated temperatures, conditions where organic acid peroxidation is enhanced (Hanson and Peterson, 1986, 1987), contain 40% more catalase activity (Zelitch, 1989, 1990). Hence catalase activity is important in protecting the plant from peroxidation reactions both within and outside of photorespiration.

Glyoxylate may also be synthesized from nonphotorespiratory sources in addition to that produced through glycolate oxidase. For example, in leaves of nodulated tropical legumes, the ureide allantoin is catabolized by way of the intermediates allantoate, ureidoglycine, and ureidoglycolate, to produce glyoxylate (Fig. 9, reaction 49). Labeling studies demonstrate that [4,5-^{14}C]allantoin is metabolized in soybean leaves, forming [^{14}C]glyoxylate, [^{14}C]glycine, and [^{14}C]serine (Winkler *et al.*, 1987), confirming that carbon enters the oxidative carbon cycle from sources other than the Calvin cycle in ureide-transporting plants. However, the subcellular localization of these reactions has not been investigated. Because peroxisomes are relatively impermeable to glyoxylate (Anderson and Butt, 1986) it may be reduced in the cytosol by an NADPH-dependent glyoxylate reductase (Fig. 9, reaction 23; Givan *et al.*, 1988b; Kleczkowski *et al.*, 1986). The resulting glycolate then enters the peroxisome. On the other hand, a recent report indicates that ureidoglycolate amidohydrolase is localized within the peroxisome (Wells and Lees, 1991). Its metabolism in the peroxisome would necessitate the uptake of either ureidoglycolate or an earlier precursor and also in the release of NH_4^+ (and CO_2) from this compartment for reassimilation elsewhere.

The glyoxylate within the peroxisome may be enzymatically oxidized to oxalate and H_2O_2 via glycolate oxidase (Fig. 6, reaction 22; Chang and Huang, 1981), reduced to glycolate by NADH-hydroxypyruvate reductase (Fig. 6, reaction 24; Husic *et al.*, 1987; Kleczkowski *et al.*, 1990), or transaminated to glycine by either of two peroxisomal aminotransferases. One of these aminotransferases uses glutamate or alanine (Figs. 6 and 7, reaction 13; Betsche, 1983; Nakamura and Tolbert, 1983), and the other (Figs. 6 and 7, reaction 46) serine (Nakamura and Tolbert, 1983) or asparagine (Ireland and Joy, 1981, 1983). Glutamine may also be involved in the synthesis of glycine (Ireland, 1986) although it is not known which of the aminotransferases mediates this reaction (Fig. 6, reactions 13 or 46). Both of these aminotransferases exhibit a property not normally associated with transaminase reactions and that is that they are essentially irreversible. This imparts direction to the metabolite flow in this portion of the photorespiratory pathway (Ireland and Joy, 1983; Nakamura and Tolbert, 1983).

Typically, serine and glutamate are portrayed as the preferred amino donors for glycine synthesis. Glutamate and 2-oxoglutarate cycle between the peroxisome and chloroplast, linking photorespiratory nitrogen cycling with C_2 metabolism. The transamination of serine is necessary for hydroxypyruvate production, which ensures that carbon is returned to the Calvin cycle. The 2-oxoglutarate produced in the presence of glyoxylate and glutamate by transamination diffuses out of the peroxisome and is transported into the chloroplast via a dicarboxylic translocator (Fig. 7, translocator A; Somerville and Ogren, 1983; Wallsgrove *et al.*, 1986;

Woo and Osmond, 1982; Woo *et al.*, 1987). There it is reaminated in the presence of glutamine in a reaction catalyzed by glutamate synthase (Fig. 7, reaction 15; Kendall *et al.*, 1986; Somerville and Ogren, 1980a; Wallsgrove *et al.*, 1983). The continual provision of glutamine for the glutamate synthase reaction is catalyzed by a chloroplastic glutamine synthetase (Fig. 7, reaction 17), which assimilates photorespiratory-derived NH_4^+ liberated during glycine cleavage (Blackwell *et al.*, 1987; Wallsgrove *et al.*, 1987). The net result of this paper chemistry is that the carbon and nitrogen moieties are potentially recycled during photorespiration (Artus *et al.*, 1986; Husic *et al.*, 1987; Keys *et al.*, 1978; Keys, 1986; Tolbert, 1980; Wallsgrove *et al.*, 1983). A major unresolved question is, however, whether this recycling actually occurs *in vivo*.

The replenishment of carbon skeletons from the C_2 cycle for glycerate and 3-phosphoglycerate synthesis necessitates a high flux of hydroxypyruvate synthesis. For every two glyoxylates transaminated one serine must also be transaminated. Indeed, a substantial flux of nitrogen from [^{15}N]serine into glycine is found with 60% of supplied serine nitrogen incorporated into glycine when glycine cleavage is inhibited with methoxylamine (Ta and Joy, 1986). This demonstrates that serine-glyoxylate aminotransferase plays a significant role in glycine synthesis. However, the flux of nitrogen from serine into glycine is fivefold greater than that from [^{15}N]glutamate (Ta and Joy, 1986), the other preferred transamination substrate (Keys *et al.*, 1978; Tolbert, 1980). Furthermore, under conditions of increasing ambient O_2, a greater proportion of supplied $^{14}CO_2$ is exported from the leaf as glutamate (Madore and Grodzinski, 1984), indicating that with elevated photorespiratory activity glutamate may not be recycled stoichiometrically through the photorespiratory nitrogen cycle. The removal of glutamate from the leaf necessitates the replenishment of 2-oxoglutarate because glutamate is no longer available for NH_4^+ assimilation through the action of glutamine synthetase. Recently two isozymes of NADP-isocitrate dehydrogenase, in addition to the NAD-isocitrate dehydrogenase found within the mitochondria, have been localized in the cytosol (Figs. 4 and 9, reaction 26), and in the chloroplast (Chen and Gadal, 1990b). It has been proposed that the cytosolic isozyme, due to its low K_m for isocitrate, is responsible for the replenishment of 2-oxoglutarate needed for glutamate synthesis and its export (Chen and Gadal, 1990a); the isocitrate required for this reaction would be supplied by a cytosolic aconitase (Figs. 4 and 9, reaction 1; Brouquisse *et al.*, 1987). Because of this removal of glutamate, other amino donors would be expected to contribute to the formation of glycine in order to maintain the appropriate flux of carbon leading to serine synthesis.

The participation of glutamate as a principal substrate for glycine synthesis is also challenged when the relative contributions of [^{15}N]alanine or [^{15}N]glutamate in the synthesis of glycine are considered. Two separate studies demonstrate that alanine is the preferred amino donor for glycine synthesis. In oat leaves the flux of nitrogen from [^{15}N]alanine exceeds that from [^{15}N]glutamate threefold (Betsche, 1983; Betsche and Eising, 1986), and in pea leaves the contribution from alanine is twofold higher (Ta and Joy, 1986). Furthermore, supplying alanine to detached leaves of glutamate synthase- or glutamine synthetase/glutamate synthase-deficient mutants of barley helps alleviate the inhibition of CO_2 fixation observed in control leaves (Blackwell *et al.*, 1988). These studies demonstrate that alanine nitrogen is able to directly contribute to the transamination of the glyoxylate pool and hence to maintain carbon flow through the photorespiratory pathway.

Even though mutants deficient in serine-glyoxylate aminotransferase activity in both *Arabidopsis* (Somerville and Ogren, 1983) and barley (Murray *et al.*, 1987) accumulate glycine and serine in air with a subsequent gradual decline in photosynthesis, they do grow under conditions of high CO_2. In wild-type barley the amount of alanine exceeds that of serine but in serine-glyoxylate aminotransferase-deficient mutants levels of serine are higher (Murray *et al.*, 1987). The reduced pool size of alanine under nonphotorespiratory conditions in this mutant, when compared with the increased amounts of alanine observed under the same conditions in wild-type plants, suggests that the serine-glyoxylate transaminase is involved in alanine synthesis (Murray *et al.*, 1987). However, in air the synthesis of alanine by this enzyme is unlikely because glyoxylate, with its lower K_m, is the preferred amino acceptor for this transaminase (Ireland and Joy, 1983; Nakamura and Tolbert, 1983). In the presence of high CO_2, alanine synthesis in the presence of glycine and serine via this transaminase would also be minimal because the synthesis of these two amino donors would be reduced. Nevertheless, alanine synthesis via this enzyme in the presence of asparagine may be of importance under conditions of both high and low CO_2 as pyruvate is a suitable amino acceptor when added along with asparagine to either crude extracts (Murray *et al.*, 1987) or purified serine-glyoxylate transaminase preparations (Ireland and Joy, 1983). In addition, substantial amounts of [^{15}N]alanine are formed *in vivo* when amino[^{15}N]asparagine is fed to detached pea leaves (Bauer *et al.*, 1977; Ta *et al.*, 1985; Ta and Joy, 1986).

Alanine is also synthesized by glutamate-glyoxylate aminotransferase in the presence of pyruvate (Figs. 3 and 6). This activity is detected in the serine-glyoxylate aminotransferase-deficient mutant at rates equivalent to those found in the wild type (Murray *et al.*, 1987). Therefore, one would expect equivalent amounts of alanine in the wild type and the deletion mutant if this enzyme were responsible for its synthesis. However, the pool size of alanine in air is greatly reduced in serine-glyoxylate deletion mutants. This indicates that the flow of nitrogen from serine to glycine via the serine-glyoxylate aminotransferase is necessary for the maintenance of reduced levels of photorespiratory intermediates, which would otherwise accumulate and perturb associated metabolic process. Because both the glutamate-glyoxylate and serine-glyoxylate aminotransferases are inhibited by glyoxylate in the presence or absence of NH_4^+ (Havir, 1986), and because the glutamate-glyoxylate aminotransferase is also inhibited by glycine (Nakamura and Tolbert, 1983), the accumulation of glycine, or glyoxylate and NH_4^+, within the serine-glyoxylate aminotransferase mutant may reduce alanine synthesis through secondary effects.

Purified serine-glyoxylate aminotransferase is also capable of using asparagine as an amino donor with glyoxylate as acceptor, indicating that it may donate nitrogen for glycine synthesis *in vivo* (Ireland and Joy, 1983). Mutants deficient in this activity also lack asparagine-glyoxylate aminotransferase activity (Murray *et al.,* 1987). The role of asparagine as a photorespiratory nitrogen donor is supported by ^{15}N feeding studies, which demonstrate that in mature pea leaves the contribution of nitrogen from asparagine in glycine synthesis (10%) is equivalent to that derived from glutamate (Ta and Joy, 1986). Furthermore, supplying asparagine to detached leaves of barley mutants deficient in glutamine synthetase allays the drop in the rate of CO_2 fixation (Lea *et al.,* 1989). The contribution of nitrogen from asparagine in the synthesis of glycine may be direct when glyoxylate is aminated, or indirect in the presence of pyruvate, depending on the concentration of oxo acid at the active site.

The involvement of asparagine in glycine formation is of interest in plants where this compound is the dominant, organically combined, nitrogen transport compound within the xylem and phloem (Pate, 1973; Sieciechowicz *et al.,* 1988). Asparagine is catabolized within developing structures and provides a major source of nitrogen for amino acid and protein synthesis (see Section II,D) and this process takes place either by hydrolysis, through the action of asparaginase (Fig. 6, reaction 5), or transamination catalyzed by serine-glyoxylate transaminase (Fig. 6, reaction 46; Sieciechowicz *et al.,* 1988). Because the amount of asparagine in peas decreases with maturation in both xylem exudates and leaves (Sieciechowicz and Joy, 1988), the [^{15}N]asparagine experiments mentioned above, which involve mature pea leaves (Ta and Joy, 1986), underestimate the flux of asparagine nitrogen into glycine. The involvement of asparagine in photorespiratory nitrogen cycling is likely to be of greater importance in developing leaves, where asparagine is the major source of nitrogen for leaf growth (Givan *et al.,* 1988a; Joy, 1988; Ta *et al.,* 1985). A similar argument can be put forth for glutamine, another dominant nitrogen transport compound that also donates its amino nitrogen to glyoxylate in the synthesis of glycine (see above; Fig. 6, reaction 46; Ireland, 1986).

The coupling of asparagine and glutamine transamination during leaf development with the C_2 cycle demonstrates that a synthetic function is associated with photorespiration leading to the formation of serine and glycine (Ta *et al.*, 1985). In fact, when $^{14}CO_2$ is supplied to leaves exposed to increasing concentrations of O_2, elevated amounts of [^{14}C]serine, [^{14}C]glycine, and [^{14}C]glutamate are exported from the leaf, indicating an association between increased photorespiration and the synthesis of metabolites for use outside of the photorespiratory pathway (Madore and Grodzinski, 1984).

Glycine leaves the peroxisome through diffusion and enters the mitochondrion either passively (Proudlove and Moore, 1982; Yu *et al.*, 1983) or in a carrier-mediated process (Fig. 7, translocator I; Day and Wiskich, 1980; Walker *et al.*, 1982). At low concentrations, glycine uptake into isolated mitochondria is inhibited by the sulfhydryl reagents mersalyl and *p*-chloromercuribenzoate (Oliver, 1987; Walker *et al.*, 1982), several glycine analogs (Walker *et al.*, 1982), and serine (Yu *et al.*, 1983). Furthermore, a deletion mutant in barley with deficient glycine metabolism has been interpreted to have impaired glycine transport (Blackwell *et al.*, 1990). However, uptake is also linear with increasing glycine concentrations (Proudlove and Moore, 1982; Yu *et al.*, 1983).

Serine is formed in the presence of NAD^+ and 2 mol of glycine through the combined activities of glycine decarboxylase (Fig. 7, reaction 21) and serine hydroxymethyltransferase (reaction 45; Oliver, 1990a); CO_2, NADH, and NH_4^+ also arise from the glycine decarboxylation reaction. Mutants lacking either glycine-dependent O_2 consumption, indicating a perturbation associated with the glycine decarboxylase complex (Somerville and Ogren, 1982a), lacking serine hydroxymethyltransferase (Somerville and Ogren, 1981), or the P or H proteins, and lacking glycine decarboxylase activity (Blackwell *et al.*, 1990) have been obtained. As with other mutants in the C_2 cycle they are conditional lethal mutants that survive only when the ambient CO_2 concentrations are high.

Serine transport across the mitochondrial membrane (Fig. 7, translocator I) resembles that of glycine, i.e., it is biphasic in nature, and is actively taken up at low concentrations in a process inhibited by glycine, and via diffusion with increasing concentration (Yu *et al.*, 1983). Because greater amounts of [^{14}C]serine are exported from leaves when the concentration of O_2 surrounding the plant is increased (Madore and Grodzinski, 1984), serine may either leave or be further metabolized within the C_2 cycle.

Serine may also be synthesized from 3-phosphoglycerate, either through the glycerate pathway (Liang *et al.*, 1984; Walton and Woolhouse, 1986) involving chloroplastic 3-phosphoglycerate phosphatase (Fig. 5, reaction 35), peroxisomal glycerate dehydrogenase (NADH-hydroxypyruvate reductase; Fig. 6, reaction 24), and glycine (asparagine, alanine): hydroxypyruvate aminotransferase (Fig. 6, reaction 46), or the phosphorylated pathway (Keys, 1980; Kleczkowski and Givan, 1988) via chloroplastic 3-phosphoglycerate dehydrogenase (Fig. 5, reaction 34), cytosolic glutamate:phosphohydroxypyruvate aminotransferase (Fig. 5, reaction 38) and chloroplastic phosphoserine phosphatase (Fig. 5, reaction 39). Serine formation via the proposed glycerate pathway may be of little physiological significance because the activity of glycerate dehydrogenase is very low (Gaudillere *et al.*, 1983), and because the equilibrium of the glycine (asparagine, alanine):hydroxypyruvate aminotransferase reaction strongly favors hydroxypyruvate, rather than serine, synthesis (Ireland and Joy, 1983; Nakamura and Tolbert, 1983). However, an uncharacterized aminotransferase may take part in this process (Kleczkowski and Givan, 1988; Liang *et al.*, 1984).

Serine crosses the peroxisomal membrane freely (Anderson and Butt, 1986). It is deaminated in the peroxisome, producing hydroxypyruvate, and donates its amino group to glyoxylate via serine-glyoxylate aminotransferase (Fig. 7, reaction 46) to form glycine. Hydroxypyruvate is reduced to glycerate by a peroxisomal NADH-dependent hydroxypyruvate reductase (Fig. 7, reaction 24), but as peroxisomes are known to readily leak hydroxypyruvate (Anderson and Butt, 1986), under conditions of limiting NADH, hydroxypyruvate may be reduced in the cytosol by an NADPH-dependent hydroxypyruvate reductase (Fig. 9, reaction 25). This cytosolic enzyme may thus function in scavenging cytosolic hydroxypyruvate (Kleczkowski *et al.*, 1988). The generation of reductant required to maintain the rates of glycerate production during photorespiration could be supplied from both mitochondrial (Fig. 4, reaction 28) and chloroplastic sources (reactions 18, 19, 28, and 29). However, Dry *et al.* (1987) have questioned the role mitochondria may play in supplying reductant for the cytosol due to the low rates of malate export from this organelle. Furthermore, Gardström and Wigge (1988) and Usuda and Edwards (1982) have indicated that the triosephosphate translocator (Fig. 5, translocator H) is less active under photorespiratory conditions. Therefore, it seems likely that chloroplast-derived NADH via malate (Fig. 5, reactions 28 and 29, translocator E) may provide the required reductant for peroxisomal glycerate production (Scheibe, 1990; also see Sections V,A and B).

Mutants lacking NADH-dependent peroxisomal hydroxypyruvate reductase (90% of the total activity), but retaining the cytosolic (NADPH-dependent) activity (10% of the total activity), exhibit 75% of wild-type rates of CO_2 fixation under conditions of ambient or elevated O_2 (Murray *et al.*, 1990). This demonstrates that total hydroxypyruvate reductase is present in excess of the amount required to metabolize the flux associated with the C_2 cycle in air. There is a substantial increase in the amount of serine in these mutants in the air (4-fold in 5 min, or 20- to 25-fold after 3 hr) relative to the levels found in wild-type plants. Thus, the conversion of serine to hydroxypyruvate is taking place at greatly reduced rates in these mutants.

Glycerate enters the chloroplast in a carrier-mediated process that involves a specific transporter (Fig. 7, translocator D) that is stimulated in the light (Robinson, 1982) and is inhibited by glycolate (Howitz and McCarty, 1983, 1985, 1987; Robinson, 1982). Glycerate is irreversibly phosphorylated, forming 3-phosphoglycerate by glycerate kinase (Fig. 7, reaction 20; Kleczkowski *et al.,* 1985).

The majority of the NH_4^+ originating from glycine decarboxylation apparently diffuses out of the mitochondrion and into the chloroplast, where it is reassimilated in the presence of glutamate by glutamine synthetase (Fig. 7, reaction 17) into glutamine. However, other enzymes may play a role in the assimilation of NH_4^+ in the mitochondria, including alanine dehydrogenase (Rowell and Stewart, 1976), glutamate dehydrogenase (Fig. 4, reaction 11; Rhodes *et al.,* 1989), or carbamoyl-phosphate synthetase (Fig. 4, reaction 8; Ludwig, 1991). The potential for alanine dehydrogenase has been previously discussed and does not appear to play a role in assimilating NH_4^+ in higher plants (Joy, 1988). Glutamate dehydrogenase may play a role in the assimilation of NH_4^+, as up to 7% the NH_4^+ produced during glycine cleavage is assimilated by mitochondrial glutamate dehydrogenase (Rhodes *et al.,* 1989; Yamaya *et al.,* 1986). Furthermore, a glutamate dehydrogenase null mutant from maize, with 15-fold less root glutamate dehydrogenase activity than wild type, exhibits a 50% decrease in the rate of ammonium assimilation (Magalhaes *et al.,* 1990). Another possible mechanism for the assimilation of photorespiratory-derived NH_4^+ is mediated by carbamoyl-phosphate synthetase in the presence of CO_2 and ATP. This protein is present in mitochondria in a range of C_3 and C_4 plants (Ludwig, 1991). Preliminary evidence suggests that chloroplasts exhibit arginine iminohydrolase (Fig. 5, reaction 4), ornithine carbamoyl-transferase (Fig. 5, reaction 33), and carbamate kinase (Fig. 5, reaction 7) activities, which liberate CO_2 and NH_4^+. This is suggestive of the operation of a series of reactions that operate to transport carbon and nitrogen from the mitochondria to the chloroplast (see Section IV,B; Ludwig, 1991). These reactions would limit the degree of NH_4^+ diffusion out of the mitochondria and would tend to concentrate both CO_2 and NH_4^+ within the chloroplast. This process would also limit the direct activation of anaplerotic reactions within the cytosol by free NH_4^+ (see Section VI,B) that would otherwise arise from photorespiratory sources.

Ammonium present in the chloroplast is assimilated by glutamine synthetase. Mutants that lack chloroplastic glutamine synthetase in barley are viable in high CO_2 but are sensitive to growth in air (Blackwell *et al.,*

1987; Wallsgrove *et al.*, 1987). Such a mutant has not been found in *Arabidopsis,* probably because it contains only the GS_2 isoform (Lea and Blackwell, 1990). The perturbation in plant growth in barley in air could be due to either the accumulation of NH_4^+ within the leaf, which has known toxic effects on cellular processes (see Section VI,B), or to the depletion of glutamine pools, which reduces the provision of amino donors for aminotransferase reactions during the C_2 cycle. Supplying glutamine to detached leaves from this mutant before they are placed in air alleviates the otherwise observed decrease in CO_2 fixation for the short term (Blackwell *et al.*, 1987). This indicates that products derived from glutamine, such as glutamate (synthesized via glutamate synthase), alanine (synthesized via glutamine-pyruvate aminotransferase), or other amino donors are needed for the operation of the C_2 cycle. As a result of feeding glutamine to these detached leaves, the pool size of NH_4^+ increases twofold, indicating that increased amounts of NH_4^+ are not responsible for decreased rates of CO_2 fixation—least not during the 10-min interval employed in these experiments. The supply of glutamine also provides carbon skeletons in the form of 2-oxoglutarate and derived products. These are catabolized at greater rates in the presence of NH_4^+ (Lawyer *et al.*, 1981; Larsen *et al.*, 1981; Platt *et al.*, 1977) and may alleviate short-term symptoms associated with elevated NH_4^+ pools. A variation of glutamine synthetase activity, from 8 to 160% of wild-type activity, is observed within the mutant population yet only those plants containing less than 38% of wild-type glutamine synthetase activity exhibit increased amounts of NH_4^+ accumulation in leaves in the air. This demonstrates that glutamine synthetase is present in twofold excess for the maintenance of wild-type amounts of NH_4^+ under photorespiratory conditions (Blackwell *et al.*, 1987).

Increased concentrations of endogenous NH_4^+ are also observed in leaves when glutamine synthetase activities are inhibited under photorespiratory conditions (Walker *et al.*, 1984a,b; Sauer *et al.*, 1987). Photosynthesis is reduced under these conditions, but the metabolic perturbations are very different from those induced by the addition of NH_4^+ alone (see Section VI,B). This is highlighted by the observation that supplying leaves with NH_4^+ results in the accumulation of more internal NH_4^+ than in leaves administered inhibitors of glutamine synthetase [e.g., methionine sulfoximine (Walker *et al.*, 1984a,b) or phosphinothricin (Sauer *et al.*, 1987)], yet the decrease in the rate of CO_2 fixation is initially similar in leaves in both treatments and thereafter declines more rapidly in the presence of methionine sulfoximine (Ikeda *et al.*, 1984; Walker *et al.*, 1984b). Even though the amount of NH_4^+ within the metabolic pool of the cell (versus compartmentalized storage pools within vacuoles) is not known when NH_4^+ is added to detached shoots there is a redirection of carbon toward amino acid synthesis (Walker *et al.*, 1984a,b), indicating that it is mediating cellular processes.

Methionine sulfoximine itself does not affect photosynthesis in isolated chloroplasts (Muhitich and Fletcher, 1983); however, supplying this inhibitor to detached leaves results in extensive catabolism of glutamate and other amino acids coupled with an increase in 4-aminobutyrate and the organic acid pool, especially malate, succinate, 2-oxoglutarate, and citrate (Walker *et al.*, 1984a,b). Similarly, a greater proportion of carbon is directed toward the synthesis of malate in glutamine synthetase-deficient mutants in air (Kendall *et al.*, 1986). Supplying either glutamine (Blackwell *et al.*, 1988; Ikeda *et al.*, 1984; Sauer *et al.*, 1987; Walker *et al.*, 1984a), glutamate, or alanine (Blackwell *et al.*, 1988) to detached leaves lacking glutamine synthetase activity temporarily alleviates the decrease in CO_2 fixation. These observations demonstrate that elevated pools of endogenous NH_4^+, produced through the inhibition of glutamine synthetase, are not solely responsible for the changes in metabolite concentrations.

Elevated concentrations of NH_4^+ do not perturb CO_2 fixation or photosynthetic electron transport for the short term under conditions of high CO_2, where a source of carbon is readily available for NH_4^+ assimilation. However, shifts in metabolism in the presence of ammonium that lead to the removal of triosephosphates from the Calvin cycle (and the regulation of enzymes involved in glycolysis, and TCA cycle; see Section VI,B), when coupled with the removal of carbon from the Calvin cycle in air as 2-phosphoglycolate, undoubtedly decrease the rate of CO_2 fixation due to depleted intermediates required for continued Calvin cycle activity when the plants are placed in air.

A continued replenishment of glutamate is required within the chloroplast for NH_4^+ assimilation and this is supplied by ferredoxin-dependent glutamate synthase (Fig. 7, reaction 15), which utilizes glutamine and 2-oxoglutarate. Deletion mutants, viable only under conditions of high CO_2, lacking plastid glutamate synthase, have been isolated (Kendall *et al.*, 1986; Somerville and Ogren, 1980a) and have been proposed to demonstrate that the primary function of this enzyme is the supply of glutamate at rates required either for the assimilation of photorespiratory-derived NH_4^+, or as amino donors for C_2-related transamination reactions. However, when these mutants are transferred to air glutamate is still present, indicating that other pathways (e.g., NADH-glutamate synthase, glutamate dehydrogenase, or transaminases) are capable of synthesizing glutamate at rates that approximate those involved in primary nitrogen assimilation (Kendall *et al.*, 1986). The pools of glycolate, glyoxylate, and citrate increase dramatically in glutamate synthase-deficient *Arabidopsis* in air (Grumbles, 1987), indicating that secondary (inhibitory) effects on metabolism associated with these metabolites cannot be dismissed. This is also supported by the observation that the rate of $^{14}CO_2$ incorporation into glutamate in the *Arabidopsis* mutant is equivalent to the wild type for only the first 6 min in air (Grumbles, 1987). The amount of 2-oxoglutarate within the leaves of the *Arabidopsis* mutant increases rapidly in air in the mutant, yet it is undetectable in the wild type and this may be due to its reduced consumption in the absence of glutamate synthase. 2-Oxoglutarate is known to regulate several metabolic process (Huault *et al.*, 1990; Schuller *et al.*, 1990b) and elevated pool sizes of 2-oxoglutarate within these mutants are likely to perturb metabolic processes through secondary effects. The rates of synthesis of glutamate by reactions other than ferredoxin-dependent glutamate synthase in the barley glutamate synthase mutant as observed by Kendall *et al.*, (1986) may therefore represent minimum values.

The import of cytosolic 2-oxoglutarate into the chloroplast for glutamate synthase activity, and the export of glutamate synthesized by this reaction for cytosolic and peroxisomal transamination reactions, take place via a dicarboxylate transporter (Fig. 7, translocator A). This transporter is detected by measuring O_2 evolution in isolated chloroplasts in the presence of NH_4^+, or glutamine, and 2-oxoglutarate (Anderson and Walker, 1983; Dry and Wiskich, 1983). The evolution of O_2 is also stimulated when aspartate, glutamate, malate, succinate, or fumarate are added along with the above reaction mixture, and the inhibition of transport of one dicarboxylic acid by another demonstrates that a translocator is involved. The flux of dicarboxylates across the chloroplast envelope via this transporter during photorespiration represents the largest transport process taking place in this organelle (Woo *et al.*, 1987). It is therefore interesting that air-sensitive mutants of barley or *Arabidopsis* have been obtained with chloroplasts that exhibit low rates of O_2 evolution in the presence of glutamine, NH_4^+, and 2-oxoglutarate. The observation indicates impairment of the dicarboxylate transporter. These chloroplasts also display reduced uptake of dicarboxylic acids (Somerville and Ogren, 1983), are missing a chloroplastic envelope protein (Somerville and Somerville, 1985), and lack the ability to synthesize or metabolize glutamate even though the required enzymes are present in these plants at wild-type levels (Wallsgrove *et al.*, 1986). Selection of these mutants demonstrates that this transporter is primarily involved with dicarboxylate fluxes across the chloroplast envelope associated with photorespiratory metabolism, and either that the shuttling of redox equivalents between the chloroplast and cytosol is of secondary importance (Somerville and Ogren, 1983) or that more than one transporter is involved in the shuttling of dicarboxylates. The transport of 2-oxoglutarate and glutamate does in fact take place through separate transporters with overlapping specificities that can be kinetically differentiated by their affinities for substrates. Two transporters have been characterized that counterexchange with malate, one preferentially involved with 2-oxoglutarate import (Fig. 5, translocator G) and the other exhibiting preferential glutamate export (Fig. 5, translocator B; Woo *et al.*, 1987). Furthermore, as chloroplasts isolated from deletion mutants lacking dicarboxylate transport still exhibit wild-type rates of glutamine uptake (Somerville and Ogren, 1983) it would appear that another transporter appears to be involved in glutamine uptake. Indeed, characterization of the glutamine transport process indicates that glutamine transport can be coupled with glutamate, but not aspartate, malate, or 2-oxoglutarate transport (Fig. 5, translocator C; Yu and Woo, 1988) and differentiates this transporter from the two described above.

B. Advantages of Compartmentation in Photorespiratory Metabolism

The metabolism of photorespiratory intermediates takes place within three organelles. This elaborate shuttling of metabolites between these compartments implies an associated functionality, yet there are key enigmas which we do not understand. Why, for example, is glycine oxidized within the mitochrondria? Why is the transamination of glyoxylate and hydroxpyruvate peroxisomal? And why is C_3 metabolism with its associated photorespiration localized in the bundle sheath cells of C_4 plants?

The flux of carbon and nitrogen through the C_2 and associated nitrogen cycles generates compounds that have demonstrated *in vitro* toxicities. These include the following.

2-Phosphoglycolate: An inhibitor of triose-phosphate isomerase (Wolfenden, 1970) and phosphofructokinase (Kelly and Latzko, 1976) activities

H_2O_2: This compound leads to the peroxidation of lipids (Kendall *et al.*, 1983), glyoxylate, and hydroxypyruvate, the latter two generate CO_2 and reduce the flux of carbon within the C_2 cycle (Grodzinski, 1979; Walton and Butt, 1981; Zelitch, 1989). H_2O_2 also inhibits the Calvin cycle, and elevates oxidative pentose phosphate pathway activities (Kaiser, 1979)

Glyoxylate: An inhibitor of glycine decarboxylase (Sarojini and Oliver, 1987) and Rubisco activities (Cook *et al.*, 1985). It also stimulates serine hydroxymethyltransferase activity (Peterson, 1982), and at low concentrations activates Rubisco by alleviating the CO_2 requirement

(Cook *et al.*, 1985). In the presence of NH_4^+, glyoxylate inhibits serine-glyoxylate, and glutamate-glyoxylate transaminase activities (Havir, 1986)

Ammonium: This compound redirects the flow of carbon from starch synthesis and carbohydrate reserves toward the synthesis of amino acids (Givan, 1979). This process would involve regulation of carbon partitioning to the synthesis or degradation of sucrose (Stitt, 1990). In the presence of NH_4^+ the incorporation of $^{14}CO_2$ into sucrose is reduced while its incorporation into glutamine, asparagine, and several other amino acids increases (Dahlbender and Strack, 1986; Mohamed and Gnanam, 1979; Lawyer *et al.*, 1981; Larsen *et al.*, 1981). The extractable activities of several enzymes increase following incubation of leaf disks in NH_4^+, including the glycolytic enzymes phosphofructokinase, enolase (Wakiuchi *et al.*, 1971), pyruvate kinase (Miller and Evans, 1957), phosphoenolpyruvate carboxylase (Dahlbender and Strack, 1986; Guy *et al.*, 1989; Platt *et al.*, 1977), and enzymes of the TCA cycle, aconitase and isocitrate dehydrogenase (Wakiuchi *et al.*, 1971). Ammonium inhibits pyruvate dehydrogenase activity by activating pyruvate dehydrogenase kinase activity (Schuller and Randall, 1989), and at low concentrations ($< 1 \text{m}M$) NH_4^+ abolishes the light activation of NADP-dependent glyceraldehyde-3-phosphate dehydrogenase and alkaline fructose-1,6-bisphosphatase (Mohamed and Gnanam, 1979; Rosa and Whatley, 1981). In the presence of glyoxylate, NH_4^+ irreversibly inhibits serine-glyoxylate and glutamate-glyoxylate transaminase activities (Havir, 1986). Furthermore, NH_4^+ dissipates the proton gradient established during photosynthetic electron transport and uncouples photosynthetic phosphorylation (Good, 1977).

It may therefore be argued that compartmentation of enzymes of the C_2 and nitrogen cycles that participate in the liberation of these compounds would limit their potential effects on cellular processes elsewhere. Thus, H_2O_2 is generated and degraded within the peroxisome by glycolate oxidase and catalase, respectively. As glyoxylate does not freely leave the peroxisome, and as glutamate-glyoxylate and serine-glyoxylate aminotransferases have a high affinity for glyoxylate, its generation and localization within the peroxisome ensures its rapid turnover. The removal of carbon from the Calvin cycle in air results in a decrease of triosephosphate export through the phosphate translocator and reduces the export of reductant and ATP into the cytosol (Gärdstrom and Wigge, 1988; Usuda and Edwards, 1982). An associated function for the decarboxylation of glycine within the mitochondria is that the reductant produced by this reaction is preferentially oxidized by the electron transport chain (Dry *et al.*, 1983,

1987) and the ATP exported into the cytosol (Gärdstrom and Wigge, 1988). Furthermore, oxidative phosphorylation within the mitochondria is resistant to high concentrations of NH_4^+ (Yamaya and Matsumoto, 1985a), in contrast with what is observed in chloroplasts (Good, 1977) and it is here that NH_4^+ is liberated through glycine decarboxylation. As NH_4^+ traverses membranes freely (Kleiner, 1981), it is currently envisaged (Fig. 7) that NH_4^+ exits out of the mitochondrion and diffuses through the cytosol en route to the chloroplast, where it is assimilated by glutamine synthetase. However, when the regulatory effects of NH_4^+ on reactions within the mitochondria, cytosol, peroxisome, and chloroplast are considered, it seems unlikely that NH_4^+ diffuses freely within an organelle or throughout the cell. To minimize the diffusion of NH_4^+ within the cytosol there may be direct transfer into the chloroplast (i.e., if mitochondria are oppressed with chloroplasts, Fig. 8), or as discussed earlier, NH_4^+ may be assimilated in the mitochondria via carbamoyl-phosphate synthetase (Fig. 4, reaction 8) and to a minor degree by glutamate dehydrogenase (Fig. 4, reaction 11), thereby limiting its effect on metabolic processes elsewhere.

It is interesting to note that several reactions lead to the synthesis of intermediates of the C_2 cycle from precursors that are not derived from the C_2 cycle itself. These reactions include the synthesis of glycolate from isocitrate (Zelitch, 1988) and hydroxypyruvate (Davies and Asker, 1985), the cytosolic reduction of hydroxypyruvate producing glycerate (Kleczkowski *et al.*, 1988), and the synthesis of serine in the chloroplast (Shah and Cossins, 1970). These reactions do not produce potentially toxic compounds and for the most part these reactions take place in the cytosol.

C. Overview

The metabolic reactions associated with the photorespiratory process in higher plants have undoubtedly evolved to minimize carbon losses, such as the glycolate excretion observed in unicellular algae. The extent of withdrawal of carbon and nitrogen from photorespiratory processes and the contribution of nonphotorespiratory sources of carbon and nitrogen to pathways associated with oxidative carbon metabolism in higher plants indicate that through time other functions have become associated with these metabolic processes. The losses and additions of carbon and nitrogen to photorespiratory metabolism are affected by several variables. These include plant age (Givan *et al.*, 1988a; Madore and Grodzinski, 1984; Ta *et al.*, 1985), temperature (Davies and Asker, 1985; Grodzinski, 1978; Hanson and Peterson, 1986, 1987; Zelitch, 1987, 1988, 1989), and the nutritional status of the plant (Lea *et al.*, 1989; Somerville and Ogren, 1982b; Ta and Joy, 1986; Winkler *et al.*, 1987). In fact, several functions have been associated with the photorespiratory process. It has been suggested that by maintaining a flux through the C_2 cycle under conditions of high light and/or low CO_2, photooxidative damage is reduced (Heber *et al.*, 1989; Powels *et al.*, 1984; Walker *et al.*, 1986). Furthermore, because photosynthesis can be limited by the availability of phosphate (Sivak and Walker, 1986; Walker *et al.*, 1986), the hydrolysis of 2-phosphoglycolate (an intermediate of the C_2 cycle), producing phosphate and glycolate, replenishes phosphate within the chloroplast and this has been shown to alleviate photoinhibitory damage (Heber *et al.*, 1989). The photorespiratory pathway also functions in a synthetic capacity, producing C_1-tetrahydrofolic acid (Cossins, 1980) either during glycine decarboxylation (Sarojini and Oliver, 1983) or formate oxidation (Grodzinski, 1979), as well as glycine and serine (Ta *et al.*, 1985) through the actions of glutamate-glyoxylate or serine-glyoxylate aminotransferase (Nakamura and Tolbert, 1983) and serine hydroxymethyltransferase (Oliver *et al.*, 1990a), respectively.

VII. Targeting of Proteins into Organelles

Enzymes involved in the biosynthesis of amino acids in the mitochondrion, chloroplast, and peroxisome are for the most part encoded by nuclear genes and translated within the cytosol. The nuclear genome encodes all of the proteins compartmentalized within the peroxisome and up to 90% of the chloroplast and mitochondrial protein complement (Ellis, 1981). This necessitates the import of proteins into the appropriate organelle for function. A typical feature of the polypeptide sequence encoding several of these enzymes is an amino-terminal transit sequence that is used to target the carboxy-terminal region of the polypeptide, or passenger protein, into the correct subcellular compartment. This transit peptide is proteolytically cleaved within the organelle following the import of the precursor protein. However, proteins that enter the peroxisome do not always contain transit sequences (Lazarow and Fujiki, 1985). Of the proteins that contain targeting information, the location of required sequences for this process are not precisely known. For example, luciferase, a protein targeted into peroxisomes of higher eukaryotes, was also found to be targeted to peroxisomes of tobacco (Gould *et al.*, 1990). This targeting requires a carboxy-terminus sequence that may be involved in the import process (Gould *et al.*, 1987). On the other hand, catalase, another peroxisomal protein, does not have a similar sequence (see White and Scandalios, 1988). Furthermore, processing may occur prior to the entry of the polypeptide into the peroxisome (Gietl and Hock, 1984).

In chloroplasts and mitochondria, where the transport process has been examined in greater detail, the transit sequence interacts with the import apparatus, a collection of proteins that span the inner and outer membrane at specific sites where the two membranes come into contact. Transit sequences interact with these protein complexes in a specific but not fully understood fashion (Lubben *et al.,* 1988; Whelan *et al.,* 1990, 1991). Comparison of transit peptides of many stromal proteins indicates that there is no primary sequence similarity (Lubben *et al.,* 1988; Keegstra, 1989). Precursor proteins destined for mitochondrial import do, however, exhibit a structural motif (an amphipathic helix) within the transit peptide that is required for protein import (Baker and Schatz, 1991; von Heijne, 1986).

It is likely that structural features of the transit peptide are recognized by the import apparatus of the chloroplast, but in some cases passenger proteins affect the uptake of chimeric precursor constructs. This indicates that the information within the transit sequence is not always sufficient for the recognition and uptake processes, and that recognition may also involve the three-dimensional structure of the entire precursor molecule (Lubben *et al.,* 1988; Keegstra, 1989). The import apparatus is best characterized in mitochondrial systems (Baker and Schatz, 1991; Hartl and Neupert, 1990; Pfanner *et al.,* 1991) and is analogous in several regards to the chloroplast apparatus. However, there are several differences between the two processes, as indicated by the different requirements in the properties of the transit peptide (noted above) and the energy requirements—chloroplast uptake does not require a potential across the inner membrane while protein uptake into mitochondria does (Keegstra *et al.,* 1989).

Both import protein complexes require cytosolic, outer and inner membrane, and matrix or stromal protein components. Initial stages of the import process involve the interaction between cytosolic 70-kDa heat shock proteins (HSP70) and precursor proteins to ensure that transit peptides adopt a conformation competent for transport. Also, unknown cytosolic proteins may participate in the release of the precursor protein from HSP70 in the presence of ATP after it has bound with a receptor on the outer membrane of the organelle. The associations between precursor molecules, heat shock proteins, and unidentified protein factors, and the requirement for ATP hydrolysis, are also required for protein uptake into pea chloroplasts (Waegemann *et al.,* 1990). It is of interest that an ATP requirement for peroxisomal protein import has also been reported (Imanaka *et al.,* 1987).

There appears to be a limited set of different receptors involved in the binding of precursors at both the chloroplast (Keegstra *et al.,* 1989) and mitochondrial (Yoshida *et al.,* 1985) outer membranes, which indicates that recognition of precursors by unique receptors does not take place. Furthermore, in mitochondria at least, these generalized surface receptor complexes do not appear essential for the uptake process. For example, partial inactivation of one protein (MAS70 in yeast, or MOM72 in *Neuro-*

spora) in isolated mitochondria by a mild protease treatment, or mutation of the gene encoding MAS70, has either no effect on the import process or reduces the rate of import, depending on the precursor examined (Baker and Schatz, 1991). However, protease pretreatment of chloroplast preparations inhibits protein import (Lubben *et al.*, 1988; Smeekens *et al.*, 1990), which suggests that this step may be important in chloroplast precursor uptake. Recently a 36-kDa import receptor from pea has been isolated (p36) and localized on the outer chloroplast membrane at sites that are in close contact with the inner membrane (Schnell *et al.*, 1990).

An integral mitochondrial membrane protein [ISP42 in yeast (Vestweber *et al.*, 1989) or MOM38 in *Neurospora* (Pfanner *et al.*, 1991)] is essential for the protein import process, and it interacts with several other as yet uncharacterized integral protein components, several receptor proteins, and a matrix-localized HSP70 (see below). Transport of the precursor across the outer and inner mitochondrial membranes occurs at sites where the two membranes come into contact and this may facilitate the formation of a transmembrane protein channel composed of these uncharacterized integral protein components. Similar contact sites have also been observed in chloroplast preparations (Pain *et al.*, 1988; Schnell *et al.*, 1990).

Once inside the mitochondrial matrix, transit peptides are cleaved through the coordinated activity of several essential proteins. These include the mitochondrial assembly polypeptides (known as MAS1 and MAS2 in yeast, or PEP and MPP in *Neurospora*) that are subunits of a matrix processing protease, and two heat shock proteins, HSP70 and HSP60. The HSP70 is involved in the initial phase of the import process within the matrix (prior to proteolytic cleavage of the precursor). It prevents premature folding or aggregation of the precursor protein (Scherer *et al.*, 1990). The other essential heat shock protein, HSP60, has a high amino acid sequence similarity with the molecular chaperonins, groEL of *E. coli* (Baker and Schatz, 1991), or the chloroplast Rubisco subunit-binding protein (Hemmingsen *et al.*, 1988). Chaperonins mediate the correct folding of imported polypeptides within an organelle either prior to, or following, proteolytic processing. The release of the folded product from HSP60 requires the hydrolysis of ATP, which provides the driving force for transport across the membranes. Association of the Rubisco binding protein (also termed cpn60, for chaperonin 60) with a range of proteins imported into the chloroplast, including glutamine synthetase, demonstrates that cpn60-mediated protein folding is widespread (Lubben *et al.*, 1989).

Stromal and thylakoid processing proteases have also been detected in chloroplasts. Proteins that are associated with the thylakoid membrane require the appropriate recognition of the amino-terminal sequence at both the chloroplast and thylakoid membranes. This requires partial processing of the N-terminal sequence after uptake into the chloroplast by a stromal processing peptidase followed by the removal of the remaining transit peptide, or thylakoid transfer domain, within the thylakoid by a thylakoid processing peptidase (Hageman *et al.,* 1990; Smeekens *et al.,* 1990).

The specificity of the precursor transport process into subcellular compartments permits targeting proteins into atypical organelles, and this allows for interesting *in vivo* manipulations of processes that are otherwise not observed in wild-type systems, for example, the targeting of a cytosolic glutamine synthetase into mitochondria (Hemmon *et al.,* 1990). The physiological question being considered in this experiment is whether glutamine synthatase can participate in the assimilation of NH_4^+ produced within this compartment during photorespiration. However, this experimental approach is not without problems, and demonstrates that several factors must be overcome when trying to establish physiological processes in perturbed systems. The chimeric precursor protein, composed of the transit peptide from β-F_1 ATPase fused onto the γ-polypeptide of a cytoslic nodule glutamine synthetase, resulted in targeting glutamine synthetase into mitochondria in transgenic tobacco. Furthermore, in *in vitro* plantlets this novel enzyme appeared to be properly processed within mitochondria, and contributed up to 25% of the total glutamine synthetase activity. However, in mature plants the protein did not assemble into an active enzyme and it was associated with the insoluble fraction of the mitochondria. The differences observed might be a result of the differences in the nutrients supplied to plantlets growing in culture versus mature growing plants, or a consequence of other metabolic differences between these two systems. Unfortunately, the question of whether or not a mitochondrial glutamine synthetase may participate in photorespiratory NH_4^+ accumulation remains unanswered.

VIII. Organization of Nitrogen Assimilation in Roots

Roots are involved in the uptake of NO_3^- and in the partitioning of nitrogen to the shoot. Because of their inaccessibility and because of the relatively low levels of enzymes the vast majority of work on roots involves analysis of the constituents of xylem sap (the exudate from the root after the shoot has been removed), and these assays may or may not include feeding with $^{15}NO_3^-$ or $^{15}NH_4^+$ (Atkins and Beevers, 1990; Pate, 1980, 1989). The assumption using this type of approach is that the xylem exudate contains products of metabolic processes that take place within the root (versus direct shoot-to-root transfer and retranslocation). An interesting finding is that all roots do not treat incoming NO_3^- in the same way (Pate, 1973; Andrews, 1986; Oaks, 1992). Temperate legumes, for example, export asparagine whereas cereal roots export NO_3^- and much smaller levels of reduced nitrogen. This means that roots from temperate legumes, of which *P. sativum* is an example, have adapted so that more NO_3^- is assimilated in the root, and relatively more photosynthate is transferred to the root, than in NO_3^- exporters. In legume roots infected with *Rhizobium,* the ability of the root–nodule complex to compete for photosynthate and the ability to export nitrogen (in the form of asparagine or allantoin) are enhanced relative to the uninfected control. It could be that the ability of legume roots to compete for photosynthate, and with that the ability to assimilate NH_4^+, has predisposed them to infection with *Rhizobium* (see Oaks, 1992, for a more extensive discussion of this point).

The reactions associated with NO_3^- reduction can be the major metabolic processes in the root system. This is in contrast to the leaf, where the flux of CO_2 reduction can be an order of magnitude greater than other reduction reactions (Lee, 1980). In addition, without photosynthesis, which has the potential of supplying excess NADPH and ATP, the root must trap these chemicals from glycolysis and electron transport in mitochondria and the enzymes of nitrogen assimilation must compete for these metabolites with other reactions in the root. The roots of maize and barley may have adapted to effectively compete for pyridine nucleotides by having an NR that can use either NADH or NADPH (Warner *et al.,* 1987; Long and Oaks, 1990) and probably in other ways that have not been examined. Vance and Griffith (1990) observed increases in glutamine synthetase, GOGAT, phosphoenolpyruvate carboxylase, and glutamic-oxaloacetic transaminase activities in the host tissues of nodules on alfalfa and soybean roots. Because the synthesis of these enzymes is turned on during the infection process these proteins may be products of genes induced during nodulation (nodulins). These proteins must in some way mediate or permit the higher levels of localized nitrogen assimilation detected in the host tissue within the nodule in infected root systems.

The basic organization of the enzymes of nitrogen assimilation is the same in roots and leaves—NR is in the cytosol, NiR and GOGAT in the plastid, and glutamine synthetase in the cytosol and plastid (Miflin, 1974; Oaks and Gadal, 1979; Suzuki *et al.,* 1981; Vézina *et al.,* 1987; Vézina and Langlois, 1989). A generalization that may be valid is that roots of *P. sativum* have relatively more glutamine synthetase in the plastid than do cereal roots (maize, rice) and that this may be related to their ability to assimilate nitrogen in the root. Emes and Bowsher (1991) have also examined plastid function in pea roots: they have established that glucose-6-phosphate and 6-phosphogluconate dehydrogenases, which supply the required NADPH, ferredoxin, and a pyridine nucleotide ferredoxin reductase are in pea root plastids. Thus the reductant for NiR and GOGAT is housed where those enzymes are localized. Oji *et al.* (1985) have also shown that these dehydrogenases are present in barley root plastids.

Enzymes, which are in both roots and leaves, are often not identical. For example, examinations made first by Gadal's group for glutamine synthetase, GOGAT, ferredoxin, and the pyridine nucleotide ferredoxin reductase showed that the root enzymes, although similar to the shoot enzymes, were not immunologically identical (Hirel and Gadal, 1980; Suzuki and Gadal, 1982; Suzuki *et al.*, 1985). These differences may have evolved due to the more competitive environment for reductant and energy faced by the root enzymes; however, we do not know at this time what advantages are attributable to these differences.

The assimilation of nitrogen is energetically more expensive in the root that in the shoot. This is due to the requirement of shoot-derived carbon for both the generation of cofactors involved in NO_3^- reduction and NH_4^+ assimilation, and for the carbon skeleton for the synthesis of transport forms of reduced nitrogen that are exported to the shoot. If NO_3^- can be assimilated in the root, does this different strategy represent an advantage or a disadvantage to the survival of the whole plant? Theoretically, NO_3^- reduction is energetically more efficient if it proceeds in the leaves, however, until recently there was no experimental evidence to support this contention.

Sinden and Durbin (1968) demonstrated that tabtoxin-β-lactam (TBL), a product synthesized by *Pseudomonas syringeae* pv. *tabacci* (tox^+), inhibits glutamine synthetase in an irreversible fashion. More recently, Knight *et al.* (1988) showed that addition of the pseudomonad or TBL to plant cultures results in the death of the host plant. They were able to select oat seedlings that were resistant to infection and unexpectedly these seedlings grew better in the presence of TBL. In the roots of these tolerant plants the glutamine synthetase was inactivated, yet the enzyme in the shoots was completely insensitive to the inhibitor. Because NO_3^- and glutamine are most likely the major forms of nitrogen exported from oat roots then the principal effect of inhibiting the root glutamine synthetase would be to force the plant to assimilate all of its nitrogen in the shoot. If this were really a more efficient way of operating, one would expect to see an increase in seedling biomass, and this is exactly what Knight *et al.* found. It will be of interest to see how general this observation is. One might predict, for example, that TBL-tolerant variants of lupine or pea should not be found because the majority of the nitrogen in these species is reduced in the root.

In a second unusual example, Knight and Langston-Unkefer (1988) showed that, although nodule-free alfalfa plants died in response to infection with *P. syringeae*, plants with well-developed active nodules grew better when they were also infected with *Pseudomonas*. This system is more complicated than the case with oat seedlings, but there is a physiologically meaningful explanation. The root glutamine synthetase is inhibited by TBL whether the alfalfa is infected with *Rhizobium* or not.

However, the nodule glutamine synthetase, which is expressed only in response to *Rhizobium,* is not inhibited. If the bacteroid nitrogenase were regulated by glutamine levels, as appears to be the case in free-living nitrogen fixers (reviewed by Vance and Griffith, 1990), changes in the size of the glutamine pool could have a profound effect on nitrogenase activity. Thus the decreased pool of glutamine might result in a higher level of nitrogenase activity and, in fact, nitrogenase activities are higher in alfalfa roots that have the double infestation (Knight and Langston-Unkefer, 1988). This response should make additional reduced nitrogen available to the plant, which in turn should allow the increased growth response that is seen in shoots, roots, and nodules of the double-infected alfalfa plants. Again this indicates that the assimilation of NH_4^+ in the nodules is more efficient due to a localized sink utilizing available photosynthate.

The efficiency inferred in these two examples is measured by enhanced yields—an agricultural measure. This kind of adaptation or specialization would be of no real advantage to wild plants, which depend for their survival on a diversity of responses to altered environmental conditions and not on higher yields.

IX. Conclusions

In this article we have focused on recent developments in the compartmentation of nitrogen metabolism in higher plants. What emerges from this analysis is that new research initiatives using genetic, biochemical, and physiological approaches are altering the understanding of key processes that have become established over the years.

Analysis using mutant and transgenic plants demonstrates that GS_1 and GS_2 have nonoverlapping functions in the assimilation of NH_4^+. The chloroplastic isozyme of glutamine synthetase is localized in mesophyll cells and appears to be responsible for the assimilation of photorespiratory-derived NH_4^+. However, the physiological role for GS_1, which is distributed within the vascular bundles, still needs to be adequately defined.

With the use of mutant methodology, inhibitors of metabolism, and the identification of new enzyme reactions, the major steps in photorespiration have been defined. Where once we thought that inputs and products of photorespiration should bear a stoichiometric relationship, it is now apparent that this pathway is open. Alternate sources of nitrogen (asparagine, glutamine and alanine) and carbon (citrate, and ureidoglycolate) may enter and exit this series of reactions. Glutamate, serine, and glycine, once thought to be integral intermediates of the photorespiratory pathway, can be exported from the leaf. The extent of CO_2 release may also be greater than previously proposed. Because of this carbon and nitrogen must be replaced if the flow of carbon is to be maintained. Mechanisms for that replacement have been proposed; however, their validation is still required. Alanine appears to be one of the major amino donors in the photorespiratory pathway, yet the mechanism for its synthesis in this capacity is not clear. It has also been proposed that the transfer of photorespiratory-derived NH_4^+ (and CO_2) from the mitochondria to the chloroplast may be mediated by citrulline, arginine, and ornithine. Although not yet substantiated, this mechanism would permit the efficient flow of NH_4^+ from mitochondrion to chloroplast that is demanded by the operation of a pathway of such magnitude as the photorespiratory pathway.

The leaves of C_4 plants are thought by some to be more efficient in nitrogen as well as carbon usage. The key to this observation may be that the enzymes of photorespiration are housed in the bundle sheath cells, where it is thought O_2 and CO_2 concentrations are maintained to reduce the flux of carbon through the photorespiratory pathway, whereas the enzymes of NO_3^- assimilation are in the mesophyll cells. Glutamine synthetase and glutamate synthase, enzymes required for the assimilation of both photorespiratory and NO_3^--derived NH_4^+, are equally distributed in both cell types. An intriguing puzzle at the moment is why glutamate dehydrogenase, an enzyme that has no established function in photorespiration, is located only in the bundle sheath cells. Indeed, in C_3 plants as well, the physiological role and the mechanism(s) involved in the regulation of this enzyme are far from clear.

Although nitrogen assimilation has not been as thoroughly investigated in roots as in leaves, it is apparent that some roots, temperate legumes in particular, assimilate relatively more nitrogen than roots of many nonlegume plants. Part of their success is the ability to direct more of the carbon assimilated by photosynthesis to the root system. When these roots become infected with *Rhizobium* species this ability is further enhanced. Interactions between enzymes in different parts of the plant, with the exception of glutamine synthetase, are a virtually unexplored area of the functioning of higher plants.

Acknowledgments

This research was supported by a Natural Sciences and Engineering Council of Canada research grant to A.O. and a Post Doctoral Fellowship to K.A.S. We wish to thank Drs. G. M. Coruzzi, J. V. Cullimore, H. W. Heldt, R. A. Ludwig, and D. Rhodes, who made manuscripts available before publication, and Dr. E. H. Newcomb for the electron micrograph.

References

Abrol, Y. P., Sawhney, S. K., and Naik, M. S. (1983). *Plant Cell Environ.* **6,** 595–599.
Amrhein, N., and Gödeke, K.H. (1977). *Plant Sci. Lett.* **8,** 313–317.
Anderson, J. W. (1980). *Biochem. Plants* **5,** 203–223.
Anderson, J. W., and Butt, V. S. (1986). *Biochem. Soc. Trans.* **104,** 106–107.
Anderson, J. W., and Done, J. (1977). *Plant Physiol.* **60,** 504–508.
Anderson, J. W., and Done, J. (1978). *Plant Physiol.* **61,** 692–697.
Anderson, J. W., and Walker, D. A. (1983). *Planta* **159,** 247–253.
Anderson, M. P., Vance, C. P., Heichel, G.H., and Miller, S.S. (1989). *Plant Physiol.* **90,** 351–358.
Andrews, M. (1986). *Plant Cell Environ.* **9,** 511–519.
Andrews, T. J., and Kane, H. J. (1991). *J. Biol. Chem.* **266,** 9447–9452.
Andrews, T. J., and Lorimer, G. H. (1987). *Biochem. Plants* **10,** 131–220.
Andrews, T.J., Lorimer, G.H., and Tolbert, N.E. (1973). *Biochemistry* **12,** 11–17.
Artus, N. N., Somerville, S. C., and Somerville, C.R. (1986). *CRC Critical Rev. Plant Sci.* **4,** 121–147.
Aslam, M., and Huffaker, R. C. (1982). *Plant Physiol.* **70,** 1009–1013.
Aslam, M., and Huffaker, R. C. (1984). *Plant Physiol.* **75,** 623–628.
Aslam, M., Huffaker, R. C., and Travis, R. L. (1973). *Plant Physiol.* **52,** 137–141.
Aslam, M., Oaks, A., and Huffaker, R. C. (1976). *Plant Physiol.* **58,** 588–591.
Atkins, C., and Beevers, L. (1990). *In* "Nitrogen in Higher Plants" (Y. P. Abrol, ed.), pp. 223–295. Res. Stud. Press, Tranton, Sommerset, England.
Atkins, C. A., Pate, J. S., and Sharkey, P. J. (1975). *Plant Physiol.* **56,** 807–812.
Avila, C., Botella, J. R., Canovas, F. M., de Castro, I. N., and Valpuesta, V. (1987). *Plant Physiol.* **85,** 1036–1039.
Awonaike, K. O., Lea, P. J., and Miflin, B. J. (1981). *Plant Sci. Lett.* **23,** 189–195.
Back, E., Burkhart, W., Meyer, M., Privalle, L., and Rothstein, S. (1988). *Mol. Gen. Genet.* **212,** 20–26.
Baker, K. P., and Schatz, G. (1991). *Science* **349,** 205–208.
Bascomb, N. F., and Schmidt, R. R. (1987). *Plant Physiol.* **83,** 75–84.
Bascomb, N. F., Plunkard, D. E., and Schmidt, R. R. (1987). *Plant Physiol.* **83,** 85–91.
Bauer, A., Joy, K. W., and Urquhart, A. A. (1977). *Plant Physiol.* **59,** 920–924.
Beevers, L., and Hageman, R. H. (1980). *Biochem. Plants* **5,** 115–168.
Betsche, T. (1983). *Plant Physiol.* **71,** 961–965.
Betsche, T., and Eising, R. (1986). *Plant Soil* **91,** 367–371.
Blackwell, R. D., Murray, A. J. S., and Lea, P. J. (1987). *J. Exp. Bot.* **38,** 1799–1809.
Blackwell, R. D., Murray, A. J. S., Lea, P. J., and Joy, K. W. (1988). *J. Expt. Bot.* **39,** 845–858.
Blackwell, R. D., Murray, A. J. S., and Lea, P. J. (1990). *Plant Physiol.* **94,** 1316–1322.
Bogness, S. R., Stewart, C. R., Aspinall, D., and Paleg, L. G. (1976). *Plant Physiol.* **58,** 398–401.
Boland, M. J., Hanks, J. F., Reynolds, P. H. S., Blevins, D. G., Tolbert, N. E., and Schubert, K. R. (1982). *Planta* **155,** 45–51.
Borchert, S., Grosse, H., and Heldt, H. W. (1989). *FEBS Lett.* **253,** 183–186.
Botella, J. R., Vervelen, J. P. and Valpuesta, V. (1988). *Plant Physiol.* **88,** 943–946.
Bouzayen, M., Latche, A., and Pech, J.-C. (1990). *Planta* **180,** 175–180.
Bowman, E. J., Ikuma, H., and Stein, H. J. (1976). *Plant Physiol.* **58,** 426–432.
Bowsher, C. G., Hucklesby, D. P., and Emes, M. J. (1989). *Planta* **175,** 334–340.
Branjeon, J., Hirel, B., and Forchoni, A. (1989). *Protoplasma* **151,** 88–97.
Brouquisse, R., Nishimura, M., Gaillard, J., and Douce, R. (1987). *Plant Physiol.* **84,** 1402–1430.

Brunswick, P., and Cresswell, K. F. (1988). *Plant Physiol.* **86,** 384–389.
Bryan, J. K. (1980). *Biochem. Plants.* **5,** 403–452.
Bryan, J. K. (1990). *Biochem. Plants.* **16,** 161–195.
Buchholz, B., and Schultz, G. (1980). *Z. Pflanzenphysiol.* **100,** 209–215.
Buchholz, B., Renpke, B., Bickel, H., and Schultz, G. (1979). *Phytochemistry* **18,** 1109–1112.
Budde, R. J. A., and Randall, D. D. (1990). *Proc. Natl. Acad. Sci. U.S.A.* **87,** 673–676.
Burdge, E. L., Mathews, B. F., Mills, W. R., Widholm, J. M., Wilson, K. G., Carlson, J. E., Debonte, L. R., and Oaks, A. (1979). *Plant Physiol.* **63,** (Suppl. 138).
Butt, V. S., and Beevers, H. (1961). *Biochem. J.* **80,** 21–27.
Butz, R. G., and Jackson, W. A. (1977). *Phytochemistry* **57,** 519–522.
Caboche, M., and Rouzé, P. (1990). *Trends Genet.* **6,** 187–191.
Cammaerts, D., and Jacobs, M. (1983). *Plant Sci. Lett.* **31,** 65–73.
Canvin, D. T., and Atkins, C. A. (1974). *Planta* **116,** 207–224.
Castric, P. A., Farnden, K. J. F., and Conn, E. E. (1972). *Arch. Biochem. Biophys.* **152,** 62–69.
Cedar, H., and Schwarz, J. H. (1969). *J. Biol. Chem.* **244,** 4112–4121.
Chang, C. C., and Huang, A. H. C. (1981). *Plant Physiol.* **67,** 1003–1006.
Chang, K. S., and Farnden, K. J. F. (1981). *Arch. Biochem. Biophys.* **208,** 49–58.
Chantarotwong, W., Huffaker, R. C., Miller, B. L., and Granstedt, R. C. (1976). *Plant Physiol.* **57,** 519–522.
Chappell, J., and Hahlbrock, K. (1984). *Nature (London)* **311,** 76–78.
Chen, F. L., and Cullimore, J. V. (1988). *Plant Physiol.* **88,** 1411–1417.
Chen, R.-D., and Gadal, P. (1990a). *Plant Physiol. Biochem.* **28,** 141–145.
Chen, R.-D., and Gadal, P. (1990b). *Plant Physiol. Biochem.* **28,** 411–427.
Chiu, J. Y., and Shargool, P. D. (1979). *Plant Physiol.* **63,** 409–415.
Chou, K.-H., and Splittstoesser, W. E. (1972). *Plant Physiol.* **49,** 550–554.
Clarkson, D. T. (1986). *In* "Fundamental, Ecological and Agricultural Aspects of Nitrogen Metabolism in Higher Plants" (H. Lambers, H. Neeleson, and I. Stulen, eds.), pp. 3–27. Nijhoff, Dordrecht, The Netherlands.
Cock, J. M., Mould, R. M., Bennett, M. J., and Cullimore, J. V. (1990). *Plant Mol. Biol.* **14,** 549–560.
Conn, E. E. (1979). *Int. Rev. Biochem.* **27,** 21–43.
Cook, C. M., Mulligan, R. M., and Tolbert, N. E. (1985). *Arch. Biochem. Biophys.* **240,** 392–401.
Coruzzi, G. M. (1991). *Plant Sci.* **74,** 145–155.
Cossins, E. A. (1980). *Biochem. Plants* **2,** 365–414.
Dahlbender, B., and Strack, D. (1986). *Planta* **169,** 382–392.
Dalling, M. J., Tolbert, N. E., and Hageman, R. H. (1972a). *Biochim. Biophys. Acta.* **283,** 505–512.
Dalling, M. J., Tolbert, N. E., and Hageman, R. H. (1972b). *Biochim. Biophys. Acta.* **283,** 513–519.
Davies, D. D., and Asker, H. (1985). *Phytochemistry* **24,** 231–234.
Davies, D. D., and Teixeira, A. N. (1975). *Phytochemistry* **14,** 647–656.
Davis, D. J., and San Pietro, A. (1977). *Arch. Biochem. Biophys.* **182,** 266–272.
Davis, R. H. (1986). *Microbiol. Rev.* **50,** 280–313.
Davis, R. H., and Weiss, R. L. (1988). *Trends Biochem Sci.* **13,** 101–104.
Day, D. A., and Wiskich, J. T. (1980). *FEBS. Lett.* **112,** 191–194.
Day, D. A., Neuburger, M., and Douce, R. (1984). *Arch. Biochem. Biophys.* **231,** 233–242.
de Duve, C. (1969). *Proc. R. Soc. London, B.* **173,** 71–83.
Deng, M.-D., Moureaux, T., Chehel, I., Boutin, J.-P., and Carboche, M. (1991). *Plant Physiol. Biochem.* **29,** 239–247.
de Vienne, D. (1983). *Can. J. Gen. Cytol.* **25,** 146–160.

Doddema, H., and Telkamp, G. P. (1979). *Physiol. Plant.* **45,** 332–338.
Doddema, H., Hofstra, J. J., and Feenstrom, W. J. (1978). *Physiol. Plant.* **43,** 343–350.
Dougall, D. K. (1974). *Biochem. Biophys. Res. Commun.* **58,** 639–646.
Dry, I. B., and Wiskich, J. T. (1983). *Plant Physiol.* **72,** 291–296.
Dry, I. B., Day, D. A., and Wiskich, J. T. (1983). *FEBS. Lett.* **158,** 154–158.
Dry, I. B., Dimitriadis, E., Ward, A. D., and Wiskich, J. T. (1987). *Biochem. J.* **245,** 669–675.
Duke, S. H., Frieden, J. W., Schrader, L. E., and Kowkkari, W. L. (1978). *Physiol. Plant.* **42,** 269–276.
Ebbighausen, H., Hatch, M. D., Lilley, R. M., Kromer, S., Stitt, M., and Heldt, H. W. (1987). *In* "Plant Mitochondria: Structural, Functional, and Physiological Aspects" (A. L. Moore and R. D. Beechy, eds.), pp. 171–180, Plenum, New York.
Edwards, J. W., and Coruzzi, G. M. (1989). *Plant Cell* **1,** 241–248.
Edwards, J. W., Walker, E. L., and Coruzzi, G. M. (1990). *Proc. Natl. Acad. Sci. U.S.A.* **87,** 3459–3463.
Ellis, R. J. (1981). *Annu. Rev. Plant Physiol.* **32,** 111–137.
Emes, M. J., and Bowsher, C. G. (1991). *In* "Compartmentation of Plant Metabolism in Non-Photosynthetic Tissues" (M. J. Emes, ed.), pp. 147–169. Cambridge Univ. Press, Cambridge, England.
Emes, M. J., and Fowler, M. W. (1979). *Planta* **144,** 249–253.
Ericson, M. (1985). *Plant Physiol.* **79,** 923–927.
Evans, H. J., and Nason, A. (1953). *Plant Physiol.* **28,** 233–254.
Fawole, M. O. (1977). *Can. J. Bot.* **55,** 1850–1856.
Ferrari, T. E., Yoder, O. C., and Filner, P. (1973). *Plant Physiol.* **51,** 423–431.
Filner, P. (1966). *Biochim. Biophys. Acta* **118,** 299–310.
Flügge, U. I., and Heldt, H. W. (1984). *Trends Biochem. Sci.* **9,** 530–533.
Forde, B. G., and Cullimore, J. V. (1989). *Oxford Surv. Plant Mol. Cell Biol.* **6,** 247–296.
Forde, B. G., Day, H. M., Turton, J. F., Shen, W.-J, Cullimore, J. V., and Oliver, J. G. (1989). *Plant Cell* **1,** 391–401.
Frederick, S. E., and Newcomb, E. H. (1969). *Science* **163,** 1353–1355.
Frieden, C. (1963). *Enzymes* **7,** 3–24.
Furuhashi, K., and Takahashi, Y. (1982). *Plant Cell Physiol.* **23,** 179–184.
Galston, A. W., and Sawhney, R. K. (1990). *Plant Physiol.* **94,** 406–410.
Gamble, J. G., and Lehninger, A. L. (1973). *J. Biol. Chem.* **248,** 610–618.
Gardeström, P., and Wigge, B. (1988). *Plant Physiol.* **88,** 69–76.
Gaudillere, M., Jolivet-Tournier, P., and Cosstes, C. (1983). *Physiol. Veg.* **21,** 1115–1124.
Genix, P., Bligny, R., Martin, J.-B., and Douce, R. (1990). *Plant Physiol.* **94,** 717–722.
Gietl, C., and Hock, B. (1984). *Planta* **162,** 261–267.
Gilchrist, D. H., and Kosuge, T. (1980). *Biochem. Plants* **5,** 507–531.
Giovanelli, J., Mudd, S. H., and Datko, A. (1980). *Biochem. Plants* **5,** 454–505.
Givan, C. V. (1979). *Phytochemistry* **18,** 375–382.
Givan, C. V. (1980). *Biochem. Plants* **5,** 329–357.
Givan, C. V., Joy, K. W., and Kleczkowski, L. A. (1988a). *Trends Biochem. Sci.* **13,** 433–437.
Givan, C. V., Tsutakawa, S., Hodgson, J. M., David, N., and Randall, D. D. (1988b). *J. Plant Physiol.* **132,** 593–599.
Glass, A. D. M. (1988). *Atlas Plant Sci., 1st* **1,** 151–156.
Glass, A. D. M., Siddiqi, M. Y., Ruth, T. J., and Rufty, T.W., Jr. (1990). *Plant Physiol.* **93,** 1585–1589.
Good, A. G., and Crosby, W. L. (1989). *Plant Physiol.* **90,** 1305–1309.
Good, N. E. (1977). *Encycl. Plant Physiol.* **5,** 429–436.
Goodchild, J. A., and Sims, A. P. (1990). *Plant Sci.* **68,** 1–7.
Goudey, J. S., Tittle, F. L., and Spencer, M. S. (1989). *Plant Physiol.* **89,** 1306–1310.

Gould, S. J., Keller, G.-A., and Subramani, S. (1987). *J. Cell Biol.* **105,** 2923–2931.
Gould, S. J., Keller, G.-A., Schneider, M., Howell, S. H., Garrard, L. J., Goodman, J. M., Distel, B., Tabak, H., and Subramani, S. (1990). *EMBO J.* **9,** 85–90.
Gowri, G., and Campbell, W. H. (1989). *Plant Physiol.* **90,** 792–798.
Granstedt, R. C., and Huffaker, R. C. (1982). *Plant Physiol.* **70,** 410–413.
Grodzinski, B. (1978). *Planta* **144,** 31–37.
Grodzinski, B. (1979). *Plant Physiol.* **63,** 289–293.
Grumbles, R. M. (1987). *J. Plant Physiol.* **130,** 363–371.
Guiz, C., Hirel, B., Shedlofsky, G., and Gadal, P. (1979). *Plant Sci. Lett.* **15,** 271–278.
Guy, R. D., Vanlerberghe, G. C., and Turpin, D. H. (1989). *Plant Physiol.* **89,** 1150–1157.
Hageman, J., Baecke, C., Ebskamp, M., Pilon, R., Smeekens, S., and Weisbeek, P. (1990). *Plant Cell* **2,** 479–494.
Hageman, R. H., and Flesher, D. (1960). *Plant Physiol.* **35,** 700–708.
Hahlbrock, K., and Scheel, D. (1989). *Annu. Rev. Plant Physiol. Plant Mol. Biol.* **40,** 347–369.
Hahlbrock, K., Sutter, A., Wellmann, E., Ortmann, R., and Grisbach, H. (1971). *Phytochemistry* **10,** 107–116.
Hall, N. P., Kendall, A. C., Lea, P. J., Turner, J. C., and Wallsgrove, R. M. (1987). *Photosynth. Res.* **11,** 89–96.
Halliwell, B. (1984). "Chloroplast Metabolism: The Structure and Function of Chloroplasts in Green Leaf Cells." Oxford Univ. Press, Oxford, England.
Hanson, K. R., and Havir, E. A. (1981). *Biochem. Plants* **7,** 577–625.
Hanson, K. R., and Peterson, R. B. (1986). *Arch. Biochem. Biophys.* **246,** 332–346.
Hanson, K. R., and Peterson, R. B. (1987). *Prog. Photosynth. Res.* **3,** 549–556.
Harper, L. A., Sharpe, R. R., Langdale, G. W., and Giddens, J. E. (1987). *Agron. J.* **79,** 965–973.
Hartl, F. U., and Neupert, W. (1990). *Science* **247,** 930–938.
Hartmann, T., and Ehmke, A. (1980). *Planta* **149,** 207–208.
Hartmann, T., Nagel, M., and Ilert, H. J. (1973). *Planta* **111,** 119–128.
Hatch, M. D., Droscher, L., and Heldt, H. W. (1984). *FEBS Lett.* **178,** 15–19.
Havir, E. A. (1986). *Plant Physiol.* **80,** 473–478.
Havir, E. A., and McHale, N. A. (1987). *Plant Physiol.* **84,** 450–455.
Havir, E. A., and McHale, N. A. (1989). *Plant Physiol.* **89,** 952–957.
Hayakawa, T., Kamachi, K., Oikawa, M., Ojima, K., and Yamaya, T. (1990). *Plant Cell Physiol.* **31,** 1071–1077.
Haynes, R. J. (1986). "Mineral Nitrogen in the Plant Soil System." Academic Press, Orlando, Florida.
Heber, U., Neimanis, S., Mimura, T., and Dietz, K. J. (1989). *Z. Naturforsch., C: Biosci.* **44c,** 524–536.
Hecht, U., Oelmuller, R., Schmidt, S., and Mohr, H. (1988). *Planta* **175,** 130–138
Heimer, Y. M., and Filner, P. (1970). *Biochim. Biophys. Acta* **215,** 152–165.
Hemmings, B. A. (1982). *Biochem. Soc. Trans.* **10,** 328–329.
Hemmingsen, S. M., Woolford, C., van der Vies, S., Tilly, K., Dennis, D. T., Georgopoulos, C. P., Hendrix, R. W., and Ellis, R. J. (1988). *Nature (London)* **333,** 330–334.
Hemmon, P., Robbins, M. P., and Cullimore, J. V. (1990). *Plant Mol. Biol.* **15,** 895–904.
Hirasawa, M., and Tamura, G. (1984). *J. Biochem. (Tokyo)* **95,** 983–994.
Hirel, B., and Gadal, P. (1980). *Plant Physiol.* **66,** 619–623.
Hirel, B., Perrot-Rechemann, C., Suzuki, A., Vidal, J., and Gadal, P. (1982). *Plant Physiol.* **69,** 983–987.
Hirel, B., Bouet, C., King, B., Layzell, D., Jacobs, F., and Verma, D. P. S. (1987). *EMBO J.* **5,** 1167–1171.
House, C. M., and Anderson, J. W. (1980). *Phytochemistry* **19,** 1925–1930.

Howitz, K. T., and McCarty, R. E. (1983). *FEBS Lett.* **153,** 339–342.
Howitz, K. T., and McCarty, R. E. (1985). *Biochemistry* **24,** 3645–3650.
Howitz, K. T., and McCarty, R. E. (1987). *Prog. Photosynth. Res.* **3,** 593–596.
Huault, C., Blondel, J.-D., and Balange, A.-P. (1990). *Plant Physiol. Biochem.* **28,** 137–140.
Humbert, R., and Simoni, R. D. (1980). *J. Bacteriol.* **142,** 212–220.
Hummelt, G., and Mora, J. (1980). *Biochem. Biophys. Res. Commun.* **96,** 1688–1694.
Hurlburt, B. K., and Garrett, R. H. (1988). *Mol. Gen. Genet.* **211,** 35–40.
Husic, D. W., Husic, H. D., and Tolbert, N. E. (1987). *CRC Crit. Rev. Plant Sci.* **5,** 45–100.
Ikeda, M., Ogren, W. L., and Hageman, R. H. (1984). *Plant Cell Physiol.* **25,** 447–452.
Imanaka, T., Small, G. M., and Lazarow, P. B. (1987). *J. Cell Biol.* **105,** 2915–2922.
Ireland, R. J. (1986). *Plant Physiol.* **80s,** 28.
Ireland, R. J., and Joy, K. W. (1981). *Planta* **151,** 289–292.
Ireland, R. J., and Joy, K. W. (1983). *Arch. Biochem. Biophys.* **223,** 291–296.
Jahnen, W., and Hahlbrock, K. (1988). *Planta* **173,** 453–458.
Jordan, B. R., and Givan, C. V. (1979). *Plant Physiol.* **64,** 1043–1047.
Journet, E. P., Bonner, W. D., and Douce, R. (1982). *Arch. Biochem. Biophys.* **214,** 366–375.
Joy, K. W. (1973). *Phytochemistry* **12,** 1031–1040.
Joy, K. W. (1988). *Can. J. Bot.* **66,** 2103–2109.
Joy, K. W., and Hageman, R. H. (1966). *Biochem. J.* **100,** 263–273.
Joy, K. W., and Prabha, C. (1986). *Plant Physiol.* **82,** 99–103.
Joy, K. W., Ireland, R. J., and Lea, P. J. (1983). *Plant Physiol.* **73,** 165–168.
Kaiser, W. M. (1979). *Planta* **145,** 377–382.
Kamachi, K., Yamaya, T., Mae, T., and Ojima, K. (1991). *Plant Physiol.* **96,** 411–417.
Kanamori, T., and Matsumoto, H. (1972). *Arch. Biochem. Biophys.* **159,** 113–122.
Kannangara, G. C., Gough, S. P., and von Weltstein, D. (1978). *In* "Chloroplast Development" (G. Akoyunoglou and J. H. Agroudi-Akoyunoglou, eds.), pp. 147–160. Elsevier/North-Holland, Amsterdam.
Keegstra, K. (1989). *Cell* **56,** 247–253.
Keegstra, K., Olsen, L. J., and Theg, S. M. (1989). *Annu. Rev. Plant Physiol. Plant Mol. Biol.* **40,** 471–501.
Kelly, G. J., and Latzko, E. (1976). *FEBS Lett.* **68,** 55–58.
Kendall, A. C., Keys, A. J., Turner, J. C., Lea, P. J., and Miflin, B. J. (1983). *Planta* **159,** 505–511.
Kendall, A. C., Wallsgrove, R. M., Hall, J. C., Turner, J. C., and Lea, P. J. (1986). *Planta* **168,** 316–323.
Kerr, M. W., and Groves, D. (1975). *Phytochemistry* **14,** 359–362.
Keys, A. J. (1980). *Biochem. Plants* **5,** 359–373.
Keys, A. J. (1986). *Philos. Trans. R. Soc. London, B* **313,** 325–336.
Keys, A. J., Bird, I. F., Cornelius, M. J., Lea, P. J., Wallgrove, R. M., and Miflin, B. J. (1978). *Nature (London)* **275,** 741–743.
Kindt, R., Pahlich, E., and Asched, I. (1980). *Eur. J. Biochem.* **112,** 533–540.
Kleczkowski, L. A., and Givan, C. V. (1988). *J. Plant Physiol.* **132,** 641–652.
Kleczkowski, L. A., Randall, D. D., and Zahler, W. L. (1985). *Arch. Biochem. Biophys.* **236,** 185–195.
Kleczkowski, L. A., Randall, D. D., and Blevins, D. G. (1986). *Biochem. J.* **239,** 653–659.
Kleczkowski, L. A., Givan, C. V., Hodgson, J. M., and Randall, D. D. (1988). *Plant Physiol.* **88,** 1182–1185.
Kleczkowski, L. A., Edwards, G. E., Blackwell, R. D., Lea, P. J., and Givan, C. V. (1990). *Plant Physiol.* **94,** 819–825.
Klein, U. (1986). *Planta* **167,** 81–86.
Kleiner, D. (1981). *Biochim. Biophys. Acta* **639,** 41–52.

Kleinhofs, A., and Warner, R. L. (1990). *Biochem. Plants* **16**, 89–120.
Klepper, L., Flesher, D., and Hageman, R. H. (1971). *Plant Physiol.* **48,** 580–590.
Knight, T. J., and Langston-Unkefer, P. J. (1988). *Science* **241,** 951–954.
Knight, T. J., Bush, D. R., and Langston-Unkefer, P. J. (1988). *Plant Physiol.* **88,** 333–339.
Kornberg, H. L. (1966). *Essays Biochem.* **2,** 1–81.
Kozaki, A., Sakamoto, A., Tanaka, K., and Takeba, G. (1991). *Plant Cell Physiol.* **32,** 353–358.
Kretovich, W. L. (1958). *Adv. Enzymol.* **20,** 319–340.
Krömer, S., and Heldt, H. W. (1991). *Plant Physiol.* **95,** 1270–1276.
Krömer, S., Stitt, M., and Heldt, H. W. (1988). *FEBS Lett.* **226,** 352–356.
Lance, C., and Rustin, P. (1984). *Physiol. Veg.* **22,** 625–641.
Larsen, P. O., Cornwell, K. L., Gee, S. L., and Bassham, J. A. (1981). *Plant Physiol.* **68,** 292–299.
Larsson, C., and Albertsson, E. (1979). *Physiol. Plant* **45,** 7–10.
Larsson, C.-M., and Ingemarsson, B. (1989). *In* "Molecular and Genetic Aspects of Nitrate Assimilation" (J. Wray and J. Kinghorn, eds.), pp. 3–14. Oxford Sci. Publ., Oxford, England.
Laurière, C., Weisman, N., and Daussant, J. (1981). *Physiol. Plant.* **52,** 146–150.
Lawyer, A., Cornwell, K. L., Larsen, P. O., and Bassham, J. A. (1981). *Plant Physiol.* **68,** 1231–1236.
Layzell, D. B., Hunt, S., Moloney, A. H. M., Fernando, S. M., and Diaz del Castillo, L. (1990). *In* "Nitrogen Fixation: Achievements and Objectives" (P. M. Gresshof, L. E. Roth, G. Stacey, and W. E. Newton, eds.). Chapman & Hall, New York. In press.
Lazarow, P. B., and Fujiki, Y. (1985). *Annu. Rev. Cell Biol.* **1,** 489–530.
Lea, P. J., and Blackwell, R. D. (1990). *In* "Perspectives in Biochemical and Genetic Regulation of Photosynthesis" (I. Zelitch, ed.), pp. 301–318. Liss, New York.
Lea, P. J., and Miflin, B. J. (1974). *Nature (London)* **251,** 614–616.
Lea, P. J., and Miflin, B. J. (1980). *Biochem. Plants* **5,** 569–607.
Lea, P. J., and Ridley, S. M. (1989). *In* "Herbicides and Plant Metabolism" (A. D. Dodge, ed.), pp. 137–167. Cambridge Univ. Press, Cambridge, England.
Lea, P. J., Blackwell, R. D., Murray, A. J. S., and Joy, K. W. (1989). *Recent Adv. Phytochem.* **23,** 157–189.
Lea, P. J., Robinson, S. A., and Stewart, G. R. (1990). *Biochem. Plants* **16**, 121–162.
Lee, R. B. (1980). *Plant Cell Environ.* **3,** 65–90.
Leech, R. M., and Kirk, P. R. (1968). *Biochem. Biophys. Res. Commun.* **32,** 685–690.
Lees, E. M., and Dennis, D. T. (1981). *Plant Physiol.* **68,** 827–830.
Lewis, O. A. M. (1986). *Inst. Biol. Stud. Biol.* **166,** 1–104 .
Liang, Z., Yu, C., and Huang, A. H. C. (1984). *Arch. Biochem. Biophys.* **233,** 393–401.
Lloyd, N. D. H., and Joy, K. W. (1978). *Biochem. Biophys. Res. Commun.* **81,** 186–192.
Long, D. M., and Oaks, A. (1990). *Plant Physiol.* **93,** 846–850.
Loulakakis, C. A., Roubelakis-Angelakis, K. A. (1990). *Plant Physiol.* **94,** 109–113.
Loyola-Vargas, V. M., and Sanchez de Jiménez, E. (1986). *J. Plant Physiol.* **124,** 147–154.
Lubben, T. H., Theg, S. M., and Keegstra, K. (1988). *Photosynth. Res.* **17,** 173–194.
Lubben, T. H., Donaldson, G. K., Viitanen, P. V., and Gatenby, A. A. (1989). *Plant Cell* **1,** 1223–1230.
Ludwig, R. A. (1991). *In* "Molecular Approaches to Compartmentation and Metabolic Regulation" (A. C. Huang and L. Taiz, eds.), pp. 99–112. Am. Soc. Plant Physiol., Beltsville, Maryland.
Madore, M., and Grodzinski, B. (1984). *Plant Physiol.* **76,** 782–786.
Magalhaes, J. R., Ju, G. C., Rich, P. J., and Rhodes, D. (1990). *Plant Physiol.* **94,** 647–656.

Mann, A. F., Fenten, P. A., and Stewart, G. R. (1979). *Biochem. Biophys. Res. Commun.* **88,** 515–521.
Mann, A. F., Fenten, P. A., and Stewart, G. R. (1980). *FEBS Lett.* **110,** 265–267.
Manning, K. (1986). *Planta* **167,** 61–66.
Marquez, A. J., Aila, C., Forde, B. G., and Wallsgrove, R. M. (1988). *Plant Physiol. Biochem.* **26,** 645–651.
Marschner, H. (1986). "Mineral Nutrition in Higher Plants." Academic Press, Orlando, Florida.
Martinoia, E., Heck, U., and Wiemken, A. (1981). *Nature (London)* **289,** 292–294.
Martinoia, E., Schramm, M. J., Kaiser, G., Kaiser, W. M., and Heber, U. (1986). *Plant Physiol.* **80,** 895–901.
Marzluf, G. A. (1981). *Microbiol. Rev.* **45,** 437–441.
Matoh, T., and Takahashi, E. (1981). *Plant Cell Physiol.* **22,** 727–731.
Matoh, T., Suzuki, E., and Ida, S. (1979). *Plant Cell Physiol.* **20,** 1329–1340.
Matoh, T., Ida, S., and Takahashi, E. (1980a). *Plant Cell Physiol.* **21,** 1461–1474.
Matoh, T., Ida, S., and Takahashi, E. (1980b). *Kyoto Daigaku Shokuryo Kagaku Kenkyusho Hokoku* **43,** 1–6.
Mazurowa, H., Ratajczak, W., and Ratajczak, L. (1980). *Acta Physiol. Plant.* **2,** 167–177.
McClure, P. R., Omholt, T. E., Pace, G. M., and Bouthyete, P. Y. (1987). *Plant Physiol.* **84** 52–57.
McKenzie, E. A., Copeland, L., and Lees, E. M. (1981). *Arch. Biochem. Biophys.* **212,** 298–305.
McNally, S., and Hirel, S. (1983). *Physiol. Veg.* **21,** 761–774.
McNally, S. F. Hirel, B., Gadal, P., Mann, A. F., and Stewart, G. R. (1983). *Plant Physiol.* **72,** 22–25.
Meister, A. (1953). *J. Biol. Chem.* **200,** 571–590.
Melzer, J. M., Kleinhofs, A., and Warner, R. L. (1989). *Mol. Gen. Genet.* **217,** 341–346.
Mestichelli, L., Gupta, J. J., and Spenser, I. D. (1979). *J. Biol. Chem.* **254,** 640–647.
Miao, G.-H., Hirel, B, Marsolier, M. C., Ridge, R. W., and Verma, D. P. S. (1991). *Plant Cell* **3,** 11–22.
Micallef, B. J., and Shelp, B. J. (1989a). *Plant Physiol.* **90,** 624–630.
Micallef, B. J., and Shelp, B. J. (1989b). *Plant Physiol.* **90,** 631–634.
Micallef, B. J., and Shelp, B. J. (1989c). *Plant Physiol.* **91,** 170–174.
Miflin, B. J. (1974). *Plant Physiol.* **54,** 550–555.
Miflin, B. J. (1980). *Biochem. Plants* **5,** 533–539.
Miflin, B. J., and Lea, P. J. (1977). *Annu. Rev. Plant Physiol.* **28,** 299–329.
Miflin, B. J., and Lea, P. J. (1982). *Encycl. Plant Physiol.* **14A,** 5–64.
Miller, G., and Evans, H. J. (1957). *Plant Physiol.* **32,** 346–354.
Miller, J. M., and Conn, E. E. (1980). *Plant Physiol.* **65,** 1199–1202.
Mills, W. R., Lea, P. J., and Miflin, B. J. (1980). *Plant Physiol.* **65,** 1166–1172.
Mohamed, A. H., and Gnanam, A. (1979). *Plant Physiol.* **64,** 263–268.
Moore, R., and Black, C. C. (1976). *Plant Physiol.* **46,** 309–313.
Morris, P. F., Layzell, D. B., and Canvin, D. T. (1989). *Plant Physiol.* **89,** 498–500.
Muhitich, M. J., and Fletcher, J. S. (1983). *Photosynth. Res.* **4,** 241–244.
Murray, A. J. S., Blackwell, R. D., Joy, K. W., and Lea, P. J. (1987). *Planta* **172,** 106–113.
Murray, A. J. S., Blackwell, R. D., and Lea, P. J. (1990). *Plant Physiol.* **91,** 395–400.
Naik, M. S., and Nicholas, D. V. D. (1986). *Phytochemistry* **25,** 571–576.
Nakamura, Y., and Tolbert, N. E. (1983). *J. Biol. Chem.* **258,** 7631–7638.
Nauen, W., and Hartmann, T. (1980). *Planta* **148,** 7–16.
Neuburger, M., Bourguignon, J., and Douce, R. (1986). *FEBS Lett.* **207,** 18–22.
Ninomiya, Y., and Sato, S. (1984). *Plant Cell Physiol.* **25,** 453–458.

Oaks, A. (1974). *Biochim. Biophys. Acta* **372,** 122–126.
Oaks, A. (1992). *BioScience* (in press).
Oaks, A., and Bidwell, R. G. S. (1970). *Annu. Rev. Plant Physiol.* **21,** 43–66.
Oaks, A., and Gadal, P. (1979). *In* "Compartmentation and Metabolic Channeling" (M. Nover, F. Lynen, and K. Mothes, eds.), pp. 245–254. Fischer-Verlag, Jena, Germany, and Elsevier/North-Holland, Amsterdam.
Oaks, A., and Hirel, B. (1985). *Annu. Rev. Plant Physiol.* **36,** 345–365.
Oaks, A., and Long, D. M. (1992). *In* "Nitrogen Metabolism in Plants" (D. J. Pilbeam and K. Mengel, eds.). Oxford Univ. Press, Oxford, England. In press.
Oaks, A., and Ross, D. W. (1984). *Can. J. Bot.* **62,** 68–73.
Oaks, A., Aslam, M., and Boesel, I. (1977). *Plant Physiol.* **59,** 391–394.
Oaks, A., Stulen, I., and Boesel, D. (1979). *Can. J. Bot.* **57,** 1824–1829.
Ogren, W. L. (1984). *Annu. Rev. Plant Physiol.* **35,** 415–442.
Oji, Y., Watanabe, M., Wakiuchi, N., and Okamoto, S. (1985). *Planta* **165,** 85–90.
Oji, Y., Kamatsu, Y., Taguchi, S., Wakiuchi, N., and Shiga, H. (1989). *Soil Sci. Plant Nutr.* **35,** 427–434.
Oliver, D. J. (1987). *In* "Plant Mitochondria: Structural, Functional, and Physiological Aspects" (A. L. Moore and R. B. Beechy, eds.), pp. 219–226. Plenum, New York.
Oliver, D. J., Neuburger, M., Bourguignon, J., and Douce, R. (1990a). *Physiol. Plant.* **80,** 487–491.
Oliver, D. J., Neuburger, M., Bourguignon, J., and Douce, R. (1990b). *Plant Physiol.* **94,** 833–839.
Omata, T. (1991). *Plant Cell Physiol.* **32,** 151–157.
Omata, T., Ohmori, M., Arai, N., and Ogawa, T. (1989). *Proc. Natl. Acad. Sci.* U.S.A. **86,** 6612–6616.
O'Neal, D., and Joy, K. W. (1973). *Nature (London)* **246,** 61–62.
Pahlich, E., and Gerlitz, C. (1980). *Phytochemistry* **19,** 11–13.
Paln, D., Kanwar, Y. S., and Blobel, B. (1988). *Nature (London)* **331,** 232–236.
Parton, W. J., Morgan, J. A., Altenhofen, J. M., and Harper, L. A. (1988). *Agron. J.* **80,** 419–425.
Pate, J. S. (1973). *Soil Biol. Biochem.* **5,** 109–119.
Pate, J. S. (1980). *Annu. Rev. Plant Physiol.* **31,** 313–340.
Pate, J. S. (1989). *Recent Adv. Phytochem.* **23,** 65–115.
Peiser, G. D., Wang, T.-T, Hoffman, N. E., Yang, S. F., Liu, H.-W., and Walsh, C. T. (1984). *Proc. Natl. Acad. Sci. U.S.A.* **81,** 3059–3063.
Peterson, R. B. (1982). *Plant Physiol.* **70,** 61–66.
Pfanner, N., Sollner, T., and Neupert, W. (1991). *Trends Biochem. Sci.* **16,** 63–67.
Platt, S. G., Plaut, Z., and Bassham, J. A. (1977). *Plant Physiol.* **60,** 739–742.
Plaxton, W.C. (1988). *Plant Physiol.* **86,** 1064–1069.
Poulson, C., and Verpoorte, R. (1991). *Phytochemistry* **30,** 377–386.
Powels, S. B., Cornic, G., and Louason, G. (1984). *Physiol. Veg.* **22,** 437–446.
Priestly, D. A., and Bruisma, J. (1982). *Physiol. Plant.* **56,** 303–311.
Proudlove, M. O., and Moore, A. L. (1982). *FEBS Lett.* **147,** 26–30.
Prunkard, D. E., Bascomb, N. F., Robinson, R. W., and Schmidt, R. R. (1986). *Plant Physiol.* **81,** 349–355.
Pryor, A. J. (1974). *Heredity* **32,** 397–419.
Pryor, A. J. (1990). *Maydica* **35,** 367–372.
Ramos, F., and Wiame, J.-M. (1980). *Eur. J. Biochem.* **108,** 373–377.
Ratajczak, L., Koroniak, D., Mazarowa, H., Ratajczak, W., and Prus-Glowacki, W. (1986). *Physiol. Plant.* **67,** 685–689.
Rathnam, C. K. M., and Edwards, G. E. (1976). *Plant Physiol.* **57,** 881–885.

Rauser, W. E. (1990). *Annu. Rev. Biochem.* **59,** 61–86.
Rhodes, D., Rendon, G. A., and Stewart, G. R. (1975). *Planta* **125,** 201–211.
Rhodes, D., Brunk, D. G., and Magalhaes, J. R. (1989). *Recent Adv. Phytochem.* **23,** 191–206.
Richardson, K. E., and Tolbert, N. E. (1961). *J. Biol. Chem.* **236,** 1285–1290.
Robinson, S. A., Slade, A. P., Fox, G. G., Phillips, R., Ratcliffe, R. G., and Stewart, G. R. (1991). *Plant Physiol.* **95,** 509–516.
Robinson, S. P. (1982). *Plant Physiol.* **70,** 1032–1038.
Rognes, S. E. (1975). *Phytochemistry* **14,** 1975–1982.
Rognes, S. E. (1980). *Phytochemistry* **19,** 2287–2293.
Rosa, R., and Whatley, F. R. (1981). *Plant Physiol.* **68,** 364–370.
Rosenfeld, S. A., and Brenchley, J. E. (1983). *In* "Amino Acids: Biosynthesis and Genetic Regulation" (K. M. Hermann, and R. L. Somerville, eds.), pp. 1–17. Addison-Wesley, London.
Rowell, P., and Stewart, W. D. P. (1976). *Arch. Microbiol.* **107,** 115–124.
Rufty, T. W., Thomas, J. F., Remmler, J., Campbell, W. H., and Volk, R. J. (1986). *Plant Physiol.* **82,** 675–680.
Russell, R. S. (1977). "Plant Root Systems: Their Function and Interactions with the Soil." McGraw Hill, London.
Sakakibara, H., Watanabe, M., Hase, T., and Sugiyama, T. (1991). *J. Biol. Chem.* **266,** 2028–2035.
Sans, N., Schindler, U., and Schröder, J. (1988). *Eur. J. Biochem.* **173,** 123–130.
Sarojini, G., and Oliver, D. J. (1983). *Plant Physiol.* **72,** 194–199.
Sarojini, G., and Oliver, D. J. (1987). *Prog. Photosynth. Res.* **3,** 569–572.
Sauer, H., Wild, A., and Ruhle, W. (1987). *Z. Naturforsch., C: Biosci.* **42c,** 270–278.
Scheibe, R. (1987). *Physiol. Plant.* **71,** 393–400.
Scheibe, R. (1990). *Bot. Acta* **103,** 327–334.
Scheid, H. W., Ehmke, A., and Hartmann, T. (1980). *Z. Naturforsch., C: Biosci.* **35C,** 213–221.
Scherer, P. E., Krieg, U. C., Hwang, S. T., Vestweber, D., and Schatz, G. (1990). *EMBO J.* **9,** 4315–4322.
Schmitt, M. R., and Edwards, G. E. (1983). *Plant Physiol.* **72,** 728–734.
Schnell, D. J., Blobel, G., and Pain, D. (1990). *J. Cell Biol.* **111,** 1825–1838.
Schubert, K. R. (1986). *Annu. Rev. Plant Physiol.* **37,** 539–574.
Schuller, K. A., and Randall, D. D. (1989). *Plant Physiol.* **89,** 1207–1212.
Schuller, K. A., Plaxton, W. C., and Turpin, D. H. (1990a). *Plant Physiol.* **93,** 1303–1311.
Schuller, K. A., Turpin, D. H., and Plaxton, W. C. (1990b). *Plant Physiol.* **94,** 1429–1435.
Scott, D. B., Farnden, K. J. F, and Robertson, J. G. (1976). *Nature (London)* **263,** 703–705.
Sechley, K. A., Oaks, A., and Bewley, J. D. (1991). *Plant Physiol.* (in press).
Sellmar, D., Lieberei, R., and Biehl, B. (1988). *Plant Physiol.* **86,** 711–716.
Sengupta-Gopalan, C., and Pitas, J. W. (1986). *Plant Mol. Biol.* **7,** 189–199.
Shah, S. P. J., and Cossins, E. A. (1970). *Phytochemistry* **9,** 1545–1551.
Shaner, D. L., and Boyer, J. S. (1976a). *Plant Physiol.* **58,** 499–504.
Shaner, D. L., and Boyer, J. S. (1976b). *Plant Physiol.* **58,** 505-509.
Shargool, P. D., Steeves, T., Weaver, M., and Russell, M. (1978). *Can. J. Biochem.* **56,** 273–279.
Shargool, P. D., Jain, J. C., and McKay, G. (1988). *Phytochemistry* **27,** 1571–1574.
Shelp, B. J., and Atkins, C. A. (1984). *Plant Sci. Lett.* **36,** 225–230.
Shelp, B. J., Atkins, C. A., Stoner, P. J., and Canvin, D. T. (1983). *Arch. Biochem. Biophys.* **224,** 429–444.
Siddiqi, M., Glass, A. D. M., Ruth, T. J., and Rufty, T. W., Jr. (1990). *Plant Physiol.* **93,** 1426–1432.

Sieciechowicz, K. A., and Joy, K. W. (1988). *Can. J. Bot.* **67,** 732–736.
Sieciechowicz, K. A. Ireland, R. J., and Joy, K. W. (1985). *Plant Physiol.* **77,** 506–508.
Sieciechowicz, K. A., Joy, K. W., and Ireland, R. J. (1988). *Phytochemistry* **27,** 663–671.
Sieciechowicz, K. A., Joy, K. W., and Ireland, R. J. (1989). *Plant Physiol.* **89,** 192–196.
Sinden, S. L., and Durbin, R. D. (1968). *Nature (London)* **219,** 379–380.
Sivak, M. N., and Walker, D. A. (1986). *New Phytol.* **102,** 499–512.
Smeekens, S., Weisbek, P., and Robinson, C. (1990). *Trends Biochem. Sci.* **15,** 73–76.
Solomonson, L., and Barber, M. J. (1990). *Annu. Rev. Plant Physiol.* **41,** 225–253.
Somers, D. A., Kao, T. K., Kleinhofs, A., Warner, R. L., and Oaks, A. (1983). *Plant Physiol.* **72,** 949–952.
Somerville, C. R. (1986). *Annu. Rev. Plant Physiol.* **37,** 467–507.
Somerville, C. R., and Ogren, W. L. (1979). *Nature (London)* **280,** 833–836.
Somerville, C. R., and Ogren, W. L. (1980a). *Proc. Natl. Acad. Sci. U.S.A.* **77,** 2684–2687.
Somerville, C. R., and Ogren, W. L. (1980b). *Nature (London)* **286,** 257–259.
Somerville, C. R., and Ogren, W. L. (1981). *Plant Physiol.* **67,** 666–671.
Somerville, C. R., and Ogren, W. L. (1982a). *Biochem. J.* **202,** 373–380.
Somerville, C. R., and Ogren, W. L. (1982b). *Trends Biochem. Sci.* **7,** 171–174.
Somerville, C. R., and Ogren, W. L. (1983). *Proc. Natl. Acad. Sci. U.S.A.* **80,** 1290–1294.
Somerville, S. C., and Somerville, C. R. (1985). *Plant Sci. Lett.* **37,** 217–220.
Srivastava, H. S., and Singh, R. P. (1987). *Phytochemistry* **26,** 597–610.
Staswick, P. E. (1990). *Plant Cell* **2,** 1–6.
Steingrover, E., Ratering, P., and Siesling, J. (1986). *Physiol. Plant.* **66,** 550–556.
Stewart, G. R., and Larher, F. (1980). *Biochem. Plants* **5,** 609–635.
Stewart, G. R., Mann, A. F., and Fenten, P. A. (1980). *Biochem. Plants* **5,** 271–327.
Stitt, M. (1990). *Annu. Rev. Plant Physiol. Plant Mol. Biol.* **41,** 153–185.
Streeter, J. G. (1973). *Arch. Biochem. Biophys.* **157,** 613–624.
Streeter, J. G. (1977). *Plant Physiol.* **60,** 235–239.
Streeter, J. G., and Thompson, J. F. (1972a). *Plant Physiol.* **49,** 572–578.
Streeter, J. G., and Thompson, J. F. (1972b). *Plant Physiol.* **49,** 579–584.
Stulen, I., Israelstam, G. F., and Oaks, A. (1979). *Planta* **146,** 237–241.
Suzuki, A., and Gadal, P. (1982). *Plant Physiol.* **69,** 848–852.
Suzuki, A., and Gadal, P. (1984). *Physiol. Veg.* **22,** 471–486.
Suzuki, A., Gadal, P., and Oaks, A. (1981). *Planta* **151,** 457–461.
Suzuki, A., Vidal, J., and Gadal, P. (1982). *Plant Physiol.* **70,** 827–832.
Suzuki, A., Jacquot, J. P., and Gadal, P. (1983). *Phytochemistry* **22,** 1543–1546.
Suzuki, A., Vidal, J., Nguyen, J., and Gadal, P. (1984). *FEBS Lett.* **173,** 204–208.
Suzuki, A., Oaks, A., Jacquot, J.-P., Vidal, J., and Gadal, P. (1985). *Plant Physiol.* **78,** 374–378.
Suzuki, A., Audet, C., and Oaks, A. (1987). *Plant Physiol.* **84,** 578–581.
Suzuki, A., Carroyal, E., Zehnacker, C., and Deroche, M. E. (1988). *Biochem. Biophys. Res. Commun.* **156,** 1130–1138.
Swarup, R., Bennett, M. J., and Cullimore, J. V. (1991). *Planta* **183,** 51–56.
Ta, T. C., and Joy, K. W. (1986). *Planta* **169,** 118–122.
Ta, T. C., Joy, K. W., and Ireland, R. J. (1984a). *Plant Physiol.* **75,** 527–530.
Ta, T. C., Joy, K. W., and Ireland, R. J. (1984b). *Plant Physiol.* **74,** 822–826.
Ta, T. C., Joy, K. W., and Ireland, R. J. (1985). *Plant Physiol.* **78,** 334–337.
Tate, S. S., and Meister, A. (1973). *In* "The Enzymes of Glutamine Metabolism" (S. Prusner and E. R. Stadtman, eds.), pp. 77–127. Academic Press, New York.
Tang, P. S., and Wu, H. Y. (1957). *Nature (London)* **179,** 1355–1356.
Taylor, A. A., and Stewart, G. R. (1981). *Biochem. Biophys. Res. Commun.* **101,** 1281–1289.
Taylor, W. C. (1989). *Annu. Rev. Plant Physiol. Plant Mol. Biol.* **40,** 211–233.
Tempest, D. W., Meers, J. L., and Brown, C. M. (1970). *Biochem. J.* **117,** 405–407.

Thibodeau, P. S., and Jaworski, E. (1975). *Planta* **127,** 133–147.
Thompson, J. F. (1980). *Biochem. Plants* **5,** 375–402.
Tingey, S. V., and Coruzzi, G. M. (1987). *Plant Physiol.* **84,** 366–373.
Tingey, S. V., Walker, E. L., and Coruzzi, G. M. (1987). *EMBO J.* **6,** 1–9.
Tittle, F. L., Goudey, J. S., and Spencer, M. S. (1990). *Plant Physiol.* **94,** 1143–1148.
Tobin, A. K., Ridley, S. M., and Stewart, G. R. (1985). *Planta* **163,** 544–548.
Tolbert, N. E. (1971). *Annu. Rev. Plant Physiol.* **22,** 45–74.
Tolbert, N. E. (1980). *Biochem. Plants* **2,** 487–523.
Tolbert, N. E. (1981). *Annu. Rev. Biochem.* **50,** 133–157.
Tolbert, N. E., Husic, H. D., Husic, D. W., Moroney, J. V., and Wilson, B. J. (1985). *In* "Inorganic Carbon Uptake by Aquatic Photosynthetic Organisms" (W. J. Lucas, and J. A. Berry, eds.), pp. 211–227. Am. Soc. Plant Physiol., Beltsville, Maryland.
Tsai, F.-Y., and Coruzzi, G. M. (1990). *EMBO J.* **9,** 323–332.
Tsushida, T., and Murai, T. (1987). *Agric. Biol. Chem.* **51,** 2865–2871.
Turpin, D. H., Botha, F. C., Smith, R. G., Feil, R., Horsey, A. K., and Vanlerberghe, G. C. (1990). *Plant Physiol.* **93,** 166–175.
Unkles, S. E., Hawka, K. L., Grieve, C., Campbell, E. I., Montagne, P., and Kinghorn, J. R. (1991). *Proc. Natl. Acad. Sci. U.S.A.* **88,** 204–208.
Uno, I., Matsumoto, K. A., Adach, K., and Ishikawa, T. (1984). *J. Biol. Chem.* **259,** 1288–1293.
Usuda, H., and Edwards, G. E. (1982). *Plant Physiol.* **69,** 469–473.
Valle, E. M., and Heldt, H. W. (1991). *Plant Physiol.* **95,** 839–845.
Vance, C. P., and Griffith, S. M. (1990). *In* "Plant Physiology, Biochemistry and Molecular Biology" (D. T. Dennis and D. H. Turpin, eds.), pp. 373–388. Longman, Singapore.
Vanetten, C. H., Miller, R. W., Wolff, I. A., and Jones, Q. (1963). *Agric. Food Chem.* **5,** 399–410.
Vanlerberghe, G. C., Schuller, K. A., Smith, R. G., Feil, R., Plaxton, W. C., and Turpin, D. H. (1990). *Plant Physiol.* **94,** 284–290.
Vanlerberghe, G. C., Joy, K. W., and Turpin, D. H. (1991). *Plant Physiol.* **95,** 655–658.
Vaughn, K. C., and Campbell, W. H. (1988). *Plant Physiol.* **88,** 1354–1357.
Vestweber, D., Brunner, J., Baker, A., and Schats, G. (1989). *Nature* (*London*) **341,** 205–209.
Vézina, L. P., and Langlois, J. R. (1989). *Plant Physiol.* **90,** 1129–1133.
Vézina, L. P., Hope, W. J., and Joy, K. W. (1987). *Plant Physiol.* **83,** 58–62.
von Heijne, G. (1986). *EMBO J.* **5,** 1335–1342.
Wada, K., Onda, M., and Matsubana, H. (1986). *Plant Cell Physiol.* **27,** 407–415.
Waegemann, K., Paulsen, H., and Soll, J. (1990). *FEBS Lett.* **261,** 89–92.
Wakiuchi, H., Matsumoto, H., and Takahashi, E. (1971). *Physiol. Plant.* **24,** 248–253.
Walker, D. A., Leegood, R. C., and Sivak, M. N. (1986). *Philos. Trans. R. Soc. London, B* **313,** 305–324.
Walker, G. H., Sarojini, G., and Oliver, D. J. (1982). *Biochem. Biophys. Res. Commun.* **107,** 856–861.
Walker, K. A., Givan, C. V., and Keys, A. J. (1984a). *Plant Physiol.* **75,** 60–66.
Walker, K. A., Keys, A. J., and Givan, C. V. (1984b). *J. Exp. Bot.* **35,** 1800–1810.
Wallace, W., Secor, J., and Shrader, I. E. (1984). *Plant Physiol.* **75,** 170–175.
Wallsgrove, R. M., Lea, P. J., and Miflin, B. J. (1982). *Planta* **154,** 473–476.
Wallsgrove, R. M., Keys, A. J., Lea, P. J., and Miflin, B. J. (1983). *Plant Cell Environ.* **6,** 301–309.
Wallsgrove, R. M., Kendall, A. C., Hall, N. P., Turner, J. C., and Lea, P. J. (1986). *Planta* **168,** 324–329.
Wallsgrove, R. M., Turner, J. C., Hall, N. P., Kendall, A. C., and Bright, S. W. J. (1987). *Plant Physiol.* **83,** 155–158.

Walton, N. J., and Butt, V. S. (1981). *Planta* **153,** 225–231.
Walton, N. J., and Woolhouse, H. W. (1986). *Planta* **167,** 119–128.
Ward, M. R., Tischner, R., and Huffaker, R. C. (1988). *Plant Physiol.* **88,** 1141–1145.
Warner, R. L., and Huffaker, R. C. (1989). *Plant Physiol.* **91,** 947–953.
Warner, R. L., Narayanan, K. R., and Kleinhofs, A. (1987). *Theor. Appl. Genet.* **74,** 714–717.
Washitani, I., and Sato, S. (1977). *Plant Cell Physiol.* **18,** 117–125.
Weger, H. G., Birch, D. G., Elrifi, I. R., and Turpin, D. H. (1988). *Plant Physiol.* **86,** 688–692.
Wells, X. E., and Lees, E. M. (1991). *Arch. Biochem. Biophys.* **287,** 151–159.
Whelan, J., Knorpp, C., and Glaser, E. (1990). *Plant Mol. Biol.* **14,** 977–982.
Whelan, J., Knorpp, C., Harmey, M.A., and Glaser, E. (1991). *Plant Mol. Biol.* **16,** 283–292.
White, J. A., and Scandalios, J. G. (1988). *Physiol. Plant.* **74,** 397–408.
Winkler, R. G., Blevins, D. G., Polacco, J. C., and Randall, D. D. (1987). *Plant Physiol.* **83,** 585–591.
Wiskich, J. T. (1980). *Biochem. Plants* **2,** 243–277.
Wiskich, J. T., Bryce, J. H., Day, D. A., and Dry, I. B. (1990). *Plant Physiol.* **93,** 611–616.
Wolfenden, R. (1970). *Biochemistry* **9,** 3404–3407.
Woo, K. C., and Canvin, D. T. (1979). *Can. J. Bot.* **58,** 17–21.
Woo, K. C., and Osmond, C. B. (1982). *Plant Physiol.* **69,** 591–596.
Woo, K. C., Morot-Gaudry, J. F., Summons, R. E., and Osmond, B. (1982). *Plant Physiol.* **70,** 1514–1517.
Woo, K. C., Flügge, U. I., and Heldt, H. W. (1987). *Plant Physiol.* **84,** 624–632.
Yamaya, T., and Matsumoto, H. (1985a). *Soil Sci. Plant Nutr.* **31,** 513–520.
Yamaya, T., and Matsumoto, H. (1985b). *Plant Cell Physiol.* **26,** 1613–1616.
Yamaya, T., Oaks, A., and Matsumoto, H. (1984). *Plant Physiol.* **75,** 773–777.
Yamaya, T., Oaks, A., Rhodes, D., and Matsumoto, H. (1986). *Plant Physiol.* **81,** 745–757.
Yoshida, Y., Hashimoto, T., Kimura, H., Sakakibara, S., and Tagawa, K. (1985). *Biochem. Biophys. Res. Commun.* **128,** 775–780.
Yu, C., Claybrook, D. L., and Huang, A. H. C. (1983). *Arch. Biochem. Biophys.* **227,** 180–187.
Yu, J., and Woo, K. C. (1988). *Plant Physiol.* **88,** 1048–1054.
Zelitch, I. (1964). *Annu. Rev. Plant Physiol.* **15,** 121–142.
Zelitch, I. (1973). *Plant Physiol.* **51,** 299–305.
Zelitch, I. (1987). *Prog. Photosynth. Res.* **3,** 621–624.
Zelitch, I. (1988). *Plant Physiol.* **86,** 463–468.
Zelitch, I. (1989). *Plant Physiol.* **90,** 1457–1464.
Zelitch, I. (1990). *Plant Physiol.* **92,** 352–357.
Zelitch, I., and Ochoa, S. (1953). *J. Biol. Chem.* **201,** 707–718.

Role of Calcium/Calmodulin-Mediated Processes in Protozoa

Leonard William Scheibel
Department of Preventive Medicine
Uniformed Services University of the Health Sciences
School of Medicine, Bethesda, Maryland 20814

I. Introduction

The physiological importance of calcium ions to eukaryote cells has been recognized for many years. The earliest studies were by Ringer (1883) and Stiles (1903) on muscle. Ringer discovered retrospectively that the saline solution used in a previous group of experiments had been inadvertently prepared with pipe water that contained considerable amounts of calcium instead of with distilled water. As a consequence, he discovered quite by accident that this solution results in an excellent artifical circulating fluid for heart muscle, resulting in contractions after 4 hr that were almost as good as those seen when the ventricle was fed with blood at the start of the experiment. If the circulating fluid was made of calcium-free saline, the ventricle grew weaker and weaker, ceasing contraction in approximately 20 min. Stiles (1903) extended these observations to frog esophagus. Later work by Heilbrunn (1940), Kamada and Kinosita (1943), and Heilbrunn and Wiercinski (1947) on skeletal muscle cells suggested the essential role of calcium for the contraction of muscle fibers. Since that time the cellular function of calcium has been extensively studied (Ebashi, 1988). Some believed the action of calcium was unique to muscle contraction but Ebashi and co-workers (Ebashi and Endo, 1968) maintained that calcium played a much wider role in cellular metabolism, regulating almost every aspect of muscle function. Two groups of investigators working independently on the calcium activation of phosphodiesterase reported on the isolation of a calcium-related activating factor in homogenates of rat brain or bovine brain (Kakiuchi and Yamazaki, 1970a,b; Kakiuchi *et al.*, 1969, 1970; Cheung, 1970, 1971, 1980). This was purified by Wang and co-workers (Teo *et al.*, 1973; Teo and Wang, 1973) and it was found to be an acidic protein of 17 kDa with the ability to bind calcium (Kakiuchi and

Yamazaki, 1970a,b; Kakiuchi *et al.,* 1972). This "calcium-modulated protein," reported to be present in a wide variety of tissues, was named calmodulin (Cheung, 1971; Walsh and Hartshorne, 1983). It has been found in all eukaryotic cells studied so far but is not present in bacteria (Kippert, 1987). Nonetheless, calmodulin-like structures (Swan *et al.,* 1987) and calmodulin-binding proteins (Iwasa *et al.,* 1981) have been identified in some prokaryotic cells.

This article gives an overview of the large body of information about the roles of calcium and calmodulin (CaM) in various protozoan systems. The presence of CaM in protozoa was first reported in *Tetrahymena* by Suzuki *et al.* (1979), Nagao *et al.* (1979), and independently by Jamieson *et al.* (1979). Subsequently CaM was reported in *Paramecium* by Satir *et al.* (1980). Two excellent reviews have been written about CaM in the ciliates (Watanabe and Nozawa, 1982; Nozawa and Nagao, 1986). *Tetrahymena* has been emphasized here because considerable work has been done on it. Also, while the structure of CaM is conserved in nature, *Tetrahymena* CaM possesses several unique structural characteristics. These appear to account for biochemical and functional properties that are peculiar to this ciliate. This is illustrative of various processes that may be important to other organisms when more is known about their biochemistry. Therefore, I will address the structure of *Tetrahymena* CaM, then the localization and cytology, followed by regulation of specific enzyme systems. A variety of mechanisms are discussed because they have been studied in some detail, usually in one or two laboratories. They also are known to be CaM mediated in mammalian systems. Mechanisms were chosen because they appear to be important to phylogenetically/structurally different organisms [e.g., ciliates, *Euglena,* slime molds (ameboid and plasmodial stages), *Amoeba,* flagellates, and sporozoa]. Considerable effort is made to present, as closely as possible, the findings and interpretations of the original investigators in chronological order. It is hoped that the creditable strides made by workers in specific organisms cited here may help in leading others to a greater understanding of less well-studied protozoa, and encourage future studies in this field. It is impossible to cover all aspects of a subject as vast as Ca^{2+}/CaM, even when it is limited to protozoa. I apologize for omitting interesting studies of countless workers on numerous systems. For instance, the considerable body of work on *Paramecium* will not be detailed here due to space limitations. However, an attempt has been made to compare specific protozoal CaM systems to that in mammals. For a more detailed treatment of mammalian CaM systems, the reader is directed to Cheung (1988) or Hartshorne (1985).

CaM is a multifunctional protein with a molecular mass of 16,680 containing 148 amino acids, a high proportion of which are acidic residues (aspartate and glutamate comprise 30% of its amino acids); this accounts for its low p*I* of 4.3 (Watterson *et al.,* 1980; Cheung, 1980). It lacks tryptophan, hydroxyproline, and cysteine, and the high phenylalanine content relative to tyrosine gives it a characteristic UV absorption spectrum (Klee *et al.,* 1980). Bovine brain CaM and CaM isolated from many other sources contains a rather unique trimethylated lysine at position 115 (Vanaman *et al.,* 1976; Anderson *et al.,* 1978). On the other hand, CaM isolated from marine organisms (e.g., *Octopus*), the slime mold *Dictyostelium discoideum,* plants, or fungi lacks this amino acid but it appears to be no less active than CaM containing trimethyllysine (Molla *et al.,* 1981; Van Eldik *et al.,* 1980a; Hartshorne, 1985). Furthermore, in CaMs where residue 115 is unmethylated lysine (or even arginine) there is a threefold increase in the level of CaM-dependent activation of NAD-kinase. There is no change in the ability of corresponding unmethylated CaMs to activate brain phosphodiesterase or gizzard myosin light chain kinase, however (Roberts *et al.,* 1986; Takeda and Yamamoto, 1987). As a whole, CaMs isolated from many sources have strikingly similar amino acid sequences, mammalin CaMs all being the same, and plant or invertebrate CaMs differing slightly in sequence (Watterson *et al.,* 1980; Takagi *et al.,* 1980). Mammalian CaMs are poorly antigenic. However, CaMs purified from a wide variety of phylogenetically distinct sources, e.g., primitive algae, slime molds, coelenterates, and advanced plants and animals, are immunologically cross-reactive (Wallace and Cheung, 1979; Chafouleas *et al.,* 1979). It appears to be one of the most highly conserved proteins studied. The major immunoreactive site lies between residues 127 and 144 in the C-terminal domain of CaMs (Van Eldik and Watterson, 1981). Van Eldik *et al.* (1983a,b) used synthetic peptide immunogens to elicit site-directed antisera reactive with native CaM. They found residues 137–143 to be the minimal immunoreactive peptide as active or more active than the entire CaM molecule, but the peptide segment must be presented in the same spatial orientation that it occupies in the intact protein to elicit antibodies reactive to intact CaM. Watterson *et al.* (1984) concluded that the side chains of all seven residues of the heptapeptide are required for full immunoreactivity with anti-CaM serum. CaM contains 4 calcium (Ca^{2+})-binding sites, each composed of 12 amino acids and flanked on each side by an 8-amino acid α helix. It appears that gross changes in structure result in loss of physiological activity. While the molecule appears relatively resistant to extremes of environmental or experimental conditions, alteration of the fundamental structure is thought to be lethal to the cell (Cheung, 1988).

The physiological mechanisms by which CaM mediates cellular activities have been the subject of intensive investigation and this is discussed by Tomlinson *et al.* (1984) and by Scheibel *et al.* (1989). Following the stimulation of a cell, there is an increase in the intracellular free Ca^{2+} concentration due to movement of Ca^{2+} through channels in the plasma membrane or to its release from intracellular organelles (Fig. 1). Each CaM molecule can bind as many as four Ca^{2+}. The binding of Ca^{2+} produces conformational changes to a more helical structure that exposes hydrophobic regions, which in turn can bind to and activate CaM-binding proteins. Conversely, lipid-soluble CaM inhibitors, if present, can also bind to CaM, inducing an overall change in conformation that prevents the activation of target enzymes. The exact sites on CaM where the various drugs bind are not presently known with certainty, but this is the subject of active investigation (Hidaka and Tanaka, 1985; Anderson *et al.*, 1985; Weiss *et al.*, 1985; Marshak *et al.*, 1985a,b; Lukas *et al.*, 1985, 1988).

CaM is known to regulate a number of cellular activities, such as contraction, transport, motility, proliferation, metabolic control, and intermediary or cyclic nucleotide metabolism. The CaM–Ca^{2+} complex can function directly and rapidly on an effector system such as the ATPase transport system or indirectly and slowly on a regulatory system such as protein kinase (Klee *et al.*, 1980).

II. *Tetrahymena* Calmodulin

A. Structural Properties

In the early 1970s protozoologists, following the lead of mammalian muscle physiologists, also recognized the existence of Ca^{2+}-binding proteins in ciliates. The presence of Ca^{2+}-dependent contractile fibers in the freshwater protozoan *Spirostomum ambiguum* (Ettienne, 1970) and ciliary reversal of *Paramecia* (Naitoh and Kaneko, 1972) were known. Normal ciliary beating (toward the rear) was seen when the concentration of Ca^{2+} was less than 10^{-6} *M*. Ciliary reversal or backward swimming of the *Paramecium* was due to cilia beating toward the front when the Ca^{2+} concentration was raised to levels greater than 10^{-6} *M*. This was attributed to Ca^{2+}-dependent membrane responses. Working with the vorticellid ciliate *Zoothamnium geniculatum* two Ca^{2+}-binding proteins of M_r 20,000 were isolated by Amos *et al.* (1975). The name "spasmin" was proposed for this class of acidic, low-molecular-weight proteins that bind Ca^{2+} with high affinity but differ from known muscle proteins in structure. It was estimated that 1.4–2.1 Ca^{2+} are bound per molecule of protein and a decrease in electrophoretic mobility on alkaline gels resulted as the Ca^{2+} concentration was raised from 10^{-8} to 10^{-6} *M*, the same concentration that induces contraction of the spasmoneme. Ca^{2+}-binding proteins were later extracted by Yamada and Asai (1982) from the contractile spasmoneme of another vorticellid ciliate, *Carchesium polypinum*. These are similar in

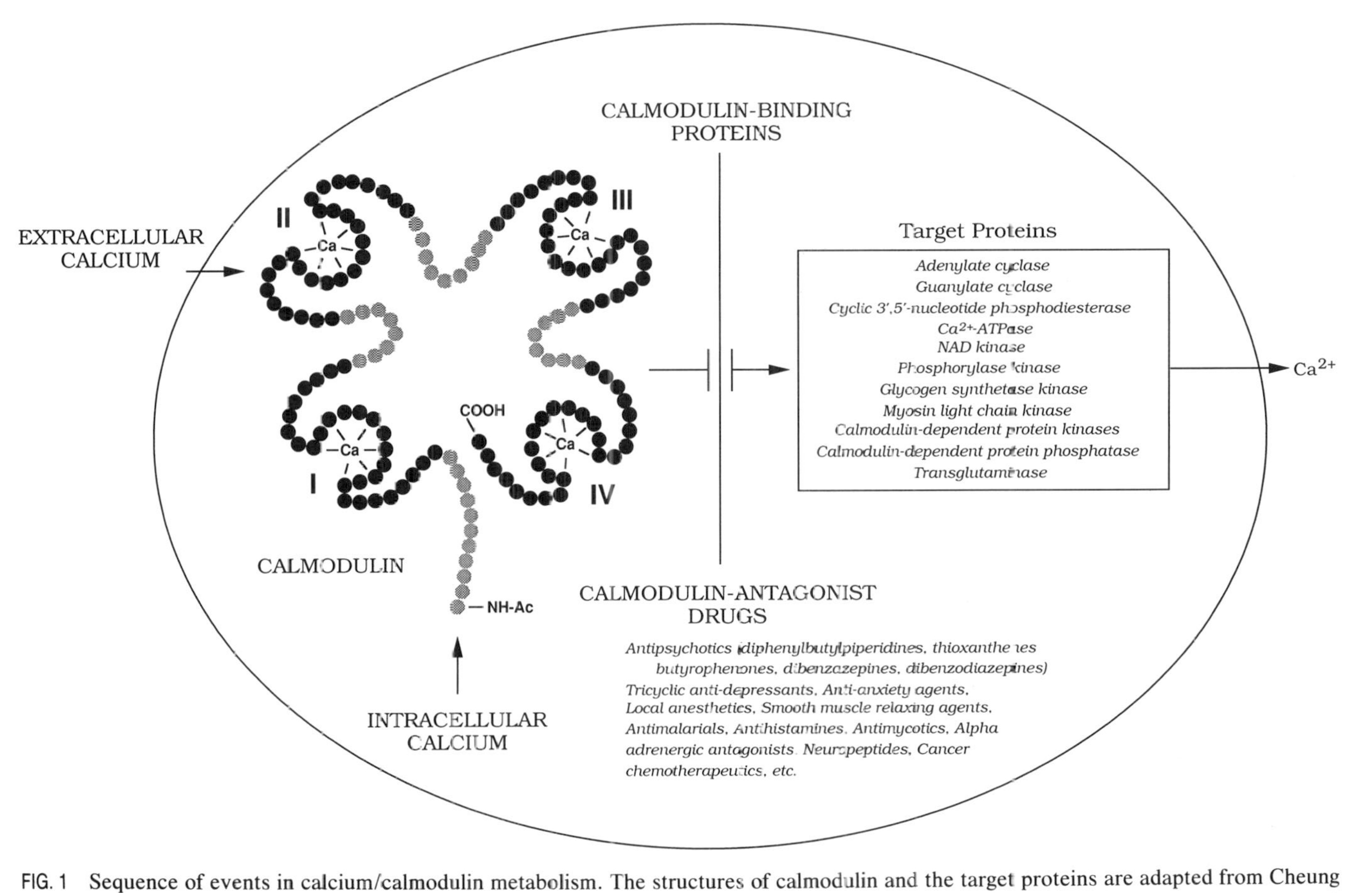

FIG. 1 Sequence of events in calcium/calmodulin metabolism. The structures of calmodulin and the target proteins are adapted from Cheung (1988) with permission.

size (molecular weight of about 16,000, 18,000, and 22,000) and other characteristics to those found by Amos *et al.* (1975) and were found not to be either CaM or troponin C. An antibody of high specificity to one of these spasmins (B) cross-reacts with spasmins of the ciliates *Stentor, Spirostomum,* and *Blepharisma* (Ochiai *et al.*, 1988). These workers suggest that the contractile Ca^{2+}-binding protein of 20 kDa is distributed widely in protozoa, and is not specific to the family Vorticellidae. In support of this, a Ca^{2+}-binding protein of approximately 22 kDa was found to be common to the cortical cytoskeleton of two rumen ciliates, *Isotricha prostoma* and *Polyplastron multivesiculatum* (Vigues and Groliere, 1985).

As a first step in understanding Ca^{2+} regulation in ciliates, Suzuki *et al.* (1979) assessed the possible existence of a "Ca^{2+}-binding protein" in the ciliate Tetrahymena (TCBP) that enhanced the activity of particulate-bound guanylate cyclase approximately 20-fold in the presence of low levels of Ca^{2+} (Nagao *et al.*, 1979). TCBP had a molecular weight of 14,000 Da and an isoelectric point at pH 4.0. It was found to bind 2 mol of Ca^{2+}/mol of protein. These are high-affinity binding sites with $K_d = 4.6 \times 10^{-6}$ *M* as determined from the Scatchard plot obtained by equilibrium dialysis. The activation of porcine brain cyclic nucleotide phosphodiesterase by TCBP in a Ca^{2+}-dependent fashion was the same as that by porcine brain CaM. The amino acid composition of TCBP was similar to that of mammalian CaM (Suzuki *et al.*, 1981). The mobility of modulator proteins isolated from *Tetrahymena,* scallop, or sea anemone in the absence of Ca^{2+} on 15% polyacrylamide gels was identical with that from brain. The addition of Ca^{2+}, however, decreased mobility such that the *Tetrahymena* CaM migrated at the same rate as that from the scallop. Brain CaM was slower than that, followed by that from sea anemone on electrophoresis (Kakiuchi *et al.*, 1981). Based on these findings Suzuki *et al.* (1981) concluded their TCBP was CaM. The phenothiazines, known CaM antagonists, were found to inhibit the TCBP-induced activation of guanylate cyclase, as does EGTA [ethylene glycol-bis(β-aminoethyl ether)-*N, N'*-tetracetic acid]. Increasing concentrations of TCBP, but not of Ca^{2+}, overcome the inhibition by trifluopenzine. This suggested to Nagao *et al.* (1981a) that trifluoperazine is a competitive inhibitor, competing for TCBP in guanylate cyclase activation as had been seen in other CaM-dependent enzyme systems from other tissues.

Independently, Jamieson *et al.* (1979) isolated a pure protein from *Tetrahymena pyriformis* and identified it as CaM. Relying on the ability of plant and animal CaMs to bind phenothiazines they utilized an immobilized chlorpromazine analog coupled to a Sepharose 4B column. This affinity-based chromatographic procedure was most valuable for purifying small amounts of starting material (Jamieson and Vanaman, 1979; Marshak *et al.* 1985a). Apparently also working independently of either Suzuki *et al.*

(1979) or Jamieson *et al.* (1979), Kumagai *et al.* (1980) used a tubulin–Sepharose 4B affinity column to isolate CaM from *T pyriformis*. They reported that it was sensitive to trifluoperazine, capable of activating porcine brain cyclic nucleotide phosphodiesterase, had a similiar electrophoretic mobility, bound to porcine brain tubulin, and had a similar isoelectric point and heat stability as brain CaM. Jamieson *et al.* (1979) recognized the importance of this protein as a primary intracellular CaM receptor regulating multiple enzyme systems in both plants and animals. They also were aware that a similar protein had been identified and partially purified from the protozoa *Euglena gracilis* and *Amoeba proteus* and the myxomycete *Physarum polycephalum* by Kuznicki *et al.* (1979), but there were no details on its physical and chemical properties. A number of investigators had documented inhibition of growth, motility, glucose utilization, and phosphate uptake by the phenothiazines in *Tetrahymena*, which were known CaM inhibitors (Jamieson *et al.*, 1979). Also, Garofalo *et al.* (Maihle *et al.*, 1979; Maihle and Satir, 1979) had attempted to demonstrate the presence of CaM in *Paramecium tetraurelia* by using a phosphodiesterase stimulation type of CaM assay, which resulted in values of 20–40 ng/μg total protein. Localization was done using the indirect immunofluorescence technique, with goat-anti-CaM as the primary label followed by rhodamine-conjugated rabbit anti-goat IgG as the secondary label. CaM was found to be concentrated in anterior and posterior areas of the cell, along the entire length of the cilia, and diffusely throughout the cytoplasm of *P. tetraurelia*. Further studies by these workers and later corroboration by Suzuki *et al.* (1982) on *Tetrahymena thermophila* revealed CaM was located also along the cilia, in a rim along the oral groove, and in a punctate pattern on the cell surface. Assay values for both *Tetrahymena* and *Paramecium* cytoplasmic CaM were estimated to be 5 ng/μg total protein (Maihle and Satir, 1980; Satir *et al.*, 1980). Jamieson *et al.* (1979) identified their *Tetrahymena* protein as being CaM by authenticating that it possessed three of the Ca^{2+}-dependent activities attributed to CaM. Their work showed that *Tetrahymena* CaM is slightly smaller (M_r 15,000) than mammalian calmodulin (e.g., bovine brain, M_r 16,680). It possesses a single trimethyllysine and lacks tryptophan and cysteine, as does mammalian CaM. Jamieson *et al.* (1979) determined the amino acid composition of *Tetrahymena* CaM by sequence analysis and recognized a greater difference between protozoal and bovine brain CaM than exists between coelenterate (*Renilla reniformis*) and brain CaM (Jones *et al.*, 1979). Yazawa *et al.* (1981) reported that there were 11 substitutions and 1 deletion of amino acid residues (total of 147) in *Tetrahymena* CaM compared to brain CaM. Ten of these changes were in the carboxyl-terminal half of the molecule near the Ca^{2+}-binding domains III and IV (Watterson *et al.*, 1980). Yazawa *et al.* (1981) and Nozawa and

Nagao (1986) proposed that the introduction of a partial positive charge, with the substitution of histidine of glutamic acid at residue 135, decreases Ca^{2+} affinity for domain IV. This, and the additional positive charge at residue 143 from the substitution of glutamine with arginine, as well as the deletion at residue 146, may alter the function of this protozoal protein and account for its specific interaction with *Tetrahymena* guanylate cyclase. (The structure of domains I and II is conserved.) Certainly *Tetrahymena* CaM appears singularly capable of stimulating *Tetrahymena* guanylate cyclase, but there is some evidence that *Paramecium caudatum* CaM can also activate *Tetrahymena* guanylate cyclase (Suzuki *et al.,* 1982; Watanabe and Nozawa, 1982). There is immunological cross-reactivity between *Tetrahymena* and *Paramecium* CaM but not CaMs from some higher organisms, e.g., porcine brain, sea urchin, sea anemone, slime mold. This may be due to structural characteristics present in CaMs of the ciliated protozoa that are different from those in higher forms of life (Suzuki *et al.,* 1982). Antibody was prepared by McCartney *et al.* (1984) against major Ca^{2+}-dependent antigenic determinants in *Tetrahymena* calmodulin. These major antigenic determinants were localized (by limited trypsin digestion of *Tetrahymena* CaM to the carboxyl-terminal half of the molecule, i.e., residues 76–147. Watterson *et al.* (1984) reported antiserum specific for vertebrate CaM residues 137–143 reacted poorly with *Tetrahymena* CaM. This lack of cross-reactivity may result, however, from differences between modified and unmodified CaMs rather than species-specific differences in primary structure. McCartney *et al.* (1984) points out that the antiserum to vertebrate CaM used by Watterson *et al.* (1984) was prepared with performic acid-oxidized protein. In any event, Watterson *et al.* (1984) suggested the substitution of the glutamine at position 143 of vertebrate CaM with a basic amino acid like arginine in *Tetrahymena* CaM would be expected to substantially reduce immunoreactivity. They go on to state that the fourth structural domain contains high sequence variability among CaMs and appears to contain a major immunoreactive region. CaMs from bovine brain or heart, human placenta, sea anemone, and scallop failed to activate the guanylate cyclase from *Tetrahymena* under those conditions where maximum activity resulted from *Tetrahymena* CaM (Yazawa *et al.,* 1981; Kakiuchi *et al.,* 1981; Watanabe and Nozawa, 1982). Despite the specific requirement of *Tetrahymena* guanylate cyclase for *Tetrahymena* CaM, brain phosphodiesterase or skeletal muscle myosin light chain kinase can be activated by the protozoan CaM but CaMs of *Tetrahymena* and either scallop or sea anemone are less potent in activating brain phosphodiesterase than is bovine brain CaM. This appears to be due to subtle structural differences between the CaMs, because *Tetrahymena* CaM has fewer hydrophobic groups than bovine brain CaM exposed to the presence of Ca^{2+}. This difference in hydrophobicity could explain the difference in brain phosphodiesterase activation, which requires larger amounts of *Tetrahymena* CaM than bovine brain CaM (Inagaki *et al.*, 1983). This suggests a graded order of activity based on structural differences, where CaMs of more primitive life forms are less able to activate mammalian proteins (Kakiuchi *et al.*, 1981; see Roberts *et al.*, 1984). The specificity of ciliate guanylate cyclase may also result, however, from its being membrane bound with an apparent requirement for Mg^{2+} instead of Mn^{2+}, the preferred metal cofactor for guanylate cyclases of many microorganisms and mammalian tissue (Kakiuchi *et al.*, 1981; Nakazawa *et al.*, 1979). The results of a ^{43}Ca nuclear magnetic resonance (NMR) spectral study done by Shimizu *et al.* (1982) suggested that the substitutions and deletions of amino acid residues do not appear to result in any significant change in Ca^{2+} mobility in *Tetrahymena* CaM compared with vertebrate CaM. The ^{43}Ca NMR of the Ca^{2+}–*Tetrahymena* CaM is markedly changed by the addition of trifluoperazine or Mg^{2+}, suggesting conformational alterations. The effects on *Tetrahymena* CaM observed in the presence of Mg^{2+} are in contrast to those observed with bovine CaM. Shimizu and colleagues (Shimizu and Hatano, 1983; Shimizu *et al.*, 1984), using ^{19}F NMR to study interactions of trifluoperazine with protozoal CaM, went on to report that 2 mol of trifluoperazine was bound to each *Tetrahymena* CaM molecule and that this bond was rather weak compared to that in porcine CaM. They suggested that the trifluoperazine is located near a hydrophobic region of the Ca^{2+}-bound *Tetrahymena* CaM molecule and near a hydrophilic region of the Ca^{2+}-free *Tetrahymena* molecule.

B. Localization within the Cell

Tetrahymena pyriformis, while in the early stationary phase of growth, was collected and fractionated according to the method of Nozawa and Thompson (1971a). CaM content was determined in fractions of cilia, ciliary supernatant, pellicles, mitochondria, microsomes, and postmicrosomal supernatant (Nagao *et al.*, 1981b). CaM content in unfractionated whole cells was 0.9 μg/mg total protein. The ciliary supernatants (43% of total) and postmicrosomal supernatant (50% of total) contain approximately 93% of total CaM, these fractions being the major CaM pools in *Tetrahymena* cells. The pool size of the ciliary supernatant is only 5.5% of the total protein, where the protein content of the postmicrosomal supernatant is 43% of the total. Therefore, CaM content in the ciliary supernatant was much higher than that in the postmicrosomal supernatant. In earlier studies Nozawa and Thompson (1971b) pulse labeled the cells with [^{14}C]palmitate or with [^{3}H]leucine. This resulted in significant differences in rates of radioactive incorporation by the lipids or proteins in these fractions, suggesting that these two pools are metabolically separated from each other to some degree. The subsequent localization of CaM in the ciliary supernatant by Nagao *et al.* (1981b) would agree with earlier observations by Maihle and Satir (1980) and by Satir *et al.* (1980), who used an indirect immunofluorescent technique to describe prominent CaM labeling along the entire length of cilia, in a rim along the oral groove, and in a punctate pattern on the surface of the cell corresponding to the ciliary basal bodies in *Tetrahymena*. Watanabe and Nozawa (1982) corroborated these observations, using the indirect fluorescent antibody technique but employing rabbit antibody to *Tetrahymena* CaM purified on an antigen affinity column [in contrast to the use of a goat antibody to rat testes CaM with no specified purification procedures, as reported by Maihle and Satir (1980) and by Satir *et al.* (1980)]. The specificity of the purified reagents used by Watanabe and Nozawa (1982) and by Suzuki *et al.* (1982) was tested with absorbed control antibodies which resulted in no fluorescence, indicating no cross-reactivity. Recent evidence suggests that control of the contractile vacuole activity is CaM mediated (Bergquist, 1989).

Both Ca^{2+} and Mg^{2+} have been recognized as being important to a variety of cell functions. *Tetrahymena pyriformis* was observed by Munk and Rosenberg (1969) to deposit Ca^{2+}, Mg^{2+}, and pyrophosphate in membrane-bound spherical cytoplasmic granules during the stationary phase of growth and in the presence of elevated Ca^{2+}, Mg^{2+}, and phosphate levels in the medium. The organism also rapidly accumulates granules if the cells were previously subjected to phosphate deprivation; conversely, the granules disappear during Ca^{2+}, Mg^{2+} or phosphate deprivation, or during rapid cell multiplication (Nilsson and Coleman, 1977). The majority of the intracellular Mg^{2+} exists in a bound form. Small changes in intracellular Mg^{2+} have a profound effect on cell activity. *Paramecium tetraurelia* appears to regulate intracellular Mg^{2+} by a Mg^{2+}-specific ion current that is dependent on changes in intracellular Ca^{2+} (Preston, 1990). Other cellular deposits of Ca^{2+} appear to be in mitochondria. In the case of *Paramecium,* intracellular deposits of Ca^{2+} appear to be on the cytoplasmic side of surface membranes, basal regions of cilia and trichocyst–pellicle fusion sites, some smooth cytomembranes, and within the ciliary axoneme. While Ca^{2+} is responsible for the orientation of the ciliary stroke, the beat frequency and the bending are regulated by an Mg^{2+}-dependent ATPase. In *Tetrahymena* the dynein ATPase attached to the doublet microtubules binds Ca^{2+} as well as Mg^{2+} (Plattner, 1975). Deposits were seen frequently facing the nine doublet microtubules on the inner side of the ciliary membrane or between the outer and central microtubules. These Ca^{2+}-dependent deposits are associated with plaques in the basal ciliary shaft membrane in *Paramacium* and *Tetrahymena* occurring with the ciliary necklace. It has been estimated that the portion of cellular Ca^{2+} contained in cilia is 30% (Kusamran *et al.*, 1980). Plattner (1975) goes on to suggest that because ciliary activity may originate at the base it would be reasonable to assume the Ca^{2+}-binding sites would be located near the ciliary base. This would suggest that Ca^{2+} may influence ciliary movement, possibly interacting with the inner or outer dynein arms (Plattner, 1975; Fisher *et al.*, 1976; Tsuchiya, 1976; see below).

Ca^{2+} has been known for some time to be involved in ciliary movement (Eckert, 1972). Bergquist and Bovee (1976) and Bergquist *et al.* (1986) recognized that cadmium inhibits ciliary activity in *T. pyriformis* and attributed this to its interacting with Ca^{2+}-binding sites. It was later suggested that cadmium could be either blocking Ca^{2+} channels or inhibiting CaM (Bergquist, 1989). The concentration of intracellular free Ca^{2+} is maintained far below the extracellular calcium concentration. In *Paramecium* an increase in the concentration of intracellular Ca^{2+} results in an increase in frequency and shift in direction of ciliary beating. This may result from the increase in Ca^{2+} conductance of the membrane permitting a strong influx of extracellular Ca^{2+}.

When a paramecium encounters an obstacle, the mechanical stimulus to the anterior end of the protozoan causes a depolarization of the membrane. This in turn activates the Ca^{2+} channels and increases the permeability of the membrane to this ion. The resulting influx of Ca^{2+} leads to reversed ciliary beating and the withdrawal of the free-swimming cell from the obstacle encountered. Accumulation of intracellular Ca^{2+} results in the inactivation of the Ca^{2+} channel. Conversely, stimulation of the posterior end produces hyperpolarization and increased ciliary beating. These "gates" are apparently not uniformly distributed throughout the surface membrane but may be located just above the ciliary base or possibly at an intramembrane particle array termed a "fusion rosette" in both *Paramecium* and *Tetrahymena*. *Paramecium caudatum* contains about 15,000 cilia and nearly all the Ca^{2+} appears to enter through the ciliary membrane (Ogura and Takahashi, 1976; Dunlap, 1977; Satir and Oberg, 1978; Eckert and Brehm, 1979). The direct role of intracellular Ca^{2+} was shown by Naitoh and Kaneko (1972), who used Triton-extracted *Paramecium* in solutions of adenosine triphosphate and magnesium ions. The cilia beat in the normal direction (toward the rear) when the Ca^{2+} concentration in the medium was less than 10^{-6} *M*. When the Ca^{2+} concentration was raised above 10^{-6} *M* the cilia beat in the "reversed" direction (toward the front). This suggested that ciliary reversal results from an increase in ciliary Ca^{2+} levels that are controlled by membrane responses to external stimuli. These data appeared to be supported by the later results of Kung *et al.* (1975) and Takahashi *et al.* (1980). These investigators worked with mutants possessing certain aberrations in locomotor behavior coupled to ciliary membrane defects in their Ca^{2+} channels as revealed by electrophysiological measurements. Triton X-100 extraction resulted in models that have no functional membrane and therefore no ion permeability barriers. These models were capable of being reactivated to swim forward in the presence of Mg^{2+}, K^+, and adenosine triphosphate and swim backward if sufficient Ca^{2+} is added to the solution. These and other studies indicated to Naitoh and Kaneko (1973) that to some degree there was Ca^{2+}-mediated membrane regulation of ciliary orientation in *Paramecium*. It is known that forward-swimming *Paramecium* cells are able to maintain concentrations of intraciliary free Ca^{2+} at less than 10^{-6} *M*. This would depend not only on the rate of influx of Ca^{2+} through voltage-sensitive Ca^{2+} channels in the ciliary membrane, which would in turn control ciliary motility (Takahashi *et al.*, 1980), but in addition this would be further controlled by a Ca^{2+} efflux pump in the ciliary membrane (Eckert, 1972). Chemical or electrical stimulation causes membrane depolarization and the opening of a Ca^{2+} gate, allowing Ca^{2+} to *passively* rush into the cell. The elevated concentration of internal Ca^{2+} would trigger ciliary reversal, and an *active* efflux of Ca^{2+} to prestimulation level (Browning and Nelson, 1976). In this regard, a Ca^{2+}-activated ATPase has been reported by Baugh *et al.* (1976) to be present in the ciliary membrane of *Tetrahymena* and it was recognized that it might be an important component in the regulation of the internal Ca^{2+} concentration of the cell, in turn influencing the behavior of cilia. Mammalian transport ATPases located in the membrane of cells have been known for some time. They hydrolyze ATP to ADP plus P_i and utilize the energy derived from this reaction to transport ions across the membrane against a gradient. One type transports Na^+ and K^+ and another less widely distributed one transports Ca^{2+} and Mg^{2+}. Examples of the latter type were described in the 1960s in muscle sarcoplasmic reticulum, which accumulate Ca^{2+} and in erythrocyte membranes, which extrude Ca^{2+}. The hydrolytic process is located on the side from which Ca^{2+} is removed to maintain a low Ca^{2+} concentration of 10^{-5} to 10^{-8} *M*. The extracellular concentration is usually about 10^{-3} *M*. Other ATPases are known that appear to transport bicarbonate, chloride, and hydrogen ions. In addition there is another Ca^{2+} transport system that consists of an Na^+–Ca^{2+} exchange carrier driven by a sodium concentration gradient but ATP does not participate directly (Stekhoven and Bonting, 1981). Recently, Ca^{2+}-dependent sodium channels have been identified in excised inside-out patches of plasma membranes from *P. tetraurelia* that are activated by CaM (Saimi and Ling, 1990).

In mammalian cells, the CaM-mediated plasma membrane Ca^{2+},Mg^{2+}-ATPase is thought to be the pump responsible for extruding cytoplasmic Ca^{2+} (Cheung, 1980; Morcos and Drummond, 1980). Increasing concentrations of Ca^{2+} in the cell activate the Ca^{2+}-ATPase, which in turn increases Ca^{2+} efflux, maintaining a steady state. However, once the gradient of transported Ca^{2+} is turned around, the overall reaction is readily reversed, unlike the Na^+,K^+-ATPase. It is this self-regulating process that is thought to maintain a low steady state level of intracellular Ca^{2+}. Muto and Nozawa (1984) recognized that this could be one of the major systems responsible for regulating cytoplasmic Ca^{2+} levels and while this enzyme has been found in higher organisms, no such system had been recognized in *Tetrahymena*. Other Ca^{2+}-stimulated ATPase activities had been reported in whole-cell homogenates of *Tetrahymena* (Conner *et al.*, 1963) and *Paramecium* (Fok *et al.*, 1981) but these had not been characterized. A Ca^{2+}-ATPase, independent of Mg^{2+}, had been reported to be present in the plasma membranes of *Trypanosoma cruzi* epimastigotes (Frasch *et al.*, 1978a), *Trypanosoma brucei* (McLaughlin, 1985) and *Entamoeba histolytica* (McLaughlin and Muller, 1979, 1981). Later work on *Entamoeba invadens* demonstrated a similar ATPase in that organism (Zarain-Herzberg and Arroyo-Begovich, 1985). While the studies on *E. histolytica* suggested CaM might be involved, inhibitor studies on *E. invadens* indicate that it is not a mitochondrial H^+-ATPase, an Na^+,K^+-ATPase (sodium pump), or mediated by sulfhydryl groups, phosphorylated intermediates, or CaM. The role of these ATPases in protozoa is far from completely understood, but will be discussed again in a later section. Muto and Nozawa (1984) described and characterized two different ATPases from microsomes and plasma membranes of *T. pyriformis* strain NT-1. One is Ca^{2+} stimulated (Ca^{2+},Mg^{2+}-ATPase) and felt to be contributory to Ca^{2+} homeostasis in *Tetrahymena* like other Ca^{2+} transport ATPases. It may be different from the ruthenium red-insensitive, Ca^{2+}-activated ATPase previously reported by Chua *et al.* (1977) from the cytosolic fraction of the E strain of this ciliate. Ruthenium red inhibits Ca^{2+},Mg^{2+}-ATPase with some degree of specificity (Stekhoven and Bonting, 1981). The other ATPase described by Muto and Nozawa (1984) is an Mg^{2+}-stimulated ATPase with a distinctly different pH profile and sensitivity to inhibitors. This Ca^{2+},Mg^{2+}-ATPase requires magnesium ions, and has a nucleotide specificity for ATP. Addition of NaCl, KCl, or ouabain (an inhibitor with almost absolute specificity for Na^+,K^+-ATPases) and inhibitors of mitochondrial ATPases have no effect. This Ca^{2+},Mg^{2+}-ATPase exhibits sensitivity to rather high doses of trifluoperazine (100 μM reduces activity to 54% of control and 10 μM to 89%). Unfortunately no attempt was made to further study this by adding CaM or Ca^{2+}, which might provide support that the agent acts by CaM binding (Nagao *et al.*, 1983; Kobayashi *et al.*, 1979). It is inhibited by vanadate to a degree, as are other Ca^{2+},Mg^{2+}-1979). It is inhibited by vanadate to a degree, as are other Ca^{2+},Mg^{2+}-ATPases. Recently, Nagao and Nozawa (1985) demonstrated a significant amount of radioiodinated *Tetrahymena*

CaM binds to microsomal membranes in a Ca^{2+}-dependent association but the biological functions of the CaM-binding proteins were not determined.

In addition to the microsomal Ca^{2+},Mg^{2+}-ATPase reported in *Tetrahymena* by Muto and Nozawa (1984) there are Ca^{2+}-stimulated ATPases of the ciliary membrane of *Paramecium*. These appear to be distinct from the axonomal ATPase proteins (dynein ATPases) because the latter are predominantly Mg^{2+} dependent, are of higher molecular weight, and are not membrane bound. It has been suggested that the membrane ATPases may function to move Ca^{2+} out of the cilium following the influx through voltage-sensitive Ca^{2+} channels during the "avoidance reaction" (Doughty, 1978a–c; Doughty and Kaneshiro, 1983). Andrivon *et al.* (1983), however, raised some questions as to whether this particular enzyme is fulfilling this role in the cell. Noguchi *et al.* (1979), recognizing the role a Ca^{2+} extrusion pump would play to keep intracellular Ca^{2+} below 1 μM, following ciliary reversal in *P. caudatum,* isolated a pellicular (surface membrane) Ca^{2+}-activated ATPase. Na^+,K^+-activated ATPase activity and dynein ATPase activity were not present in their pellicle fraction. Bilinski *et al.* (1981) isolated a Ca^{2+}-ATPase from surface membranes of *Paramecium* freed of cilia. Their preparation was 10 times more active than that isolated by Noguchi *et al.* (1979), due to procedural changes to maximize ATPase activity. This CaM-stimulated Ca^{2+}-ATPase, free of secretory proteins and ciliary tubulin, was sensitive to inhibition by the Ca^{2+},Mg^{2+}-ATPase inhibitors ruthenium red and vanadate as well as the CaM inhibitors, trifluoperazine and compound R24571. The authors attribute its activity to exocytosis regulation (Tiggemann and Plattner, 1982).

Riddle *et al.* (1982) pointed out that there are a number of reports of divalent cation-dependent ATPases detected in either the cilia or surface of *Paramecium*. All are clearly different from dynein ATPase in cilia. They suggest published reports are compatible with two ATPases being present in cilia: one an Mg^{2+}-dependent activity of dynein and another being the membrane-bound Ca^{2+}-ATPase reported by Doughty (1978a), Andrivon *et al.* (1983), and Noguchi *et al.* (1979). However, the amount of Ca^{2+}-ATPase associated with isolated *Paramecium* cilia depends on the method of deciliation utilized by the investigator. The Ca^{2+} shock procedure results in three times more Ca^{2+}-ATPase being associated with cilia than when they are prepared by Mn^{2+} shock. Several energy-dependent cell functions are probably controlled by the Ca^{2+}-ATPase, which appears to originate at the cell surface. These cell functions, as stated previously, include direction (ciliary reversal) and orientation of ciliary beat. Another Ca^{2+}-dependent acitivity is the explosive discharge of trichocysts (secretory vesicles) that lie just below the cell surface in *Paramecium*. Following stimulation of the protozoal external surface, the Ca^{2+} channels ("rosette particles") open. Histochemical evidence points to a Ca^{2+}-ATPase activity at the site of trichocyst fusion with the surface membrane and subsequent extrusion of contents exocytotically. The secretory product is easily visible by light microscopy and can be seen as long, thin needles surrounding the entire cell. One of the major proteins of extruded trichocysts (1–10% of total) is CaM (Matt *et al.*, 1978; Satir and Oberg, 1978; Plattner *et al.*, 1977; Rauh and Nelson, 1981; Riddle *et al.*, 1982).

In any event, Watanabe and Nozawa (1982) recognized that the free Ca^{2+} level necessary for ciliary reversal is about 10^{-6} *M*, which nearly coincides with the K_d of Ca^{2+} binding to *Tetrahymena* CaM reported by Suzuki *et al.* (1981). Therefore, to investigate the probable role CaM may play in ciliary reversal, they utilized gel electrophoresis and immunoelectron microscopy to localize CaM and CaM-binding proteins within the cilia. It was this that prompted Suzuki and co-workers (1982) to prepare antiserum directed against *Tetrahymena* CaM to attempt its localization intracellularly by immunofluorescence. Trifluoperazine caused *Tetrahymena* to stop forward swimming, and in some instances initiate backward swimming (Rauh *et al.*, 1980; Suzuki *et al.*, 1982). This drug also inhibited the uptake of food into the food vacuole from the oral apparatus (endocytosis) in a dose-dependent relationship, as measured by the ability of this ciliate to ingest India ink. In addition, this drug also inhibited excretion of contractile vacoule contents, (suggesting CaM regulation of osmotic pressure, i.e., exocytosis, and fluid secretion). Ohnishi *et al.* (1982) succeeded in isolating CaM from *Tetrahymena* cilia. There were earlier reports by Gitelman and Witman (1980) and by Van Eldik *et al.* (1980b) that they had isolated such a protein from flagellar fractions of the green alga, *Chlamydomonas*. The protein isolated by Gitelman and Witman (1980) on a fluphenazine–Sepharose 4B column was heat stable and activated cyclic nucleotide phosphodiesterase in a Ca^{2+}-dependent manner, but was not associated with the 12S and 18S dynein ATPase fractions. The protein isolation by Van Eldik *et al.* (1980b) was done on a phenothiazine–Sepharose affinity column, a procedure that has been used successfully to isolate CaMs from various sources. While this protein had properties characteristic of CaM, the protein did not activate brain phosphodiesterase. It was concluded not to be CaM. The material isolated by Ohnishi *et al.* (1982), on the other hand, from cilia, cell bodies, and whole cells of *T. pyriformis* strain W and *T. thermophila* strain B1868 met several criteria strongly suggesting it to be CaM. The cilia were fractionated into a membrane plus matrix fraction and an axoneme fraction. The term axoneme or "axial filament complex" includes all components limited by the accessory microtubules and coarse fibers without membranous or mitochondrial structures (Warner, 1970). This axoneme fraction was further separated into an outer doublet microtubule fraction and a crude dynein fraction. Alkali-glycerol gel electrophoresis of these fractions revealed

CaM in all but the crude dynein fraction. The greatest percentage of CaM in cilia is localized on the interdoublet or nexin links (Ohnishi *et al.,* 1982). These structures are related to the conversion of active sliding in the outer-doublet microtubules to bending, resulting in change in ciliary beat direction. Ohnishi *et al.* (1982), however, acknowledged that their findings are in contrast to observations made by Jamieson *et al.* (1979) and Blum *et al.* (1980), who reported that much of the ciliary *Tetrahymena* CaM could be extracted by Tris/EDTA with the crude dynein fraction of demembranated cilia. These crude dyneins were resolved by sucrose density gradient centrifugation and CaM appeared to be associated with the 14S dynein ATPase. Only the CaM bands of the axoneme and outer-doublet microtubule fractions found by Ohnishi *et al.* (1982) showed electrophoretic differences in the presence or absence of Ca^{2+}, but they concede the CaM–dynein interaction seen by Blum *et al.* (1980) may be weaker than that seen in other Ca^{2+}-dependent CaM complexes. These differences may be the result of variation between methodologies [i.e., in the absence of a large supply of CaM from *Tetrahymena,* Blum *et al.* (1980) used bovine brain CaM in most of the experiments (see Hirano and Watanabe, 1985)]. Immunoelectron microscopy by Ohnishi *et al.* (1982) localized CaM at regular 90-nm intervals along the long axis of the outer-doublet microtubules by using monospecific rabbit anti-*Tetrahymena* CaM IgG and ferritin-conjugated goat anti-rabbit IgG. They point out that this 90-nm interval corresponds to that of the radial spokes and interdoublet links. The CaM-binding site was suggested to be interdoublet links in the ciliary axoneme. This specific localization was thought to explain how a Ca^{2+}-dependent regulator might be involved in ciliary reversal (Ohnishi *et al.,* 1982). In addition, the sliding–bending mechanism appears to be a result of molecular changes that occur uniformly along the axoneme (Warner, 1976). The reader is strongly encouraged to refer to the report by Doughty (1979) for a comprehensive discussion of this subject with detailed refereces. Ohnishi and Watanabe (1983) later demonstrated that cilia of *T. thermophila* strain B contain another Ca^{2+}-binding protein of about 10kDa, TCBP-10, in addition to CaM. TCBP-10 is in the CaM family, related to vitamin D-dependent Ca^{2+}-binding proteins and PAP I-b protein, which are present in higher vertebrates. TCBP-10 is present in both the cilia and cell body. While it has a subunit molecular weight of 10,000, it exists as a dimer in the native state. Partially purified TCBP-10 failed to activate adenylate cyclase, guanylate cyclase, and phosphodiesterase of *Tetrahymena* as well as porcine brain phosphodiesterase. They conceded at that time that the role played by TCBP-10 in ciliary reversal was not known. In addition, at least 36 kinds of CaM-binding proteins (CaMBPs), which are dependent on Ca^{2+} and specific for CaM, were ultimately detected in the cilia using an improved [^{125}I]CaM overlay method (Hirano and Watanabe, 1985). The major CaM-binding proteins appear to Hirano-Ohnishi and Watanabe (1988) to be localized in the outer-doublet microtubule fraction. (At least six kinds of CaMBPs are associated with microtubules as suggested by their study.) Presumably motility would result from ATP-induced microtubule sliding. This is modulated by cAMP and a Ca^{2+}/CaM-dependent phosphorylation of β-tubulin by protein kinase (Hirano-Ohnishi and Watanabe, 1989). They speculate that this phosphorylation may affect interaction between dynein arms and microtubules or interdoublet links and spokes with doublet microtubules. However, the increase or decrease in frequency of ciliary beating did not always occur at Ca^{2+} concentrations of 10^{-6} M (the K_d of CaM for Ca^{2+}). Therefore, Takemasa *et al.* (1989) deduced that the presence of a Ca^{2+}-binding protein other than CaM might explain this. Such a protein might be TCBP-10 [which may or may not correspond to the Van Eldick *et al.* (1980b) CaM-like protein in *Chlamydomonas* flagella]. Further investigation by Takamasa *et al.* (1989) revealed that TCBP-10 is in reality a degradation product of a 25-kDa protein, TCBP-25, which exists in addition to CaM in *Tetrahymena* cilia.

Movement of a cilium or a flagellum has been well described by Alberts *et al.* (1989) and their text should be read for details on the subject. Briefly, they state that movement is produced by bending of the core or axoneme. Here nine doublet microtubules are arranged in a ring around a pair of single microtubules known as a "9 + 2" array (Fig. 2). Each member of the central pair is a complete microtubule. Each of the outer doublets is made up of one complete microtubule (A tubule) fused to one partial microtubule (B tubule) in such a way that they share a common tubule wall. Radial spokes project inward from each doublet to the inner sheath, which envelopes the central pair of microtubules.

Force is generated to produce wavelike movements by the arms that join adjacent doublets in the outer ring of 9 doublets. Pairs of inner and outer arms, made up of a protein known as dynein and capped with globular heads containing ATPase activity, are spaced every 24 nm along the A tubule. Nexin links are more widely spaced and extend between adjacent doublets to resist sliding. Bending is produced by the sliding of microtubules, which results from movement of dynein heads along the adjacent microtubule resulting from ATP binding and hydrolysis. Alberts *et al.* (1989) suggested that the sliding of microtubules in a cilium is similar to the operation of myosin heads in muscle. In addition to the comprehensive treatment of the subject by Alberts *et al.* (1989), the reader is encouraged to refer to original papers by Gibbons (1966, 1982), Warner (1970, 1976), Warner *et al.* (1977), Pitelka (1974), Allen (1968), Witman *et al.* (1976, 1978), Gibbons and Grimstone (1960), Ringo (1967) and Reed and Satir (1980).

The ATPase activity of 14S and 30S dynein can be activated by either

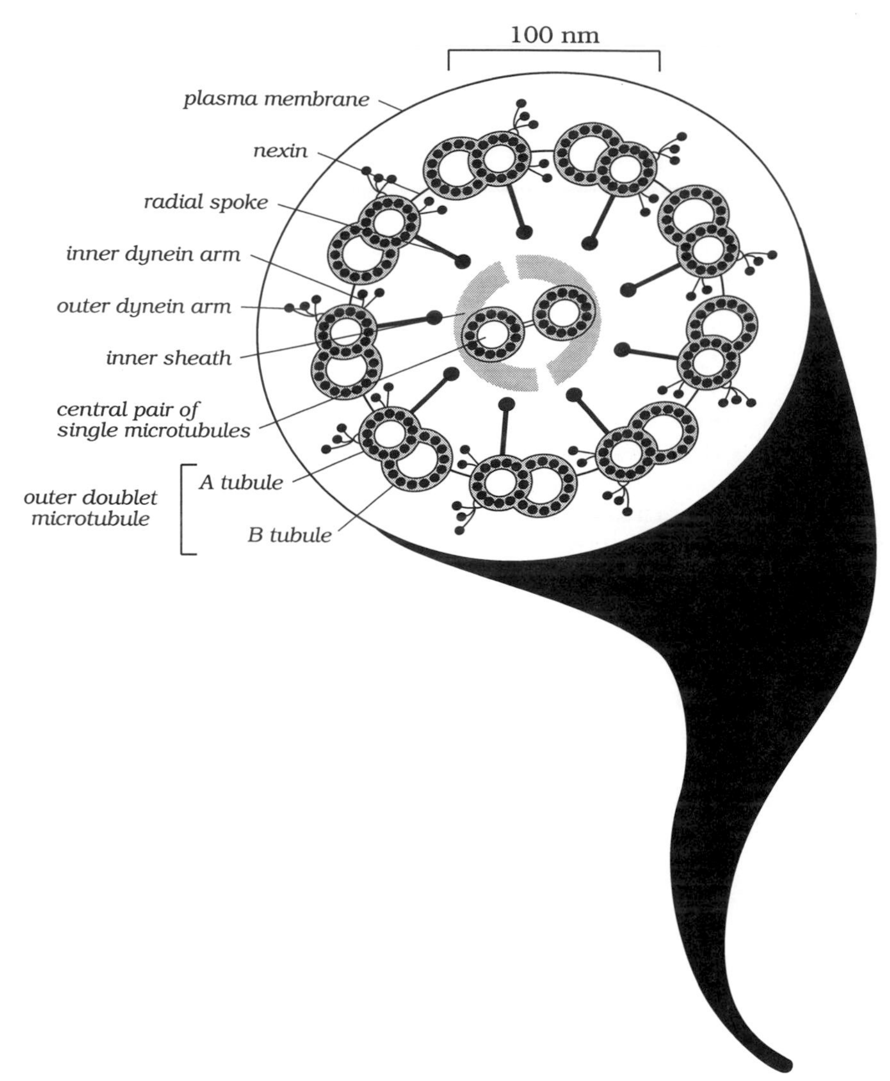

FIG. 2 A cilium shown in cross-section. [Adapted from Alberts *et al.* (1989) with permission of publisher.]

Mg^{2+} or Ca^{2+} (Gibbons, 1966). However, experiments by Naitoh and Kaneko (1972) on Triton-extracted models of *Paramecium* suggested that the presence of Mg^{2+} plus ATP is required for ciliary motion. Concentrations of Ca^{2+} in the millimolar range or greater cause a reversal in ciliary motion. This suggested to Doughty (1979) that Ca^{2+} alters the Mg^{2+}-

dependent dynein ATPase activity. He was able to solubilize dynein proteins from demembranated cilia of *Paramecium*. Micromolar concentrations of Ca^{2+} inhibited the Mg^{2+}-dependent ATPase activity of the crude dynein extracts. In addition, the presence of Ca^{2+} (4 μM) reduced electrophoretic mobility of the crude dynein extract. Sepharose 4B chromatography of these crude extracts resulted in three dynein fractions with magnesium-dependent ATPase activity. One dynein fraction showed Ca^{2+}-stimulated ATPase activity that in the absence of Ca^{2+}, resolved into three bands on native gels and, in the presence of Ca^{2+}, migrated as a single band. The second fraction appeared to be a Ca^{2+}-insensitive Mg^{2+}-dependent ATPase and resolved as a single band in the presence or absence of Ca^{2+}. The third dynein fraction was a Ca^{2+}-inhibited, Mg^{2+}-dependent ATPase resolving into two bands in the absence of Ca^{2+} and combining into a single, slower moving band at micromolar levels of Ca^{2+}. This third dynein fraction could be further resolved into two subfractions with Mg^{2+}-ATPase activities, opposing Ca^{2+} sensitivities, and differing specific activities. These data suggested that there is a Ca^{2+}-dependent alteration in the molecular and functional state of differing classes of dyneins within the ciliary axoneme. Doughty (1979) pointed out this gradation of Ca^{2+} interaction at various sites could theoretically affect interdoublet sliding, interdoublet cross-bridge activity, doublet arms, and/or radial spoke attachment to the central sheath or spoke heads.

Jamieson *et al.* (1979) conducted isolation studies by extraction of demembranated cilia with Tris/EDTA, and reported that much of the ciliary CaM could be isolated and the crude dyneins were resolved by sucrose density gradient centrifugation. In *Tetrahymena* ciliary axonemes, the peak activity for ATPase is associated with 14S dynein. While the mechanism of these Ca^{2+} effects on the axoneme was not understood, this suggested that CaM might endow Ca^{2+} sensitivity on the dynein ATPase of 14S and 30S dyneins. Jamieson *et al.* (1980) and Blum *et al.* (1980) went on to report that their initial studies on Tris/EDTA-extracted crude dynein from *Tetrahymena* cilia was enhanced by Ca^{2+} to a small degree in the presence of additional bovine brain CaM. Doughty (1979) had utilized KCl extraction to observe the effects of low levels of Ca^{2+} on Mg^{2+}-dependent dynein ATPase. They hoped this procedure would improve CaM extraction from the dyneins, yielding only those dyneins that were Ca^{2+} sensitive in the presence of CaM. They did indeed observe a greater CaM-induced Ca^{2+} sensitivity by dynein ATPase if the crude dyneins were obtained by KCl extraction, especially (up to 10-fold) with CaM supplementation (either from bovine brain or *Tetrahymena*). The 14S dynein ATPase extracted by either the KCl or Tris/EDTA procedure is more sensitive to Ca^{2+}-CaM activation than is the 30S fraction. On the other hand, troponin C (a Ca^{2+} regulatory protein of actomyosin ATPase similar in structure to CaM) plus Ca^{2+} activated the 30S dynein and it did so to a greater extent than the 14S dynein. This activation was less than that achieved by CaM. Also, gentle heat treatment or *N*-ethylmaleimide treatment produce major differences on the activation of the 14S dynein versus the 30S dynein (Blum *et al.*, 1981, 1982; Blum and Hayes, 1984). These observations suggest that the calmodulin-binding sites differ between the two dyneins.

Because the 14S and the 30S dyneins both bound to CaM–Sepharose 4B columns in the presence of Ca^{2+} and were eluted in the presence of EGTA, Blum *et al.* (1980; Blum and Hayes, 1984) suggested that both dyneins had specific binding sites for CaM. The 14S and 30S dyneins consisted of different high-molecular-weight species (Jamieson *et al.*, 1980; Blum *et al.*, 1980). In addition, photoaffinity labeling studies indicated the heavy chains of the 14S dyneins had substantially more labeling than those of the 30S, which was in accord with the greater CaM activation of the 14S dynein (Blum *et al.*, 1980; Blum and Hayes, 1984). It remains to be seen what physiological roles these play in the protozoa (Piperno and Luck, 1982).

C. Regulation of Enzymes

1. Cyclic Nucleotide Metabolism

Cyclic nucleotides play an important part in cell growth and proliferation. In general, levels of cyclic nucleotides are regulated by cyclases, enzymes responsible for synthesis, and phosphodiestterases, enzymes responsible for degradation (Kudo *et al.*, 1981a,b; Goldberg and Haddox, 1977; Murad *et al.*, 1979; Ross and Gilman, 1980).

Guanylate cyclase is the enzyme that catalyzes the conversion of GTP to cyclic 3′, 5′-guanosine monophosphate (cGMP).

$$\text{Guanosine triphosphate (GTP)} \xrightarrow[\text{cyclase}]{\text{guanylate}} \text{cyclic GMP} + PP_i$$

It is present in most mammalian tissues, birds, fish, mollusks, insects, and bacteria. Guanylate cyclase activity is distributed between soluble and particulate fractions of the cell, differing quantitatively and qualitatively from tissue to tissue. There also may be a selective distribution of guanylate cyclase in discrete structures in the cell, or within specific cell types of an organ. In addition, changes occur in the total activity as well as proportion of enzymatic activity found in particulate versus soluble fractions through the growth cycle of the cell and in times of abnormal growth.

Cyclic GMP in turn may activate protein kinase, which stimulates the incorporation of phosphate into cellular proteins, in so doing activating secondary enzymes. Cyclic GMP may also stimulate other enzymes, including a cyclic nucleotide phosphodiesterase (PDE) that catalyzes hydrolysis of cyclic AMP (cAMP).

$$\text{ATP} \xrightarrow{\text{adenylate cyclase}} \text{cyclic AMP} + \text{PP}_i$$

$$\text{cAMP} + \text{H}_2\text{O} \xrightarrow{\text{phosphodiesterase}} \text{AMP} + \text{P}_i$$

Tetrahymena pyriformis contains cGMP and cAMP as well as their related enzymes (Wolfe, 1973; Ramanathan and Chou, 1973a; Nakazawa *et al.*, 1979; Kudo and Nozawa, 1983). These cyclic nucleotides are reported by some investigators to vary during the cell cycle (Voichick *et al.*, 1973; Kariya *et al.*, 1974; Dickinson *et al.*, 1976; Graves *et al.*, 1976; Gray *et al.*, 1977). This presumably reflects the role of these components in cell growth and proliferation because there appeared to be a temporal relationship of elevated adenylate cyclase activity followed by high cAMP and low PDE activity coincident with cell division of *T. pyriformis* strain W (Kariya *et al.*, 1974; Dickinson *et al.*, 1976) or strain E (Voichick *et al.*, 1973). Low cAMP was observed by these workers when adenylate cyclase was at a minimum and PDE had increased. There also appeared to be an increase in guanylate cyclase activity associated with an increase in cGMP and a decrease in cyclic GMP PDE, the enzyme responsible for degradation of this cyclic nucleotide (Graves *et al.*, 1976) in synchronized cultures of *T. pyriformis* strain W (Gray *et al.*, 1977). Kudo *et al.* (1981a,b) found essentially the same results with respect to guanylate cyclase activity in strain GL of *T. pyriformis*. However, no significant change was observed in either adenylate cyclase or cyclic AMP PDE activities during growth. This contrasts with the findings of Voichick *et al.* (1973), who reported adenylate cyclase and cyclic AMP PDE activities are dependent on the stage of growth of *T.pyriformis* strain E. Except for the difference in the strain used by these two groups, there is no obvious explanation for this discrepancy. Adenylate cyclase from the NT-1 strain has a higher activity than that of WH-14 and W strains (Shimonaka and Nozawa, 1977). In addition, enzymatic and morphological studies suggest the activity of adenylate cyclase in the pellicle membrane changes as a function of temperature, presumably as a result of altered physical states (fluidity). Further work by Kudo and Nozawa (1983) with the NT-1 strain of *T. pyriformis* reported on a protein that specifically binds cAMP. Fractions having this binding activity eluted in a single peak coincident with protein kinase activity. While total cAMP-binding activity did not change at various stages of cell growth, these workers suggest that the regulation of cAMP concentration in the cell may be due to a redistribution of cAMP-binding proteins. These results are consistent with earlier work by Majumder *et al.* (1975), who reported that a cytosolic protein kinase from strain E of *Tetrahymena* is activated by cAMP, which fluctuates markedly during the growth cycle. They suggest that protein kinase as well as cAMP play a key role in the regulation of cell growth.

Guanylate cyclase in *T. pyriformis* strain NT-1 is associated in large part with the plasma membranes (pellicles): it is particulate bound to a 105,000 *g* precipitate and is not part of the soluble fraction [as is seen in most mammalian tissues (Nagao and Nozawa, 1987; Kudo *et al.*, 1981a)]. Small amounts are also associated with cilia, mitochondria, and microsomes (Schultz and Klumpp, 1984). Nagao *et al.* (1979) and later Muto *et al.* (1983) reported that this guanylate cyclase activity was enhanced approximately 20 to 40-fold in the presence of Ca^{2+} by a *Tetrahymena* Ca^{2+}-binding protein (TCBP). This protein has the same mode of activation as mammalian CaM and except for 11 substitutions and 1 deletion it has essentially the same amino acid composition as bovine brain CaM (Suzuki *et al.*, 1981). Kudo *et al.* (1982b) also showed that bovine brain CaM is very similar to *Tetrahymena* CaM in heat stability, isoelectric point, and Ca^{2+}-dependent electrophoretic mobility. No other CaM (i.e., bovine brain, sea anemone, scallop, *R. reniformis, Chlamydomonas, Dictyostelium,* spinach, English cucumber) activated this cyclase enzyme from *Tetrahymena* to the extent of *Tetrahymena* CaM or *Paramecium* CaM (Yazawa *et al.*, 1981; Kakiuchi *et al.*, 1981; Kudo *et al.*, 1982a,b; Nagao and Nozawa, 1987). This may be related to the fact that the sequence of amino acids in *Tetrahymena* CaM that activates guanylate cyclase differs more from that of brain CaM than any other reported CaM. *Tetrahymena* CaM migrates significantly slower than bovine brain CaM on 15% polyacrylamide gel electrophoresis (Kudo *et al.*, 1983, 1985a). While it is almost as effective as brain CaM in activating brain or heart cyclic nucleotide PDE, adenylate cyclase, myosin light chain kinase, erythrocyte (Ca^{2+},Mg^{2+}) ATPase, and plant NAD kinase, it appears to have no effect on the activity of guanylate cyclases from rat brain, rat lung, or human platelets (Nagao and Nozawa, 1987). Kudo *et al.* (1985a) tested the effects of brain CaM and other structurally similar Ca^{2+}-binding proteins on *Tetrahymena* guanylate cyclase. Bovine brain CaM (B-CaM) inhibits the Ca^{2+}-dependent *Tetrahymena* CaM (T-CaM)-mediated activation of *Tetrahymena* guanylate cyclase in a concentration-dependent relationship. This antagonism between CaM and B-CaM was not observed in the activation of Ca^{2+}-dependent cyclic nucleotide PDE of bovine brain or heart. Kudo *et al.* (1983, 1985a) suggested that the B-CaM or S-100, based on similarities in their amino acid sequence, have similar binding domains for *Tetrahymena* guanylate cyclase that prevent T-CaM from activating the cyclase.

Guanylate cyclase from the particulate fraction of rat lung and mouse parotid required Mn^{2+} as the sole cofactor. Mg^{2+} can be substituted in a variety of particulate mammalian guanylate cyclases, but with a substantial loss of activity (Nagao and Nozawa, 1987). *Tetrahymena* guanylate cyclase is stimulated by Mg^{2+} to a greater extent than by Mn^{2+} in the absence of CaM (Nakazawa *et al.*, 1979; Kudo *et al.*, 1982a). Maximum activities were obtained at greater than 5 m*M* for both cations, but 5 m*M* Mn^{2+} stimulated activity almost three times more than did 5m*M* Mg^{2+} in digitonin-solubilized enzyme from plasma membranes (Nagao and Nozawa, 1987). Triton X-100, Lubrol PX, deoxycholate, or trypsin did not stimulate the enzyme. Unexpectedly these inhibited, even though detergents generally increase the activities of particulate guanylate cyclase fractions from 3 to 12 times (Nagao and Nozawa, 1987; Nakazawa *et al.*, 1979; Murad *et al.*, 1979).

Classical CaM antagonists such as the phenothiazine derivatives, as well as *N*-(6-aminohexyl)-5-chloro-1-naphthalene sulfonamide (W-7), local anesthetics, and EGTA inhibited this Ca^{2+}–CaM-mediated activation of guanylate cyclase (Nagao *et al.*, 1981a, 1983; Muto *et al.*, 1983). A common feature of mammalian guanylate cyclase is the activation of the enzyme by nitroso compounds. However, unlike what is observed with pig heart guanylate cyclase, agents such as sodium azide, $NaNO_2$, or *N*-methyl-*N'*-nitro-*N*-nitroguanidine (MNNG) failed to activate *Tetrahymena* guanylate cyclase in the presence of Mg^{2+} or Mn^{2+} (Kudo *et al.*, 1985b).

Kudo *et al.* (1985b) recognized that changes in the intracellular levels of cGMP, Ca^{2+}, and guanylate cyclase activity correlate to various stages of cell growth and differentiation. The biochemical nature of the stimuli affecting these changes remains to be elucidated, however. Using synchronized cultures (by a single hypoxic shock or heat shock) of *T. pyriformis* strains W and GL, Gray *et al.* (1977) and Kudo *et al.* (1981a) found guanylate cyclase activity decreases to the lowest level at an early stationary phase of growth about 75 min after the cells enter the first synchronous division. There is then a rise in activity at the middle of the exponential phase of growth before the second cell division. A consistent correlation exists between these changes in guanylate cyclase activity in the particulate fraction and CaM concentration in the soluble fraction of the synchronized cell. Further work by Singh and Chatterjee (1988) showed that the level of CaM did vary at different phases of the cell cycle in heat shock-synchronized *T. pyriformis* strain W. The average CaM level was 0.742 $\mu g/mg$ protein during the time taken for cells in mitosis to form two daughter cells (mid-M phase). This was about 85 min after the end of heat shock and lasted 20 min. The concentration of CaM then fell over the next 60 min (mid-G_1 phase) to 0.323 $\mu g/mg$ protein, with a gradual increase in the uptake of radiolabeled thymidine. Over the next 40 min there was a maximum rate of incorporation of this radiolabeled nucleotide precursor and CaM levels rose to 0.790 μg/mg (mid-S phase). The concentration of CaM levels remained at 0.776 μg/mg (mig-G_2 phase) over the next 60 min but there was a steep decline of radiolabeled thymidine incorporation at the end of S and the beginning of G_2 before the onset of the next synchronous division. The CaM inhibitor, trifluoperazine (at 18 μM), and 2 mM EGTA (the Ca^{2+} chelator), were found to inhibit DNA synthesis about 10 min prior to the end of G_1 and the beginning of S, a period in which the cell is known to require Ca^{2+}. The uptake of radiolabeled thymidine was not affected by these agents, however. Trifluoperazine also did not inhibit the synchronous mitotic division of *Tetrahymena*. This suggested to these workers that CaM is involved in the replication of nuclear DNA and would agree with the earlier findings of Dickens *et al.* (1975). They found cGMP, its dibutyrl derivative and caffeine, but not cAMP, caused the induction of TMP synthetase in strain W of *Tetrahymena*. Cyclic AMP decreased both the cGMP and caffeine mediated increases in TMP synthetase activity. They interpreted this to mean that cGMP but not cAMP acts as the signal for induction of cell proliferation

Ca^{2+} and Mg^{2+} levels were reported by London *et al.* (1979) to increase by 50% about 15 min prior to the burst of cell division (75 min after heat shock). This increase in CaM, observed using atomic absorption spectroscopy, takes place over a 10-min time period preceding a 2-fold decrease in cAMP and a 100-fold increase of cGMP at the onset of cell division. Ca^{2+} and Mg^{2+} levels drop at the peak of cell division and reach predivision levels by the end of cell division. London *et al.* (1979) interpreted this to indicate that Ca^{2+} was the "trigger" for cell division and responsible for concomitant changes in cyclic nucleotide level (i.e., elevating cGMP and lowering cAMP). The possibility that Ca^{2+} or Mg^{2+} functions as a "trigger" for cell division, however, was disputed by Walker and Zeuthen (1980), who observed that synchronous cell division continued in buffer alone or even distilled water after heat shock. While no net uptake of Ca^{2+} or Mg^{2+} occurs, the cells continue to divide. These authors do concede, however, that intracellular stores of bound Ca^{2+} or Mg^{2+} may be released as free ions into the cytosol, where they could function as modulators of cell division either through microtubule assembly mechanisms or the control of intracellular cyclic nucleotide levels. This agrees with later work by Kim *et al.* (1984, 1985) that significant amounts of Ca^{2+} can be released from mitochondria, which in *Tetrahymena* can occupy approximately 15% of the total volume of the cell (there are approximately 4000–6000 mitochondria/cell). This Ca^{2+} can significantly increase the Ca^{2+} concentration in the cytoplasm, stimulating changes in the lag phase of the cells. In addition, the conclusion of London *et al.* (1979) that Ca^{2+} was the "trigger'' for lowered cAMP is somewhat at odds with those of Nandini-Kishore and Thompson (1979), who reported Ca^{2+} release was followed by a rise of cAMP.

In an effort to understand the role of structure and function in *Tetrahymena* biomembranes Kameyama *et al.* (1982) attempted to modify their lipid composition and compare biological functions. They modified the phospholipid composition by treatment with phospholipase A_2 and measured the activities of the two membrane-bound cyclases (guanylate cyclase and adenylate cyclase). Approximately 45% of the total phosphatidylethanolamine was converted to its lysolipid (lysophosphatidylethanolamine) with a slight increase in membrane fluidity occurring at 25°C compared to untreated control membranes. There was a decrease in guanylate cyclase activity by 50%, with little effect on adenylate cyclase activity. This suggests that the former enzyme is more dependent on phospholipid environment than the latter. It remains to be seen whether agents such as Ca^{2+}, norepinephine, acetylcholine, carbachol, or serotonin (known to stimulate phospholipase A_2 activity) would have any effect on intracellular guanylate cyclase and cGMP levels (Kameyama *et al.*, 1982; Murad *et al.*, 1979). In mammalian tissues lysophosphatidylcholine stimulated, but lysophosphatidylethanolamine was ineffective at stimulating, guanylate cyclase. The free fatty acids that are cleaved from phospholipids by phospholipase A_2 hydrolysis, as well as phospholipase A_2 itself, are known to enhance guanylate cyclase.

Adenylate cyclase activity in *Tetrahymena* is reported to be in the plasma membrane rather than intracellular membrane (Kudo *et al.*, 1985b). It requires Mg^{2+} or Mn^{2+} in the presence of ATP for activity. It is well known that *Tetrahymena* CaM has no effect on *Tetrahymena* adenylate cyclase (with or without Ca^{2+}) or cyclic nucleotide PDE activity (Kudo *et al.*, 1980, 1982a; Nagao *et al.*, 1979, 1983; Kameyama *et al.*, 1982). Adenylate cyclase is inhibited in a dose-dependent manner by trifluoperazine, local anesthetics, and W-7, but this inhibition is not Ca^{2+}–CaM dependent (Nagao *et al.*, 1983; Muto *et al.*, 1983). It is argued by Nagao *et al.* (1983) that the CaM antagonists may act nonspecifically on the lipid bilayer, resulting in the inhibition of adenylate cyclase. This contrasts with guanylate cyclase inhibition, where the antipsychotic CaM antagonists inhibit the Ca^{2+}-dependent activation of the enzyme by competing with CaM. In the case of the local anesthetics competitive antagonism explains only part of the inhibition. In addition to differing from mammalian adenylate cyclase in its requirement for CaM and Ca^{2+} (Klee *et al.*, 1980; Ross and Gilman, 1980), activators of mammalian cell adenylate cyclase activity (adrenalin, histamine, carbamylcholine, prostaglandin E, etc.) failed to activate the *Tetrahymena* enzyme (Kameyama *et al.*, 1982). *Tetrahymena pyriformis* strain NT-1 adenylate cyclase was not stimulated by fluoride,

5-guanyl-imidodiphosphate (GppNHp), or forskolin, unlike the same enzyme from rat heart (Kudo *et al.*, 1985b). Also, hormones such as epinephrine, norepinephrine, and glucagon or the β-adrenergic agonist, isoproterenol (at concentrations of 50 μ*M*), did not activate the adenylate cyclase in this strain of *Tetrahymena*. Dopamine, histamine, serotonin, and prostaglandin E_2 elicited little effect (even at 100 μ*M*). Adenylate cyclase was activated by low levels of Ca^{2+} and inhibited by high levels (1.0 m*M*). In mammalian cells, cAMP is reported to antagonize Ca^{2+} influx through the plasma membrane and its release from internal pools (Tsien *et al.*, 1984). Therefore, this may play a role in control in *Tetrahymena*. These results are in contrast to earlier work by Rozensweig and Kindler (1972) and by Kassis and Kindler (1975), who, working with strain W of *Tetrahymena*, reported activation of this enzyme by epinephrine (1 m*M*), serotonin (1 m*M*), or NaF (1–10 m*M*) but not norepinephrine (1 m*M*). The stimulation by epinephrine could be antagonized by the β-adrenergic blocker, propranolol, at a 10 μ*M* concentration. These differences may reflect the difference between strains of ciliate or differing experimental conditions. The earlier studies utilized a different adenyl cyclase assay and activity was dependent on addition of either caffeine or aminophylline to inhibit a potent PDE present in their preparations.

Early work by Janakidevi *et al.* (1966a,b), reporting that the W strain of *Tetrahymena* contained serotonin and catecholamines (apparently synthesizing both epinephrine and norepinephrine), led Blum (1967) to the observation that the growth of this ciliate (strain HSM) was inhibited by a variety of adrenergic and/or serotonergic drugs (i.e., α- and β-adrenergic blocking agents and other drugs known to interact with mammalian catecholamine systems). Janakidevi *et al.* (1966b) speculated that the role of these catecholamines was to regulate the large stores of glycogen found in *Tetrahymena*, because adenyl cyclase was known to affect production of 3′, 5′-AMP, a cofactor in the conversion of mammalian phosphorylase *b* to phosphorylase *a*, the enzyme responsible for glycogen breakdown. Blum (1970) found that *Tetrahymena* has a 3′, 5′-cyclic AMP PDE sensitive to methyl xanthine inhibition (K_i caffeine = 5.3 m*M*, K_i theophylline = 1.7 m*M*), and speculated that the increase seen in glycogen synthetase activity (and concomitant increase in glycogen content) and decrease in phosphorylase activity by theophylline (at a concentration of ~0.2–0.6 m*M*) resulted from this phosphodiesterase inhibition. This agrees with findings by Voichick *et al.* (1973), who reported a reciprocal relationship between cAMP and glycogen synthesis during growth of strain E of *Tetrahymena*. They found a rise in cAMP when glycogen is not synthesized, and conversely decreased cAMP during active gluconeogenesis. They recognized that this corresponds with what takes place in mammals where cAMP activates protein kinase, which in turn phophorylates or activates phosphorylase used in glycogenolysis. High levels of cAMP would also inhibit glycogen synthetase, which in turn inhibits net glycogen formation. Later work by Muto *et al.* (1984) confirmed that glycogen phosphorylase of NT-1 *Tetrahymena* is activated by cAMP in the presence of Mg^{2+}, ATP, and a 120,000 *g* supernatant. The latter cytosolic fraction may contain a cAMP-dependent protein kinase (Majumder *et al.,* 1975) and cAMP-binding protein (Kudo and Nozawa, 1983). This binding protein is distinct from CaM. A redistribution of this protein may take place during cell growth, such that cAMP-binding activity increases in the ciliary, mitochondrial, and microsomal fractions during transition from exponential to stationary phase of growth. In this way this protein may participate in modulating cAMP in response to cell functions. Rozensweig and Kindler (1972) hypothesized that epinephrine and serotonin are activators of adenyl cyclase and cAMP synthesis in *Tetrahymena*. The growth inhibition by the α- and β-adrenergic blocking agents (Blum, 1967) as well as abolition of epinephrine stimulation of *Tetrahymena* adenyl cyclase may reflect the role of a primative β-adrenergic-mediated control of metabolism in this ciliate.

In support of this view, Nandini-Kishore and Thompson (1979) reported that the addition of glucose (0.5–1.0%) to *T. pyriformis* NT-1 in the stationary phase of growth was followed by a "small but consistent increase in Ca^{2+} uptake." This coincided with a marked increase in cell motility and increased cellular cAMP, which reached 8-fold higher than initial values. This increase of cAMP was decreased to control values by 100 μM dichloroisoproterenol, a potent epinephrine antagonist. EGTA-treated cells did not increase cAMP following glucose stimulation but they did increase cAMP within 30 min of exposure to 100 μM epinephrine, which was prevented by dichloroisoproterenol. These authors propose that the glucose-enhanced Ca^{2+} influx results from a transient change in membrane conductance caused by glucose. They pose epinephrine as the agent coupling the glucose-enhanced depolarization resulting in Ca^{2+} influx with the increase in cAMP, especially because *Tetrahymena* was reported to synthesize epinephrine, and both norepinephrine and serotonin have been reported in this ciliate (Janakidevi *et al.,* 1966a,b; Blum, 1967, 1970). These workers and Wolfe (1973) attribute cAMP as playing a role in carbohydrate metabolism of this protozoan. While the levels of epinephrine and dichloroisoproterenol used in these experiments (100 μM) may be argued to be high, these investigators note that there is a remarkable dependence of cAMP on cell density. This results in variation from experiment to experiment. In addition, changing cGMP may also participate in this sequence of reactions.

The primary action of the catecholamines, epinephrine and norepinephrine, is on the adenyl cyclase system, which affects the production of 3′,

5′-AMP, a cofactor in the conversion of phosphorylase *b* to phosphorylase *a*, the enzyme responsible for the breakdown of glycogen (Janakidevi *et al.*, 1966b). *Tetrahymena* has large stores of glycogen. It is possible, therefore, that the role of the catecholamines in this organism is to regulate the amounts of polysaccharide reserve, in a manner similar to mammalian systems. Work by Blum (1967) suggests that *T. pyriformis* (strain HSM) possesses a fully developed catecholamine metabolic control system. In *T. pyriformis* stain GL-C, cAMP is increased by caffeine. Dibutyryl cAMP potentiates the effects of caffeine. The metabolic effects of caffeine and theophylline are generally believed to result from inhibition of 3′, 5′-cAMP PDE. Theophylline increased glycogen synthetase activity and decreased phosphorylase of *Tetrahymena* grown on glucose. *Tetrahymena* possesses a theophylline-sensitive cAMP PDE. It seems likely that effects on glycogen synthetase and phosphorylase are mediated through cAMP (Blum, 1967, 1970). Both dibutyryl cyclic AMP and caffeine cause a partial inhibition of cell division and ciliary regeneration in a concentration-dependent manner. Therefore cAMP may regulate growth and motility in *Tetrahymena* and low concentrations of α-adrenergic blockers appear to promote growth and β blockers inhibit it (Kariya *et al.*, 1974).

Schultz *et al.* (1983) reported on the presence of an active guanylate cyclase in the highly purified ciliary membranes of *Tetrahymena*, which was stimulated by other protozoan CaMs. Greater than 90% of activity was reported to be associated with the ciliary membrane fraction, with little in that of the cell body (Schultz *et al.*, 1983; Schultz and Klumpp, 1984; Kudo *et al.*, 1985a). As in the case of pellicular guanylate cyclase, essentially all activity was associated with the particulate fraction; no activity was found in the soluble fraction. Activity of the ciliate cyclase was maximal with Mg^{2+} or Mn^{2+}. The addition of *Tetrahymena* or *Paramecium* CaM increased the cyclase activity in whole cilia about 1.3 times. Pig brain or soybean CaM was incapable of enhancing activity, however, beyond that level attained in the absence of added CaM. The same was also seen with *Dictyostelium* (slime mold) CaM, which lacks trimethyllysine, and the Ca^{2+}-binding proteins S100, troponin C, or parvalbumin. The authors speculate that *Tetrahymena* guanylate cyclase may have a portion of CaM associated with it as if it were a subunit. Examples of this are guanylate cyclase of *Paramecium*, PDE of rabbit lung, or phosphorylase kinase (Schultz *et al.*, 1983). These workers speculate that there may be several reasons why protozoan CaM differs from mammalian CaM in activating *Tetrahymena* guanylate cyclase. The first reason is that *Tetrahymena* guanylate cyclase may contain two binding sites, one being activated by unique protozoan CaMs, the other by various other CaMs (Klee *et al.*, 1980). Alternatively, the other possibility may be that the CaM-binding site is subtly altered or modified during enzyme cyclase preparation. Optimum activation may depend on the critical features of *Tetrahymena* CaM, and other activators may not possess these for optimum binding. They further postulated that CaM may be a critical subunit at the active site of the enzyme as observed in other systems. In any event, they suggest that this CaM-dependent guanylate cyclase may play an important role in ciliary function (Schultz and Klumpp, 1984).

Muto *et al.* (1985) determined the activities of guanylate cyclase and adenylate cyclase in cilia and cell bodies of *Tetrahymena* during their growth from log phase to stationary phase. The guanylate cyclase activity in the early log phase (18 hr) was relatively low in both cilia and cell bodies. By the mid-log phase (24 hr) there was an increase of over fourfold and at the stationary phase the cell bodies showed an elevation of activity of about sevenfold. The corresponding activity in cilia was almost unchanged. Adenylate cyclase activity in cell bodies was 15-fold higher than that in cilia and did not change with the stage of growth. This increase of guanylate cyclase activity was attributed to an increased level of enzyme but not CaM. The authors speculate this CaM-dependent guanylate cyclase may modulate the cGMP level, which may play a substantial role in the growth of this organism. This agrees with the earlier observations by Quinones and Renaud (1975), who found cGMP stimulated the rate of cilia regeneration in *Tetrahymena* where cAMP, dibutyryl cAMP, caffeine, or theophylline inhibited cilia regeneration.

The enzyme catalyzing the degradation of the 3′, 5′-cyclic nucleotides is cyclic nucleotide PDE. It hydrolytically cleaves the 3′-phosphodiester bond to 5′-nucleotide. This enzyme is found in mammalian, insect, bacterial, and plant sources (Appleman *et al.,* 1973). Cyclic nucleotide PDE was isolated from *T. pyriformis* strain GL by Ramanathan and Chou (1973a,b). They reported that it resembled the mammalian enzyme in possessing multiple molecular forms and in its response to the methylxanthine inhibitors and being markedly stimulated by imidazole. These are major differences from this enzyme as isolated from other microorganisms.

Further work by Kudo *et al.* (1980) on this enzyme from the NT-1 strain in the early stationary growth phase showed optimum activity in both the soluble and 105,000 *g* particulate fractions at pH 8.0–9.0. It was enhanced by addition of Mg^{2+} or Mn^{2+} but, in contrast to the cAMP form, most of the cGMP PDE was located in the soluble fraction. PDEs may be either soluble or particulate. Mg^{2+} is the preferred cation, especially for guanosine PDE, but high Mn^{2+} concentrations will do equally well (Appleman *et al.,* 1973).

A Ca^{2+}-dependent activating factor was extracted by Kudo *et al.* (1980) from *Tetrahymena* cells that increased PDE activities, but this factor was not the same activator that stimulated guanylate cyclase. Conversely, the Ca^{2+}-binding protein from *Tetrahymena* that stimulated guanylate cyclase had no effect on the PDE activities of this protozoan, suggesting these Ca^{2+}-dependent modulator proteins are different.

On the other hand the enzymatic hydrolysis of cGMP in mammals is stimulated by Ca^{2+} and an activating factor or modulator protein. However, this factor, CaM, also stimulates adenylate cyclase and modifies cGMP hydrolysis to a greater or lesser extent than cAMP, depending on the tissue and conditions within the cell (Watterson and Vanaman, 1976; Appleman *et al.*, 1973; Klee *et al.*, 1980). In many cells CaM appears to act more specifically on cGMP degradation. Cyclic GMP metabolism depends in large part on the availability of Ca^{2+}, which controls the presence of the Ca^{2+}–CaM complex. This is turn regulates the PDE, which controls cGMP.

Protein kinase was extracted by Murofushi (1973) from the axoneme fraction of *T. pyriformis* (strain W) cilia separated from the cell bodies. Activity was independent of cAMP, inhibited by Ca^{2+}, and capable of phosphorylating tubulin. They later modified the extraction procedure to utilize Triton X-100 and purified the enzyme on a DEAE-cellulose column (Murofushi, 1974). Four peaks of protein kinase were eluted, three of which were cGMP dependent and one that was selectively enhanced by cAMP. Two of the cGMP dependent protein kinases preferred casein and histone. The remaining cGMP dependent protein kinase and the cAMP-dependent protein kinase showed a higher reaction rate when protamine or histone was used as a substrate. The molecular weights of the cGMP-dependent kinases were about 100,000 while that of the cAMP-dependent kinases was about 60,000. The location was thought to be in the soluble fraction or the membrane of the cilia, and its function was thought to be regulation of ciliary motile processes in the protozoa. Cyclic AMP has been recognized as playing an important role in axonemal motility. Tash and Means (1982) later reported on a cAMP-dependent phosphoprotein in sperm axokinin, which when phosphorylated was used in movement.

Subsequently a different cAMP-dependent protein kinase was isolated from the cytosol *T. pyriformis* strain E (Majumder *et al.*, 1975). These investigators suggest the enzyme (RC) consists of a regulatory (R) and a catalytic (C) subunit. An increase in cellular cAMP causes an increase in the rate of formation of the functionally active catalytic subunit of protein kinase C by specific interaction of cAMP with cyclic AMP-dependent inactive protein kinase, which is the metabolically stable storage form of the enzyme in the cell. Catalytically active protein kinase acts in turn on protein substrates, phosphorylating them to multiple organelle-specific phosphoproteins thought to be important in the different phases of cell growth. This would account for the changing specific activity of the protein kinase as the growth of *Tetrahymena* progresses. This cAMP-dependent protein kinase of Majumder *et al.* (1975) may correspond to the cAMP-dependent protein kinase later reported in strain NT-1 (Kudo and Nozawa, 1983).

2. Dolichol Kinase

Initial investigations by Gandhi and Keenan (1983a) reported on the existence of a Ca^{2+}-requiring CTP-dependent kinase in particulate preparations of *T. pyriformis*. This and a report by Kakiuchi *et al.* (1981) on the presence of a Ca^{2+}-CaM-regulated guanylate cyclase in the same organism prompted Gandhi and Keenan to investigate the possibility that the Ca^{2+}-binding protein, CaM, might also mediate activity in their kinase system. This enzyme was dolichol kinase, which catalyzes the phosphorylation of dolichol to dolichol monophospate, and is thought to regulate glycoprotein synthesis (Hemming, 1974). Keenan's group had been interested in glycoprotein synthesis for some time in *Tetrahymena* (for references see Adrian and Keenan, 1979). They recognized that glycoproteins serve as receptors on cell membranes for bacteria, hormones, or neighboring cells and were important as enzymes, antigens, and immunoglobulins (Waechter and Lennarz, 1976; Sairam, 1989). Dolichol kinase from *Tetrahymena* and from rat brain is regulated by CaM. When these enzyme preparations were washed with EGTA by Gandhi and Keenan (1983a) to deplete them of Ca^{2+} and CaM, both *Tetrahymena* and rat brain dolichol kinase required Ca^{2+} and CaM to restore their activity to original level. These authors conceded that EGTA washing probably did not remove all of the CaM, however, because the enzyme preparation retained about 25% of the control activity after addition of exogenous Ca^{2+}. This may reflect a strong bond between CaM and the rat brain or *Tetrahymena* dolichol kinase. [A similar phenomenon had been observed in original studies by Nagao *et al.* (1979) with a particulate-bound guanylate cyclase of *Tetrahymena*.] Washing with EGTA had little if any effect on rat liver dolichol kinase in Gandhi and Keenan's (1983a) studies.

Rat brain and *Tetrahymena* enzyme preparations require the same amount of Ca^{2+} and CaM for full activation. In addition, both rat brain and *Tetrahymena* dolichol kinase are strongly inhibited by the CaM inhibitors, trifluoperazine and chlorpromazine. The rat liver enzyme is unaffected by such treatment. Gandhi and Keenan (1983a) presented a final piece of evidence to suggest that the dolichol kinase enzyme from rat brain is similar to that from *Tetrahymena*. This was the total inactivation of both rat brain and *Tetrahymena* enzymes, which resulted from CaM antiserum with no effect on the dolichol kinase of rat liver. Unfortunately the source of CaM from which the antiserum was prepared was not identified by the authors but it may have been beef brain.

Tetrahymena dolichol phosphokinase is very specific in its requirement for CTP (cytidine triphosphate) as the phosphoryl donor. This enzyme has two pH optima, one at about pH 7.4 and the other at about pH 9.0 (Gandhi and Keenan, 1983b). The activity at both pH values is Ca^{2+}–CaM dependent. The enzyme is also stimulated to a lesser degree by Mg^{2+} in a calmodulin-independent mechanism. The preferred substrate in *Tetrahymena* is dolichol-17 or dolichol-18 and the product is identified as dolichyl monophosphate by thin-layer chromatography in three different solvent systems and by high performance liquid chromatography. Dolichyl phosphate phosphatase activity was also present in this ciliate. This enzyme was inhibited up to 58% in the presence of Ca^{2+} concentrations routinely used to assay the kinase. It is also inhibited by 10 m*M* NaF as well as 20 m*M* UTP (uridine triphosphate) (Gandhi and Keenan, 1983b).

III. *Euglena* Calmodulin

CaM has been reported to be present in the algal flagellate *E gracilis* (Kuznicki *et al.*, 1979). Lonergan (1984) observed the Klebs strain Z of this organism to exhibit biological clock-controlled shape changes and further observed that (1) alterations in free cytoplasmic Ca^{2+} concentration, (2) addition of the Ca^{2+} ionophore A23187 (which would be expected to increase intracellular Ca^{2+}), or (3) addition of CaM antagonists (i.e., trifluoperazine or chlorpromazine), induces changes in cell shape. This was interpreted to mean that Ca^{2+} regulates the contractile process of the shape change. In further work, Lonergran (1985) focused on immunofluorescence patterns of microfilaments, myosin, and CaM during the biological clock-controlled shape changes. These led him to suggest that *Euglena* may possess a CaM-controlled actomyosin contractile system.

There have been many attempts to explain the periodic cycle of physiological events that ultimately leads to mitosis in a variety of cells. One such possible regulator of cell cycle variation is cAMP and this has been discussed by us in Section II, C. There are difficulties with all of the hypotheses put forth. One mechanism has been proposed by Goto, Laval-Martin, and Edmunds, in which mitochondrial Ca^{2+} and CaM play a crucial part (Goto *et al.*, 1985; Edmunds *et al.*, 1987). They measured NAD^+, NADH, $NADP^+$, and NADPH levels in synchronous slowly dividing cultures of *E. gracilis*, Klebs (Z) strain. There was a circadian rhythm with NAD^+ oscillating 180° out of phase with NADP(H). There was also a circadian rhythm in NAD^+ kinase activity such that it could control the level of NAD^+. Presumably this oscillation could result in NAD^+ being converted to $NADP^+$ or NADH to NADPH through the participation of NAD^+ kinase or $NADP^+$ phosphatase. Oxidation–reduction reactions between NAD^+

and NADH do not participate in this mechanism. This mechanism would constitute a closed loop self-sustaining circadian oscillator and according to Goto *et al.* (1985) and Edmunds *et al.* (1987) encompasses three steps.

Step 1: NAD^+ would augment the net Ca^{2+} efflux rate from the mitochondria. This would result in maximum concentrations of cytosolic Ca^{2+} in 6 hr. Another possibility, in cells with a photoreceptor stimulated by light, is that there would be enhancement of net Ca^{2+} efflux from the mitochondria or net Ca^{2+} influx across the plasmalemma into the cytoplasm.

Step 2: Ca^{2+} would activate CaM, resulting in the presence of active Ca^{2+}/CaM complex in cell cytoplasm.

Step 3: The active Ca^{2+}/CaM complex would in turn activate NAD^+ kinase (rate becoming maximal in 12 hr) and inhibit $NADP^+$ phosphatase in the cytoplasm. After 12 hr Ca^{2+}/CaM would be minimal. Following six more hours there would be a decrease of net NAD^+ production; NAD^+ level would be at a minimum. At this time the regulatory process just described could repeat itself.

This, however, does not exclude a number of other mechanisms for cellular regulation of eukaryotic cells (Kippert, 1987). An example of such a system is present in mammals and also appears to be functional with subtle differences in *T. pyriformis* (Kim *et al.,* 1984) and the acellular slime mold, *P. polycephalum* (Holmes and Stewart, 1979). This is the mitochondrial Ca^{2+} cycle. In this mechanism, influx and efflux of Ca^{2+} across mitochondrial membranes does not take place at exactly the same rate. This results in a net flux in one direction during various time periods. The Ca^{2+} concentration of the cytoplasm will be inversely proportional to that in the matrix. It is believed that cellular processes are timed by cation concentration resulting from these transmembrane fluxes (Kippert, 1987).

IV. *Amoeba* Calmodulin

A. *Physarum polycephalum*

Actomyosin is responsible for muscle contraction in higher animals. It also appears to explain the cytoplasmic streaming and ameboid movement of various kinds of cells in lower organisms (Kohama and Ebashi, 1986; Ebashi, 1988). The slime mold *P. polycephalum* is unique, exhibiting both rapid cytoplasmic streaming in the multinucleate diploid plasmodium stage like some plants, and slow ameboid movement in the uninuclear haploid amebal stage. Some consider that this organism bridges the gap between the plant kingdom and animal kingdom (Ebashi, 1988). Ca^{2+} exhibits a regulatory role only in rare instances of prokaryotic physiology (Swan *et al.*, 1987; Iwasa *et al.*, 1981) and, unlike eukaryotic cells, there is no convincing evidence of CaM in bacteria (Kippert, 1987). On the other hand, *P. polycephalum* (plasmodial stage), *Amoeba proteus* (Kuznicki *et al.*, 1979) *Dictyostelium*, and *Acanthamoeba* (Wylie and Vanaman, 1987) have CaMs, all of which differ from each other and from mammalian CaM significantly. In addition, 10 μM concentrations of the phenothiazines were amebastatic, 50 μM was either amebastatic or amebicidal, and 100 μM was amebicidal to several species of *Acanthamoeba* and *Naegleria fowleri* (Schuster and Mandel, 1984). Kohama and Takano-Ohmuro (1984) proposed that the difference in motility between the stages of the same organism could be attributed to a difference in contractile proteins. Coelectrophoresis of amebal and plasmodial myosin B by these workers revealed that the heavy chain of the plasmodial stage migrated slower in the electric field. Digestion of the amebal myosin heavy chain showed a few peptides were missing when compared with the plasmodial heavy chain. Therefore it would appear that the heavy chains of these two stages were different in amino acid sequence and that this was a stage-specific protein. Actin–myosin–ATP interaction is presumed to account for cytoplasmic streaming in plasmodia of *P. polycephalum*. Ca^{2+} (at micromolar levels) inhibits cytoplasmic streaming, causing contraction apparently through repressing the actin–myosin–ATP interaction in this organism (Kohama and Kohama, 1984; Kohama *et al.*, 1985; Kohama and Shimmen, 1985; Kohama and Kendrick-Jones, 1986). Ca^{2+} and actin are known to reversibly inhibit the phosphorylation of an actin-binding protein. It consists of two distinct subunits of 42,000 Da and apparently caps the fast-growing end of actin filaments, blocking actin polymerization at this end and causing fragmentation of actin filaments, which in turn reduces low shear viscosity (Maruta *et al.*, 1983; Maruta and Isenberg, 1983). The Ca^{2+} inhibition of actin–myosin–ATP interaction is in sharp contrast with the activating effect of Ca^{2+} on this interaction in muscle. The extensive work by Kohama *et al.* (1980); Kohama, 1981) has provided the biochemical explanation for these observations. They isolated a light chain of M_r 14,000 that confers the inhibitory sensitivity of actin–myosin–ATP interaction by binding to actin under high Mg^{2+} conditions. This they called a "Ca^{2+} dependent inhibitory factor," and it appears to be different from CaM.

Myosin isolated from *Physarum* ameba is designated α-myosin and that from the plasmodial stage is p-myosin. The interaction of these myosins with actin is inhibited by Ca^{2+}, essentially to the same degree. They each consist of two 250K heavy chains as well as 18K and 14K light chains on SDS-PAGE, and have similar shape. They are two headed and long tailed in structure. The peptide maps of the heavy chain and the 18K light chain are quite different when comparing those of α-myosin to those of p-myosin. The 18K light chains appear to be phosphorylatable (as was seen in *Dictyostelium*) (Flicker *et al.*, 1985; Kuczmarski and Spudich, 1980). Comparing α-myosin with p-myosin by reacting them with a specific antibody revealed the heavy chains to be distinctly different. The same is true for the 18K light chains by two-dimensional gel electrophoresis. The 14K light chain of the amoebal stage cannot be differentiated from that of the plasmodial stage by either two-dimensional gel electrophoresis, by its Ca^{2+}-binding activity, peptide map, or reactivity to antibodies (Kohama *et al.*, 1986; Kohama and Ebashi, 1986; Ebashi, 1988).

The commonality of the 14K light chain in the two stages of *Physarum* agrees with its crucial role in Ca^{2+} regulation. It explains how Ca^{2+} inhibits the activity of actin-activated ATPase at micromolar levels. In 1988 this light chain was completely sequenced by Kobayashi *et al.* (reviewed by Korn and Hammer, 1988) and it was found to be very similar to brain CaM. Only two of the Ca^{2+}-binding sites (I and III) are retained, however.

Electron microscopy reveals the α-myosin is made up of a 185-nm fibrous tail with two globular heads and a diameter of approximately 13 nm. This α-myosin appears to contain three major polypeptide bands with a heavy chain (M_r 220,000) and two light chains (M_r 18,000 and 14,000) in a molar ratio of 1:1:1 (Kohama *et al.*, 1986).

The p-myosin heavy chain could be phosphorylated such that 1 mol of heavy chain contained 1 mol of phosphate. This phosphorylation is independent of the presence of Ca^{2+} or (porcine brain) CaM, unlike smooth muscle myosin light chain, where Ca^{2+}-dependent myosin light chain phosphorylation results from a light chain kinase–CaM system (Ogihara *et al.*, 1983; Kohama and Kendrick-Jones, 1986). One might wonder about Ca^{2+}/CaM dependence of phosphorylation if *Physarum* CaM had been available to these workers. The presence of F-actin stimulated the Mg^{2+}-ATPase activity of phosphorylated myosin seven or eight times. In the absence of Ca^{2+} this was about 30-fold. There was no corresponding activation of the dephosphorylated form (Ogihara *et al.*, 1983; Kohama and Kendrick-Jones, 1986). Therefore, phosphorylation (and the absence of Ca^{2+}) are indispensable to the activation of Mg^{2+}-ATPase by F-actin. This is similar to what is seen with *Acanthamoeba* "myosin I," where the phosphorylation of the heavy chain also activates actomyosin ATPase activity. This contrasts with the actin activation and formation of thick filaments seen in *Acanthamoeba* "myosin II" and *Dictyostelium* "conventional myosin" on dephosphorylation. The myosin molecule from the plasmodium of *P. polycephalum,* on the other hand, resembles *Acanthamoeba* myosin II and conventional myosin of *Dictyostelium* in several ways. These are its double-headed and long-tailed shape (Takahashi *et al.*, 1983) and the Ca^{2+}-ATPase activity is higher than its K^+–EDTA–ATPase activity (Kohama *et al.*, 1986).

B. *Acanthamoeba castellanii*

Pioneering work by Pollard and Korn (1973a,b) reported a myosin-like protein in *Acanthamoeba castellanii* (Neff strain). This enzyme has properties in common with skeletal muscle myosin. In muscle, the mechanism of contraction appears to involve the cyclic interaction of actin (which activates myosin ATPase), myosin (a group of related ATPases believed to generate energy for movement or contraction), and ATP. They therefore suggested that myosin in amebae might act with other proteins to generate the force for movement similar to the mechanism utilized by higher organisms for muscle contraction. This protein, however, differed significantly from muscle myosin in that it was a single-headed enzyme of about 180,000 kDa, incapable of forming filaments, but resembling the S1 subunit of muscle myosin, composed of a single heavy chain of an approximate molecular weight of 140,000, and two different light chains with molecular weights of 16,000 and 14,000. This was considerably less than the myosin of muscle (460,000) and that of the slime mold (*Dictyostelium* conventional myosin) or platelet (500,000) but there was no evidence of proteolytic clevage during the purification of the ameba protein. In contrast to the myosins of higher animals, the *Acanthamoeba* enzyme is reported to be soluble at physiological (low) ionic strength and like *Physarum* myosin it lacks cysteine (Pollard and Korn, 1973a; Warrick and Spudich, 1987). There is a 110-kDa protein in the brush border of chicken intestine that has many characteristics of this "myosin I." It binds to F-actin with three- to fourfold increases in Mg^{2+}-ATPase activity but, in contrast to myosin I of protozoa, it is activated by Ca^{2+} in the presence and absence of actin. The light chains in brush border myosin are CaM-like. There is structural similarity between the 110-kDa protein and myosin S1, including the ATP/actin-binding site (Carboni *et al.*, 1988; Warrick and Spudich, 1987; Korn and Hammer, 1988). This suggests a vertebrate counterpart of myosin I. Under physiological conditions, the activation (but not the binding) of Mg^{2+}-ATPase by F-actin depends on an M_r 97,000 activator protein isolated from *Acanthamoeba*. The early studies by Pollard and Korn (1973a,b) employed rabbit muscle actin because *Acanthamoeba* actin was difficult to obtain in adequate quantities. There are some differences between actins of different species (Korn, 1982). This amebal activator protein is not analogous to the Ca^{2+}-sensitive control protein seen in

Physarum and other systems (Pollard and Korn, 1973b). Subsequently, Maruta and Korn (1977a) reported that this "cofactor" protein is a kinase that phosphorylates the single heavy chain but not the two light chains. In contrast, vertebrate smooth muscle and nonmuscle myosins are regulated by a phosphorylation of their M_r 20,000 light chains (P-light chains) by specific light chain kinases. Further purification to near homogeneity revealed the globular myosin I kinase to be a single polypeptide of M_r 107,000. This enzyme catalyzes the incorporation of 0.9 to 1.0 mol of phosphate into a single site on the heavy chain of both myosin IA or IB. This site is a serine residue, at least in the case of myosin IB. It requires Mg^{2+} but is not dependent on Ca^{2+}, Ca^{2+}/CaM (in contrast with muscle and nonmuscle light chain kinase), or cAMP for activity (Hammer *et al.*, 1983). It is specific for the heavy chain of myosin I from *Acanthamoeba* or *Dictyostelium* but not for the *Acanthamoeba* myosin II heavy chain (Korn and Hammer, 1988). It should be kept in mind that while myosin I heavy chain kinase was studied for possible dependence on CaM, the source of CaM was bovine brain rather than *Acanthamoeba* CaM (Hammer *et al.*, 1983). Because *Acanthamoeba* CaM has rather striking differences to mammalian CaM there may be a strong species specificity to *Acanthamoeba* CaM (Wylie and Vanaman, 1987), as is seen in the case of *T. pyriformis* guanylate cyclase (Kakiuchi *et al.*, 1981) or plant NAD kinase (Roberts *et al.*, 1984) as described in this article. Failure to demonstrate a CaM requirement may rest on using heterologous CaM as a result of the unavailability of autologous CaM. The actin-activated Mg^{2+}-ATPase activity is lost when the heavy chain is dephosphorylated by a phosphatase. The phosphorylation or dephosphorylation had no effect on either the K^+-EDTA-ATPase or Ca^{2+}-ATPase activities of this myosin from *Acanthamoeba*.

Subsequently a second myosin-like enzyme was isolated and purified from *Acanthamoeba* (Maruta and Korn 1977b,c; Pollard *et al.*, 1978). It was named myosin II to distinguish it from the myosin ATPase that they previously described and which they termed myosin I. *Acanthamoeba* myosin II, with a molecular weight of about 350,000 to 400,000, contains two heavy chains with molecular weights of approximately 170,000 and two pairs of light chains with molecular weights of about 17,500 and 17,000, respectively. It has two globular heads and a tail about 90 nm long. The two myosin proteins (I and II) differ in their enzymatic properties. Myosin II is more similar to other muscle and nonmuscle myosins. [A "conventional myosin" similar to myosin II in *Acanthamoeba* had been previously reported in the cellular slime mold *D. discoideum* ameba (Clarke and Spudich, 1974; Kuczmarski and Spudich, 1980; Peltz *et al.*, 1981; Korn and Hammer, 1988). It has a heavy chain of 243 kDa and a tail of

180–186 nm.] This is based on composition, size, distinct head and tail regions, and propensity to form filaments on the addition of Mg^{2+} and under conditions of low ionic strength (Hammer *et al.*, 1983).

The K^+-EDTA-ATPase activity is greater than the Ca^{2+}-ATPase activity in *Acanthamoeba* myosin I, where the reverse is true for the myosin II form. Both myosin I and myosin II have low levels of Mg^{2+}-ATPase that are slightly activated by F-actin. Maruta and Korn (1977a) found only the Mg^{2+}-ATPase of myosin I was activated to a significant degree by the heavy chain kinase. Later work by Collins and Korn (1980), using different purification procedures for myosin II, showed a 25-fold activation by F-actin.

Work by Maruta *et al.* (1978, 1979) demonstrated that there were three forms of myosin I, all with molecular weights of about 180,000. The single-headed M_r 159,000 *Acanthamoeba* myosin IA has a heavy chain of M_r 130,000 and light chains of M_r 17,000 and 14,000. The single-headed M_r 150,000 IB form of myosin has a heavy chain of M_r 125,000 and light chains of M_r 27,000 and 14,000. Both IA and IB myosins are globular proteins with little if any tail segment (only 35 kDa and totally distinct from the tail of muscle myosin) and remain monomolecular under conditions in which other myosins self-associate into bipolar filaments (Albanesi *et al.*, 1985a; Warrick and Spudich, 1987). Both myosin I isozymes contain variable amounts of the M_r 14,000 peptide (Hammer *et al.*, 1983). The heavy chains alone of *Acanthamoeba* myosin IA and IB contain the ATPase catalytic site, the actin-binding site, and the phosphorylation site. Each is fully active enzymatically in the absence of light chains (Maruta *et al.*, 1978; Maruta and Korn, 1981a,b). [A low-molecular-weight myosin has been purified from the slime mold *D. discoideum* that is similar to those single-headed myosins IA and IB of *Acanthomoeba* (Cote *et al.*, 1985).] This differs from the double-headed *Acanthamoeba* myosin II, where both catalytic and phosphorylation sites are also on the heavy chains but there are three phosphorylation sites per heavy chain. Localization of these sites by chymotryptic digestion reveals the structure of *Acanthamoeba* myosin II to be similar to *Dictyostelium* amoeba conventional myosin (Peltz *et al.*, 1981). The catalytic site where actin binds when activating Mg^{2+}-ATPase is at the amino-terminal globular head region and the phosphorylation sites on each heavy chain of myosin II are within a 3000-Da peptide near the C terminus (tail) of the molecule (Cote *et al.*, 1981; Collins *et al.*, 1982a). It is the serine residue that is phosphorylated to phosphoserine. More than one site per heavy chain must be phosphorylated before the actin-activated Mg^{2+}-ATPase of myosin II is inhibited. Collins *et al.* (1982a) commented that it is surprising that phosphorylation sites so near the carboxy-terminal tail of the *Acanthamoeba* myosin heavy chain can completely inhibit the enzymatic activity of the ATPase site in the heads. The phosphorylation site is separated from the catalytic site a linear distance of the equivalent of 100,000 Da and the tail is presumed to be a rigid α-helical rod made up of the two coiled heavy chains. Enzymatic activity may depend on formation of myosin dimers with the phosphorylation site at the tail of one molecule lying near the head of an oppositely oriented myosin monomer, thereby regulating the actin-activated ATPase of the second molecule. For further details on the structure of myosin IA and IB refer to the report by Korn and Hammer (1988). Correlating substructure to function is complex. The amino acid sequence, while more conserved in the head than the tail, between species, suggests only a one-dimensional array. Function may depend on the three-dimensional structure (Warrick and Spudich, 1987).

Another feature where myosin II contrasts with myosin I is that dephosphorylation instead of phosphorylation activates the Mg^{2+}-ATPase of *Acanthamoeba* myosin II and the conventional myosin of *Dictyostelium* ameba (Maruta and Korn, 1981a,b; Collins *et al.*, 1982b; Collins and Korn, 1980, 1981). Both the phosphorylated and dephosphorylated myosin form bipolar filaments in $MgCl_2$/KCl but the filaments of the dephosphorylated form were larger and less readily dissociated in physiological levels of Mg^{2+}-ATP. This suggests dephosphorylation of heavy chains increases self-association, binding to F-actin, and actin-activated ATPase activity (Collins *et al.*, 1982b). As these workers point out, different enzymatic activities may be the result of subtle effects on the state of phosphorylation and interaction of myosin molecules in the filaments. *Acanthamoeba* myosin II Mg^{2+}-ATPase is activated to a greater extent by *Acanthamoeba* actin than by muscle actin. This actomyosin ATPase of *Acanthamoeba* is similar to the conventional myosin of *D. discoideum* (Clarke and Spudich, 1974) and is independently regulated by Ca^{2+}. It is inhibited 80–90% by 1 mM concentrations of the Ca^{2+} chelator, EGTA. It is reactivated in free Ca^{2+} of less than 1 μM with appropriate Mg^{2+} concentrations. This Ca^{2+} regulation of the myosin itself may be similar to the regulation of molluskan myosins. The kinase that phosphorylates all three sites of the myosin II heavy chain has been isolated and partially purified (Cote *et al.*, 1981). This enzyme is specific for only the myosin II heavy chain, not the light chains of myosin II or myosin IA or IB. Activity is dependent on the presence of Mg^{2+}, not Mn^{2+}, Ca^{2+}, CaM, or cAMP. Work has been published to suggest that cAMP stimulates phosphorylation of a *Dictyostelium* polypeptide that comigrates with the heavy chain myosin of the organism (Rahmsdorf *et al.*, 1978). Later work shows a fourfold increase in phosphorylation of the 18-kDa myosin light chain *in vivo* by cAMP (Warrick and Spudich, 1987). The protein phosphatase that dephosphorylates the *Acanthamoeba* myosin II has also been highly purified (see Hammer *et al.*, 1983).

The myosin referred to by some as IC appears to be the same as

Acanthamoeba myosin IA plus a 20,000-Da regulatory component. In support of this, the 130,000-Da heavy chains of *Acanthamoeba* myosin IA and IC are identical. Heavy chains of *Acanthamoeba* myosins IA, IB, and II are different and appear to be products of different genes (Gadasi *et al.*, 1979; Gadasi and Korn, 1979; Hammer *et al.*, 1984). Gadasi and Korn (1980) speculated that these different myosin species may serve different motility functions because myosin IA and IB isozymes appear to be localized near the plasma membrane while myosin II appears to be deeper in the cytoplasm. In addition, these two very different myosins being oppositely regulated may have different functions. In summary, Collins and Korn (1981) suggested that actomyosin ATPase activity is regulated in four ways.

1. Extent of phosphorylation of myosin II heavy chains
2. Mg^{2+} concentration
3. Ca^{2+} concentration
4. Ratio of myosin II to F-actin

Further information on this subject is contained in articles by Albanesi *et al.* (1983, 1985b,c), Fujisaki *et al.* (1985), and Pantaloni (1985).

C. *Naegleria gruberi*

Naegleria gruberi is a cosmopolitan, nonpathogenic soil and freshwater ameba. It is closely related to the pathogen, *Naegleria fowleri,* which causes meningoencephalitis. While classed with the amebae it is not related to *Acanthamoeba*. Both organisms, however, do share characteristics with slime molds. When *N. gruberi* are transferred from growth medium to a nutrient-free aqueous solution they transform within an hour or so into streamlined, rapidly swimming flagellates with two anterior flagella, which "swim" about 100 times faster than amebae "crawl." Much of our knowledge about the physiology of this transformation results from the efforts of Fulton and colleagues (Fulton, 1977a,b).

The development of these ameboflagellates from the amebal stage to the flagellated stage involves the production of microtubules as well as a number of other processes. All components of the flagellar apparatus are rapidly assembled during differentiation because no microtubules are found in the cytoplasm of *Naegleria*. Fulton (1977a,b) suggested that amebae with a functioning actin-based motility system and no cytoplasmic microtubules alternate with flagellates that have a microtubule cytoskeleton and no functional actin-based motility system. The actin-mediated motility system of amebae depends on critical levels of free intracellular Ca^{2+}. At the time of differentiation the principal protein of amebae, actin, stops being produced. While there is a certain commonality between actin of various forms of life, *Naegleria* actin is unusual in several ways when compared to that of the ameboid forms of *Physarum, Dictyostelium,* or *Acanthamoeba*. As the synthesis of actin declines, translatable actin mRNA rapidly disappears. This occurs within 7 min after the initiation of differentiation (Sussman *et al.*, 1984). Conversely, as the tubulin is synthesized for flagella the translatable tubulin mRNA is first detected in 20 min, increases parallel to the rate of flagellar tubulin synthesis to a maximum in 50 to 60 min, then decreases (Lai *et al.*, 1979).

Ca^{2+} is compartmentalized into Ca^{2+} reservoirs at differentiation, and free Ca^{2+} becomes limiting, thus stopping ameboid-type movement. At this time, the microtubular cytoskeleton is assembled in an environment of low free Ca^{2+} and the flagellated shape is formed. Reversal of the process to ameboid motility occurs with release of Ca^{2+} from local intracellular stores. This results in disassembly of the flagellate cytoskeleton and resumption of ameboid movement (Fulton, 1977a,b, 1983). Only the amebal stage feeds and undergoes reproduction. They have a diameter of about 15 μm and move about 50–100 μm/min. When amebae reach the stationary phase, they may form cysts that can survive extremes in temperature, desiccation, and pH. Excystment occurs when amebae are confronted with bacteria, their usual food.

In the span of minutes (it takes an average time of 57 min for visible flagella to form) basal bodies are assembled, the flagella elongate, and rootlets are assembled, presumably to anchor the flagellar apparatus. This appears to depend on the synthesis of new RNA (presumably mRNA) until about halfway (40 min) through the period from initiation of transformation to assembly of flagellum, and synthesis of specific new proteins (protein synthesis is required until three-quarters of the way through the same period). Actinomycin D or daunomycin (inhibitors of RNA synthesis) and cyclohexamide (inhibitor of protein synthesis) prevent differentiation if they are added soon after amebae are transferred to the nonnutrient media (Fulton and Walsh, 1980).

The α and β subunits (M_r 55,000) that make up the tubulins of both outer doublets as well as the central pair of microtubules are synthesized *de novo* during differentiation, beginning about 10 to 20 min after the initiation of differentiation and stopping with production of full-length flagella (Fulton, 1983). Three of four mRNAs specific for the differentiation of *Naegleria* amebal forms into flagellated forms code for flagellar proteins: two are tubulin subunits, and one of these is the flagellar CaM (Mar *et al.*, 1986; Shea and Walsh, 1987). Rootlet protein is also synthesized to the same extent as outer-doublet tubulin during differentiation.

There are two CaM-like Ca^{2+}-binding proteins not present in the amebal stage that are also synthesized during differentiation. They have molecular weights of 15,500–16,000 and 15,000–15,300 and they meet several criteria of CaM. They are smaller than CaM of mammalian tissue but are the same size as that of *Dictyostelium* and *Tetrahymena*. The larger CaM (CaM-1) is localized in flagella and the smaller (CaM-2) is localized in the cell bodies (Fulton and Lai, 1980; Fulton *et al.*, 1986). At the present time, the available evidence suggests that most other eukaryotic cells contain only a single CaM. Fulton *et al.* (1986), however, present compelling data to indicate that *Naegleria* has two CaMs.

The mRNAs of these Ca^{2+}-binding proteins of the flagellated stage are not present in the amebal stage. They appear, reaching peak abundance in about 60 min of differentiation (the period when flagella are being assembled), and then decline precipitously by 100 min. They then disappear congruent with the mRNAs of the α- and β-tubulins. The complete mechanism of regulation remains to be elucidated, however (Fulton, 1983).

V. *Trypanosoma* Calmodulin

The regulation of intracellular Ca^{2+} is also known to be important to another group of protozoan parasites of the family Trypanosomatidae. These hemoflagellates include two genera of human parasites, *Leishmania* and *Trypanosoma*. While the importance of this cation is well recognized, the mechanisms by which it acts in the trypanosomes are far from well understood. Two different Ca^{2+}-dependent ribonucleases, an endoribonuclease and an exoribonuclease, have been isolated from *Trypanosoma brucei* cytoplasm, which are possibly involved in degration of mRNA (Gbenle and Akinrimisi, 1982; Gbenle, 1985; Gbenle *et al.*, 1986). Concentrations of Ca^{2+} are known to activate adenylate cyclase in plasma membranes of bloodstream forms of *T. brucei,* resulting in the increased production of cAMP (Voorheis and Martin, 1981). This Ca^{2+}-stimulated activity is not inhibited by trifluoperazine (50 μM), however, suggesting to these workers that the Ca^{2+} receptor may not be CaM. They suggest that host antibody binding to the trypanosomal surface coat may trigger a Ca^{2+}-dependent transformation of the cells to produce a different surface antigen. Further work by Voorheis *et al.* (1982) presented evidence that only Ca^{2+} (at 0.5–1.0 mM) and the Ca^{2+} ionophore, A23187, were involved in the release of surface coat protein during antigenic variation or in the transformation from the bloodstream form into the insect gut form. Unfortunately, some concessions were made in the experimental procedures employed in this study. Conditions had to be chosen by Voorheis *et al.* (1982) to speed up an event requiring 8–9 days to take place *in vivo* into an *in vitro* period of approximately only 10–15 min due to complications involved in long-term cultivation of this organism. It has been suggested that the control of variant surface glycoprotein (VSG) release from the plasma membrane results from its phosphorylation and dephosphorylation (deAlmeida and Turner, 1983).[1]

In support of the hypothesis *T. brucei* does have two protein kinases, a suramin-sensitive form and a nucleoside-stimulated form (Walter and Opperdoes, 1982). In addition, protein kinase C activity had been identified in *T. brucei* homogenates (Keith *et al.*, 1990). The phosphorylation pattern of the organism appears to change at different stages of the life cycle (Mancini and Patton, 1983). Conversely, just as Ca^{2+} can function to the benefit of the cell, Ca^{2+} can also interact with a variety of toxic substances that act to the detriment of the cell (Schanne *et al.*, 1979). Ca^{2+} has been reported to be essential for trypanocidal activity of normal human serum (D'hondt *et al.*, 1979), probably by synergizing inhibitory effects of high-density lipoprotein (Clarkson and Amole, 1982). Ca^{2+} has also been shown to synergize the antitrypanosomal activity of salicylhydroxamic acid/glycerol (Clarkson and Amole, 1982) as well as potentiating melarsoprol (the drug of choice for treating African trypanosomiasis with central nervous system involvement).

CaM has been identified in promastigotes of *Leishmania braziliensis*, *Leishmania mexicana* (Walter and Opperdoes, 1982; Benaim *et al.*, 1987), epimastigotes of *T. cruzi* (Goncalves *et al.*, 1980; Tellez-Inon *et al.*, 1985), the African trypanosomes, *Trypanosoma brucei rhodesiense*, *Trypanosoma congolense*, and *Trypanosoma vivax* (Walter and Opperdoes, 1982; Ruben *et al.*, 1984; Ruben and Patton, 1985a,b), and the enteric flagellate *Giardia lamblia* (Munoz *et al.*, 1987).

CaM antagonists inhibit growth of *G. lamblia* trophozoites. There are minor differences, however, between the response to these antagonists in the activation of heart cyclic phosphodiesterase by mammalian versus *G. lamblia* CaM. The apparent molecular weight of the protozoan protein is 16,700 and that of brain is 16,800. There are some distinct differences in amino acid composition [e.g., the *G. lamblia* CaM contains equal amounts of threonine and serine whereas that from most other sources has an excess of threonine; the parasite CaM has more glycine than others and a low content of glutamate (Munoz *et al.*, 1987)]. As discussed earlier, it is

[1] In 1983, however, it was just becoming known that VSG contained covalently bound phosphate. Subsequently it was shown that VSG could not be labeled with [γ-^{32}P]ATP but could be metabolically labeled with $^{32}PO_4$. Eventually it was recognized that a phosphodiester bond was used to link carboxyl-terminal ethanolamine to a mannose containing glycosylphosphatidyl inositol, the so-called GPI linkage (reviewed by Boothroyd, 1985). Consequently, cycles of phosphorylation are not likely to regulate VSG release. Instead, a cation-independent phospholipase C was shown to specifically break the GPI linkage and release soluble VSG.

very important that cytosolic Ca^{2+} be regulated within well-defined limits. Because Ca^{2+} passively enters the cell, regulation frequently results from a Ca^{2+}-ATPase extrusion pump. A Ca^{2+} transport system was described by Munoz *et al.* (1988) in *Giardia,* in which Ca^{2+} influx was blocked by micromolar concentrations of the CaM antagonists, W-7 and fluphenazine, and high concentrations of antimalarial drugs chloroquine (1.0–3.0 mM) and quinacrine (0.1–1.0 mM) (which are also known to inhibit a variety of CaM-mediated systems). *Giardia lamblia* has two types of ATPases. One is a cytosolic (soluble) enzyme activated by Mg^{2+} alone and the other is a membrane-associated (particulate) Ca^{2+}-activated enzyme. Activity of the latter in the presence of Ca^{2+} alone is only 36% of the total when Mg^{2+} is also present. This is therefore designated as a Ca^{2+},Mg^{2+}-ATPase and it is inhibited by a variety of CaM antagonists, some of which are neuroleptic drugs. The authors point out that this latter enzyme is different from the ATPase (described earlier) present in *E. histolytica,* which had a Ca^{2+} requirement but in which Mg^{2+} was without effect (McLaughlin and Muller, 1981). A similar Ca^{2+}-dependent, Mg^{2+}-independent ATPase was reported by McLaughlin (1985) to be in the surface membrane fraction of *T. rhodesiense*. Addition of bovine brain CaM was not stimulatory and chlorpromazine or trifluoperazine (50 μM) was not inhibitory. A brief report described the presence of a Ca^{2+},Mg^{2+}-ATPase in *T. brucei brucei* (bloodstream forms), which was partially purified on a CaM-affinity chromatography column. Unfortunately, insufficient details were provided to allow evaluation of the measured activity (Bababunmi *et al.,* 1982). Older work by Frasch *et al.* (1978a) describes an oligomycin-insensitive Ca^{2+}-ATPase in cultures *T. cruzi* epimastigotes. In addition, there is an Mg^{2+}-activated oligomycin-sensitive membrane-bound ATPase similar to that found in mammalian mitochondria (Sastre and Stoppani, 1973; Frasch *et al.,* 1978b).

CaM from *L. braziliensis* and *L. mexicana,* the cause of cutaneous leishmaniasis, migrates on polyacrylamide gel electrophoresis in SDS in an identical manner to that from bovine brain (Benaim *et al.,* 1987). The molecular weight of 16,700 appears similar to bovine brain CaM. The UV absorption spectra of CaM from *L. mexicana* show a diminution in the major peak at 277 nm, which is also characteristic of bovine brain CaM. The authors point out that this change is more typical of invertebrate and plant CaM, which possess only one tyrosine residue instead of two, similar to that reported by Ruben *et al.* (1983) for *T. brucei.* CaM from either *L. braziliensis* or *L. mexicana* is quantitatively equal in potency to that from bovine brain in stimulating the Ca^{2+},Mg^{2+}-ATPase of human red cells. It remains to be seen if differences could have been measured between the protozoal versus mammalian CaMs if the Ca^{2+},Mg^{2+}-ATPase of *Leishmania* (autogenous enzyme) had been available. A recent report describes a plasma membrane (Ca^{2+},Mg^{2+}-ATPase from *L. braziliensis* that is stimulated 1.6-fold by bovine brain CaM (Benaim and Romero, 1990). The same caveat was raised by Benaim *et al.* (1987) when they tested the sensitivity of the activation of red cell Ca^{2+},Mg^{2+}-ATPase by leishmanial CaM to inhibition by trifluoperazine. They found it equal to that observed for bovine brain. *Leishmania donovani,* the cause of visceral leishmaniasis, was also found to be sensitive to the phenothiazine drugs (Pearson *et al.,* 1982). Both promastigotes and extracellular amastigotes are particularly sensitive to chlorpromazine and trifluoperazine. Of particular interest, however, is the observation that chlorpromazine also kills amastigotes inside human monocyte-derived macrophages. The authors argue that serum concentrations of phenothiazine adequate to kill *Leishmania* could be achievable in the clinical setting. This remains to be conclusively determined.

CaM from the African trypanosomes was elegantly studied by Ruben and colleagues (Ruben *et al.,* 1983, 1984; Ruben and Patton, 1985a, 1987) and found to be more closely related to that from *Tetrahymena* than CaM from host tissues. The CaM of *Trypanosoma brucei gambiense* is encoded by three identical tandemly repeated genes. Transcription of the CaM genes generates polygene transcripts resulting in mature CaM mRNAs (Tschudi *et al.,* 1985; Tschudi and Ullu, 1988). There is a rather dramatic decrease and redistribution of CaM as *T. brucei rhodesiense* change from slender bloodstream trypomastigotes (approximately 50% : 50% cytosolic : particulate) to slow-growing procyclics (30% : 70% cytosolic : particulate).

The molecular weight of *Tetrahymena* and *Trypanosome* CaM is 13,500 on SDS-polyacrylamide gels while rat erythrocyte or bovine brain CaM has an apparent molecular weight of about 16,500 when analyzed under identical conditions (Ruben *et al.,* 1983). CNBr-cleavage fragments and Western procedures suggest the two protozoan proteins, while related, are distinct from each other. Trypanosome CaM moved faster electrophoretically and differed antigenically from bovine brain CaM, suggesting structural differences between them (Ruben and Patton, 1985a,b, 1987). Structural differences between host and trypanosome CaMs notwithstanding, both are capable of activating brain phosphodiesterase or erythrocyte Ca^{2+},Mg^{2+}-ATPase to a similar degree and are inhibited to the same extent by the phenothiazines. Trifluoperazine was trypanosomacidal *in vitro,* having an ED_{50} of 15 μM. Seebeck and Gehr (1983) extended these observations by reporting that a number of the neuroleptic phenothiazines at micromolar concentrations rapidly inhibited *in vitro* growth as well as [^{3}H]uridine incorporation of *T. brucei.* The authors suggest that because the phenothiazines can cross the blood–brain barrier, these drugs may show promise in treating trypanosomal infections of the CNS. Unfortunately, mice infected with *T. b. brucei* were not protected by trifluoperazine or thioridazine (50 mg/kg/day) given on days 21 to 25 (Rice *et al.*, 1987). Examination by electron microscopy revealed treatment with these drugs selectively caused disintegration of the pellicular microtubules with the complete exclusion of any effect on flagellar microtubules. This selective effect by CaM antagonists in *T. brucei* would seem to agree in some ways with the more recent work by Dolan *et al.* (1986), in which Ca^{2+} (100 μM) initiated selective depolarization of pellicular but not flagellar microtubules. There was no effect on either the central pair of flagellar microtubules or the nine outer-doublet microtubules, even after a 30-min incubation with Ca^{2+}. It is well known that Ca^{2+} regulates microtubular assembly and disassembly in mammalian tissues. This effect by Ca^{2+} is also greatly enhanced by CaM in host cells. However, in the case of *T. brucei* pellicular microtubules, the presence of bovine brain CaM (50 mg/ml) had no effect on tubulin release (one β-tubulin and two α-tubulins). It may be alleged that the presence of tightly bound *Trypanosoma* CaM in the preparations might obscure any stimulation that might be seen with added mammalian CaM. Dolan *et al.* (1986) conceded that it is also possible that *T. brucei* CaM is structurally distinct enough from mammalian CaM (which they used in their study) to make the added bovine brain CaM ineffective in the protozoal system. In support of this hypothesis, they go on to remind the reader that the cAMP phosphodiesterases from *T. cruzi* and *T. brucei* are not sensitive to activation by mammalian CaM (Dolan *et al.*, 1986; Goncalves *et al.*, 1980). Actually, later work by Tellez-Inon *et al.* (1985) demonstrated that *T. cruzi* phosphodiesterase is activated best in the presence of Ca^{2+} and in the presence of *T. cruzi* CaM. In any event, it remains to be seen whether or not autologous *T. brucei* CaM would result in an increase in stimulation of tubulin release in African trypanosomes. The CaM inhibitors trifluoperazine at 10^{-5} M or 48/80 (10^{-5} g/ml) had no effect on the release of pellicular tubulin at Ca^{2+} concentrations of 10 or 100 μM in *T. brucei,* but these drugs also fail to inhibit both cold-sensitive and cold-stable microtubules in mammalian cells (Dolan *et al.*, 1986). The amino acid sequences of β-tubulins from *Trypanosoma, Leishmania, Euglena,* and *Chlamydomonas* are very similar (93–95%). That of human tubulin is distinct [only 80% similarity (Fong and Lee, 1988)]. These workers go on to point out leishmanial tubulins are developmentally regulated. There is a rapid increase in their biosynthesis during the transformation of the amastigote to the promastigote stage. This is followed by a decrease in biosynthesis during the change from promastigote to amastigote stage. Chan and Fong (1990) reported that the herbicide, trifluralin, at micromolar concentrations selectively binds to the tubulin of *Leishmania mexicana amazonensis,* interfering with growth of the amastigotes. There was no binding to mammalian tubulin and no inhibition of mammalian cell growth.

Ruben *et al.* (1984) isolated CaM from both particulate and soluble fractions of 3 species of African trypanosomes, and of the 66 amino acids sequenced they reported 6 substitutions in amino acids at residues 77, 79, 97, 99, and 108; at residue 115 there is most notably a substitution of lysine for trimethyllysine compared to bovine brain CaM. There are eight substitutions when comparing trypanosomal CaM to *Tetrahymena* CaM (see Fig. 3). Later work by Strickler, Ruben, and Patton in 1986 (Ruben, personal communication) showed 17 amino acid substitutions out of 148 in the primary structure of *Trypanosoma* CaM compared to that of bovine brain and 22 when compared to *Tetrahymena* CaM (Ruben and Patton, 1985a; Ruben and Haghighat, 1990). The large share of substitutions are in the carboxy-terminal half of the molecule. The substitutions at the carboxyl-terminal end are typical of invertebrate or protozoan CaM and the absence of trimethyllysine is typical of yeast and mold CaM (unmethylated lysine at position 115 contributes to maximum activation of NAD kinase) (Ruben *et al.*, 1984; Roberts *et al.*, 1985; Takeda and Yamamato, 1987). The amino-terminal region before domain I (residues 1–11) contains 3 substitutions. The connecting regions between domains II and III (residues 76–84) and between domains III and IV (residues 113–120) have, respectively, 2 and 1 substitution. Domain II and III each contain three substitutions while domain IV contains five substitutions (L. Ruben, personal communication). This differs slightly from the version reported earlier (Nozawa and Nagao, 1986), in that three substitutions originally described in domain I have now been correctly placed in the amino terminus and the number of substitutions in domain IV has now been correctly listed as 5 and not 6. This difference in sequence of amino acids results in an increase in Ca^{2+} requirement (threefold greater) for the activation of bovine brain cyclic nucleotide phosphodiesterase by *Trypanosoma* CaM compared to the Ca^{2+} requirement with the autologous CaM. This is due to decreased affinity of the protozoal CaM for Ca^{2+} (Ruben *et al.*, 1984). This is similar to what is seen with *Tetrahymena* CaM, where potency to activate bovine brain cyclic nucleotide phosphodiesterase is only 55% of bovine brain CaM, as mentioned earlier (Nozawa and Nagao, 1986; Kakiuchi *et al.*, 1981). These unusual structural features of trypanosome CaM may contribute to its interacting with proteins or pathways unique to the protozoa.

Localization of CaM throughout the length of the flagellum of procyclic stages of *T. b. rhodesiense* or *T. congolense* by immunofluorescence (see Fig. 4) suggests a role for CaM in *Trypanosome* motility. Ruben *et al.* (1984), however, also identified a cytosolic 65-kDa CaM-binding protein in

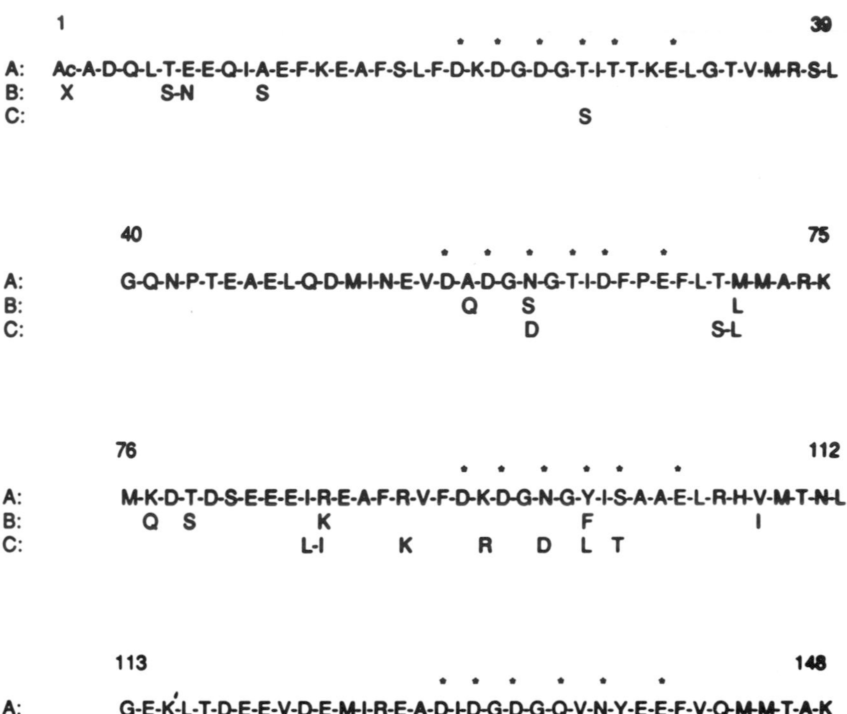

FIG. 3 Comparison of calmodulin amino acid sequences. Calmodulins are from (A) bovine brain (Watterson *et al.*, 1980), (B) *T. brucei* (J. E. Strickler, L. Ruben, and C. L. Patton; unpublished observations, and (C) *Tetrahymena* (Yazawa *et al.*, 1981). The single-letter abbreviations for amino acids are used: A, alanine; D, aspartic acid; E, glutamic acid; F, phenylalanine; G, glycine; H, histidine; I, isoleucine; K, lysine; L, leucine; M, methionine; N, asparagine; P, proline; Q, glutamine; R, arginine; S, serine; T, threonine; V, valine; Y, tyrosine. K′, M^{ε},N^{ε},N^{ε}-Trimethyllysine; X, unknown blocking group; #, space inserted to give maximal alignment; *, proposed calcium-coordinating residues. (Courtesy of L. Ruben.)

the 100,000 *g* supernatant, suggesting to them that CaM fulfills other functions in addition to motility. Subsequently, CaM-binding proteins of M_r 40,000, 38,000, 36,000, and 32,000 were found, the latter being completely unrelated in amino acid sequence to any known protein (Ruben and Haghighat, 1990). This cluster of proteins is capable of binding CaM from *T. brucei* and that of bovine brain. They are cysteine protease-dependent fragments of a 58-kDa parent protein that corresponds to the variable surface glycoprotein (VSG). Ruben *et al.* (1991) suggested that CaM association with VSG may serve as a CaM-sensitive signal. In addition, Ruben *et al.* (1990) identified particulate CaM-binding proteins with molecular

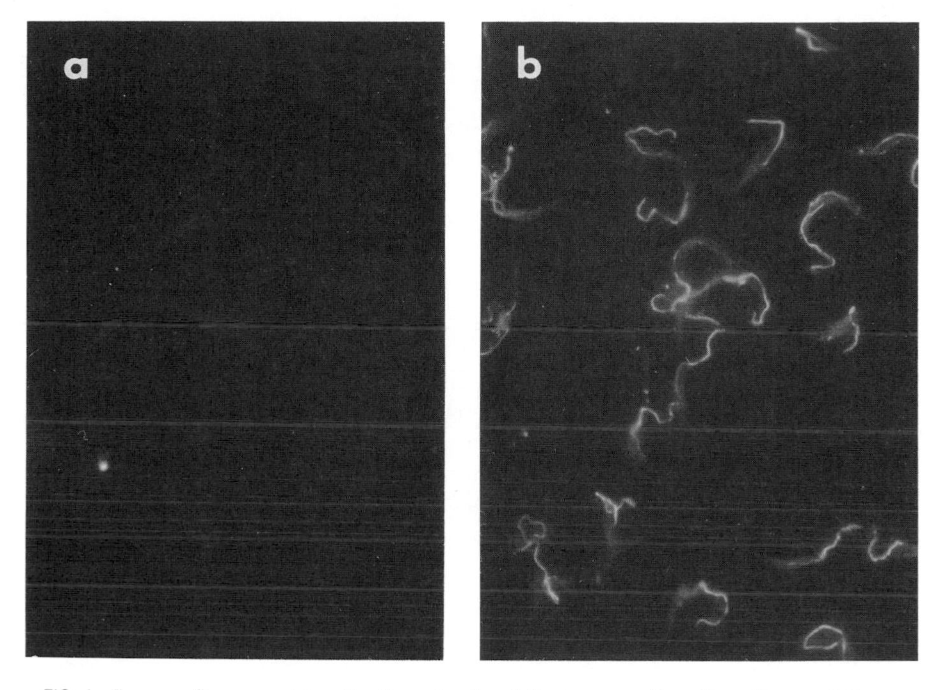

FIG. 4 Immunofluorescent localization of calmodulin in procyclic culture forms of *Trypanosoma brucei*. Cells were fixed in 95% ethanol (5 min) followed by acetone (10 sec). Cells were incubated with (a) a 1 : 50 dilution of preimmune serum or (b) a 1 : 50 dilution of antiserum. Calmodulin distribution was determined following addition of fluorescein isothiocyanate-conjugated goat anti-rabbit IgG. (Courtesy of L. Ruben and C. L. Patton.)

weights of 126,000 and 106,000. These are developmentally regulated, being present in slender bloodstream forms and absent in procyclic culture forms. They are also not found in host tissue. Ruben (1987) went on to identify eight Ca^{2+}-binding proteins in *T. brucei,* demonstrating this protozoan has a variety of Ca^{2+}-binding proteins to respond to different Ca^{2+} signals. Five of these were cytosolic and had molecular weights of 84, 70, 64, 22, and 15 kDa. The latter appeared to be trypanosome CaM. The three remaining were particulate with molecular weights of 55, 46, and 29 kDa. The 46-kDa protein bound three times more Ca^{2+} in cultured procyclic forms than it did in the slender bloodstream forms. It was not present in *L. mexicana,* however, suggesting that the cyclical differentiation of various stages of the life cycle or forms of these related organisms may result in differing response to Ca^{2+} cues. The reader is strongly urged to read a recent review by Ruben and Haghighat (1990), which deals with the biochemistry and Ca^{2+} regulatory pathways of *T. brucei.* It details areas of importance that were omitted here due to space limitations.

CaM has been purified from *T. cruzi* epimastigotes (Tellez-Inon *et al.*, 1985) and found to have an apparent molecular weight of 16,000. *Trypanosoma cruzi* CaM in the presence of Ca^{2+} was capable of activating either homologous cyclic AMP phosphodiesterase (PDE) or brain PDE. The highest activity resulted from *T. cruzi* CaM activating *T. cruzi* PDE in the presence of Ca^{2+}. This reaction, as well as cell motility, was blocked by micromolar concentrations of EGTA, chlorpromazine, fluphenazine, and compound 48/80. The results found by Tellez-Inon *et al.* (1985) differ somewhat from those of Goncalves *et al.* (1980), insofar as the latter workers describe a *T. cruzi* soluble cAMP PDE that is not Ca^{2+} dependent or activated by *T. cruzi* CaM. It also is not inhibited to a significant degree by methylxanthines such as theophylline and caffeine. The authors state that PDE from some microorganisms, like that from *T. cruzi,* is also resistant to these inhibitors, in contrast to the PDE of mammals. Goncalves *et al.* (1980) estimated the molecular weight of *T. cruzi* CaM as 30,000 by gel filtration. Tellez-Inon *et al.* (1985) attributed the differences between their results and those of Goncalves *et al.* (1980) to differences in purification of CaM and PDE. For a discussion of other enzymes (e.g., adenylate cyclase, cAMP PDE, and protein kinases) in the trypanosomes, one should refer to Walter and Opperdoes (1982). While others have reported the existence of protein kinase activities in trypanosomes that are not apparently activated by cAMP, Ulloa *et al.* (1988) rather convincingly established the presence of a cAMP-dependent protein kinase in *T. cruzi.* Cyclic AMP, adenylate cyclase activators, and inhibitors of cAMP PDE appear to stimulate transformation of *T. cruzi* epimastigotes to infective metacyclic trypomastigotes. In addition, activators of cAMP PDE and inhibitors of CaM inhibit this process (Gonzales-Perdomo *et al.*, 1988). These results are in agreement with those of Goncalves *et al.* (1980), Oliveira *et al.* (1984), and Rangel-Aldao *et al.* (1987), who found changing levels of cAMP during growth of *T. cruzi* epimastigotes *in vitro*. There is an initial increase in cAMP coincident with adenyl cyclase peaking about the stationary phase of growth. Furthermore, this adenylate cyclase is stimulated by isoproterenol, which in turn has an effect on DNA synthesis as measured by labeled thymidine uptake (Oliveira *et al.*, 1984). Trypomastigotes exhibit a cAMP content four times that found in epimastigotes, suggesting cAMP plays a role in the differentiation of *T. cruzi* as reported in *Trypanosoma lewisi, T. brucei,* and *Leishmania* (Rangel-Aldao *et al.*, 1987).

Verapamil has been demonstrated to have a favorable effect on the murine model of Chagas' disease (Tanowitz *et al.*, 1989). It appears, however, that the major role of this Ca^{2+} antagonist is to prevent the microvascular abnormalities leading to vasospasm and hypoxia, facilitating vascular perfusion, and not through a direct effect on the parasite.

VI. Calmodulin in Sporozoa (Baker, 1977)

Keogh and Shaw observed in 1944 that Ca^{2+} added to an isolated muscle preparation caused relaxation. However, if the intestine is previously treated with quinine or other antimalarials the muscle contracts. Based on these observations, these workers suggested that quinine interferes with growth and development of the malaria parasite by reducing the amount of Ca^{2+} available in the red cell for reproduction. Subsequently, Johnson *et al.* (1980) reported that EDTA or EGTA (2.5 m*M*) greatly reduced or abolished attachment and invasion of erythrocytes by *Plasmodium knowlesi* merozoites. Further work by Wasserman *et al.* (1982) demonstrated that EGTA (1 m*M*) blocked the asexual cell cycle of synchronized cultures of *Plasmodium falciparum* at two points. The first is 20–26 hr following invasion of the red cell. The second point of inhibition was the invasion of the erythrocyte by the merozoite. In addition, cytoadherence of *P. falciparum*-infected erythrocytes to thrombospondin, an adhesive glycoprotein involved in a number of cell-to-cell interactions, is Ca^{2+} dependent (Barnwell *et al.*, 1989; Barnwell, 1989). The binding of Ca^{2+}, presumably to a 22-kDa protein, is also important in regulating surface interactions resulting from host cell entry by *Toxoplasma* (Sibley *et al.*, 1986), as well as the spore discharge system of the microsporidian intracellular parasite *Spraguea lophii*, which is Ca^{2+}/CaM mediated (Pleshinger and Weidner, 1985). As early as 1981 it was recognized that malaria parasites accumulate Ca^{2+} levels that are up to 30 times greater than that found in uninfected red cells. This Ca^{2+} is located almost exclusively within the parasite with an increased content in late stage parasites. The rate of influx into infected cells is at least 7 times that which takes place in normal erythrocytes. Efflux of Ca^{2+} is also decreased in infected erythrocytes. Both influx and efflux are presumed to be energy dependent (Leida *et al.*, 1981; Bookchin *et al.*, 1981; Tanabe *et al.*, 1982a; Krungkrai and Yuthavong, 1983; Mikkelsen *et al.*, 1984; Krishna and Ng, 1989). A class of acidic phosphoproteins of plasmodial (murine) origin was described by Wiser *et al.* (1982, 1983b, 1988). These are associated with the host red cell membrane and thought by these workers to have a Ca^{2+}-binding function even though cAMP, Ca^{2+}, and EGTA fail to affect phosphorylation. Unfortunately no direct data on Ca^{2+} binding were provided in their studies. Protein kinase activity has been reported to be exclusively associated with the parasite and specific activity of this enzyme increases with parasite maturity. Phosphorylation appears not to be regulated by cAMP, cGMP, CaM, or Ca^{2+} but optimum kinase activity occurs with ATP, acidic substrates, spermine, or spermidine (Wiser *et al.*, 1983a; Wiser and Schweiger, 1985).

Protein phosphorylation also occurs in *Babesia bovis* and several phosphoproteins have been reported (Willadsen, 1984); Ray *et al.* (1990) described a Ca^{2+}/CaM-dependent protein kinase and a protein kinase C in the membrane fraction of *B. bovis*. The former enzyme appears to phosphorylate a 40-kDa protein but its function remains to be elucidated in *Babesia*. Normal erythrocytes are almost impermeable to Ca^{2+} and the very low intraerythrocytic Ca^{2+} content is maintained through active extrusion by a Ca^{2+}-ATPase (Shalev *et al.*, 1981). Ca^{2+},Mg^{2+}-ATPase was assayed by Tanabe *et al.* (1982b) in *Plasmodium chabaudi*-infected erythrocytes and uninfected cells under identical conditions. They found enzyme activities of both membranes were stimulated to the same extent by bovine brain CaM, but the lowered pH and decreased ATP concentration in the infected erythrocytes suppressed the activity of Ca^{2+},Mg^{2+}-ATPase to 15–20% normal values. Tanabe *et al.* (1982b, 1983) speculated that the decreased Ca^{2+},Mg^{2+}-ATPase as well as the low intracellular pH and ATP concentrations in *P. chabaudi*-infected rat red cells account for the observed decrease in Ca^{2+} efflux. These workers (Tanabe *et al.*, 1983) did report a 30% reduction in CaM levels in schizont-infected cells in contrast to later findings by Scheibel *et al.* (1987, 1989), who reported amounts of CaM increasing with parasite maturity. In any event, Tanabe *et al.* (1983) conceded that even with the decreased levels of CaM in schizonts that they observed, there is an approximately 100-fold excess of what is required to activate the Ca^{2+} efflux pump.

Clinically effective antimalarials, quinine, and quinacrine (Mepacrine) were known for years to inhibit CaM (Asano and Hidaka, 1984; Roufogalis, 1985), and chloroquine was known to inhibit Ca^{2+}-dependent phospholipase A_2, which is considered to be a Ca^{2+}/CaM-regulated enzyme (Loffler *et al.*, 1985). Chloroquine also competitively inhibits the CaM activation of rat brain PDE and CaM stimulation of Ca^{2+},Mg^{2+}-ATPase in erythrocytes (Nagai *et al.*, 1987). It was this anti-CaM activity of the antimalarials, the suspected importance of CaM in *Plasmodia,* the apparent obligate requirement for Ca^{2+} at specific points of the plasmodial life cycle, and the presence of a Ca^{2+} efflux pump that appeared to be CaM mediated in this parasite that caused Scheibel *et al.* (1987, 1989) to investigate Ca^{2+}CaM functions in *Plasmodia*. Following stimulation of mammalian cells there is movement of Ca^{2+} through channels in the plasma membrane or from the intracellular membranes or organelles, resulting in the increase in concentration of intracellular free Ca^{2+} (Tomlinson *et al.*, 1984). Each molecule of CaM can bind as many as four Ca^{2+}. This produces conformational changes regulating CaM activity and facilitating CaM binding to either CaM-binding proteins or lipid-soluble drugs. Because the activity of CaM is controlled by changes in the concentration of intracellular Ca^{2+}, Scheibel *et al.* (1987, 1989) proposed an attempt to inhibit these sequential reactions in malaria-parasitized red cells by utilizing two types of drugs. First, Ca^{2+} channel blockers were to be employed to attenuate the Ca^{2+} flux throught the plasma membrane or from intracellular sites and prevent the activation of CaM. The subsequent step in this process, the binding of the active Ca^{2+}/CaM complex to target proteins or enzymes, was to be blocked by the utilization of various prototype CaM inhibitors. Precedent for this approach was reversal of drug resistance in tumor cells, which Scheibel *et al.* (1987, 1989) theorized was due, in some cases, to a drug with Ca^{2+} channel-blocking activity potentiating a classical CaM antagonist. The anti-CaM agent may or may not be inhibiting a CaM-mediated efflux pump.

Initial studies by Scheibel *et al.* (1987, 1989) demonstrated the presence of CaM in *P. falciparum*. Normal erythrocytes contained only 10 to 11 ng CaM/10^6 cells, whereas ring stage parasites contained about 18 ng/10^6 cells. In addition, it appeared that CaM content increased with parasite maturity such that schizonts contained up to twice the value found in rings.[2] Similar levels of CaM were later reported in *B. bovis* by Ray *et al.* (1990). CaM appeared by immunoelectron microscopy to be scattered throughout the cytoplasm of the malaria parasite (Fig. 5). This tended to agree with the fluorescence microscopy and electron microscopic autoradiographic studies done by Scheibel *et al.* (1987, 1989). Previous studies in their laboratory showed the immunosuppressive drug, cyclosporine A (CsA) (at micromolar concentrations), binds to CaM (in a Ca^{2+}-dependent interaction) in T cells and appears to inhibit CaM-dependent activation of PDE (Scheibel *et al.*, 1989; Hess *et al.*, 1985, 1986; LeGrue *et al.*, 1986; Colombani *et al.*, 1985, 1986). Their results showed CsA at micromolar levels, which was known to kill malaria parasites *in vitro* and *in vivo*,[3] concentrates initially in the food vacuole (Fig. 6), as do the clinically effective antimalarials mepacrine, chloroquine, and mefloquine (Aikawa, 1972; Warhurst and Thomas, 1975; Jacobs *et al.*, 1987). CsA then distributes within the cytoplasm of mature *P. falciparum* parasites as confirmed by fluorescence microscopy.

Binding studies by flow cytometric analysis showed increased binding of micromolar concentrations of CsA to schizont-infected erythrocytes compared to normal erythrocytes. There was competition for binding to infected erythrocytes for free parasites between CsA and either the clas-

[2] Robson and Jennings (1991) analyzed the calmodulin gene in *P. falciparum* and found it to be unique among other known calmodulin genes.

[3] A combination of CsA and quinine has been reportedly used to treat cerebral malaria in humans (see page 43 of Scheibel *et al.* (1989). However, a double blind placebo-controlled trial carried out in Ho Chi Minh City showed no effect on mortality (see page 16 of Warrell *et al.* (1990).

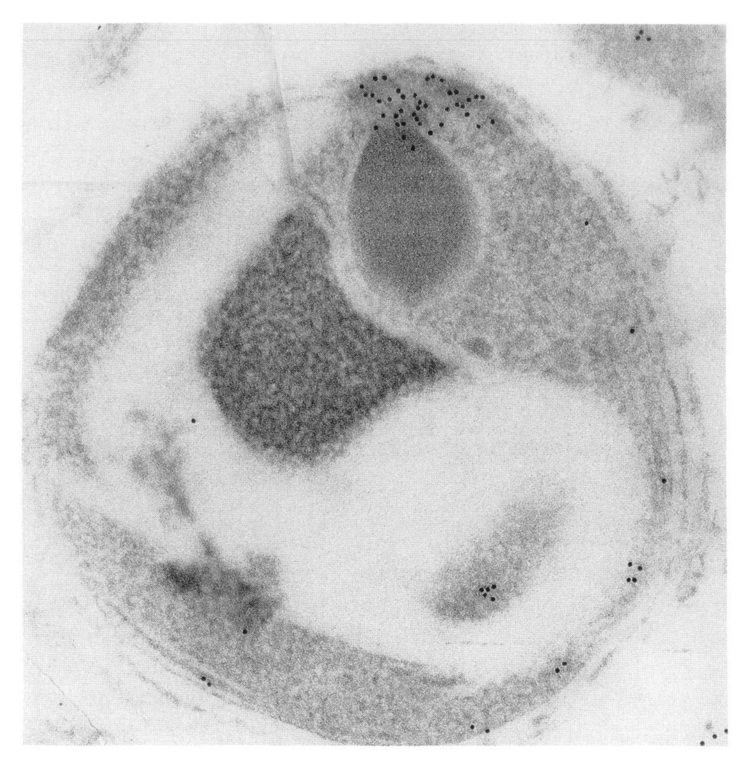

FIG. 5 Immunoelectron micrograph of *P. falciparum* merozoite (Brazilian strain ZG8, drug sensitive) showing localization of calmodulin (see Scheibel *et al.*, 1987, 1989, for methods). An area of intense immunoreactivity is located on the apical end of the merozoite. (Courtesy of M. Aikawa.)

sic CaM inhibitor, W-7, or the clinically effective antimalarial drug, chloroquine.

The *in vitro* growth of *P. falciparum* was inhibited to varying degrees by the Ca^{2+} antagonists, as would be expected from the obligate dependence of the parasite on Ca^{2+}. It is unlikely that binding to CaM or affinity of this binding by the Ca^{2+} antagonists contributes to channel blockade in mammalian tissues. The reported affinity of binding that exists between the channel blockers and CaM (Roufogalis, 1985; Johnson and Wittenauer, 1983) does appear to correlate with antimalarial potency, however. The growth of the human malaria parasite was also sensitive to CaM antagonists, in many instances to a greater degree than to the Ca^{2+} channel blockers. The degree of inhibition of the parasite was directly proportional to the known anti-CaM potency or degree of binding to CaM in mammalian tissues by these Ca^{2+} antagonists and CaM antagonists (Asano and

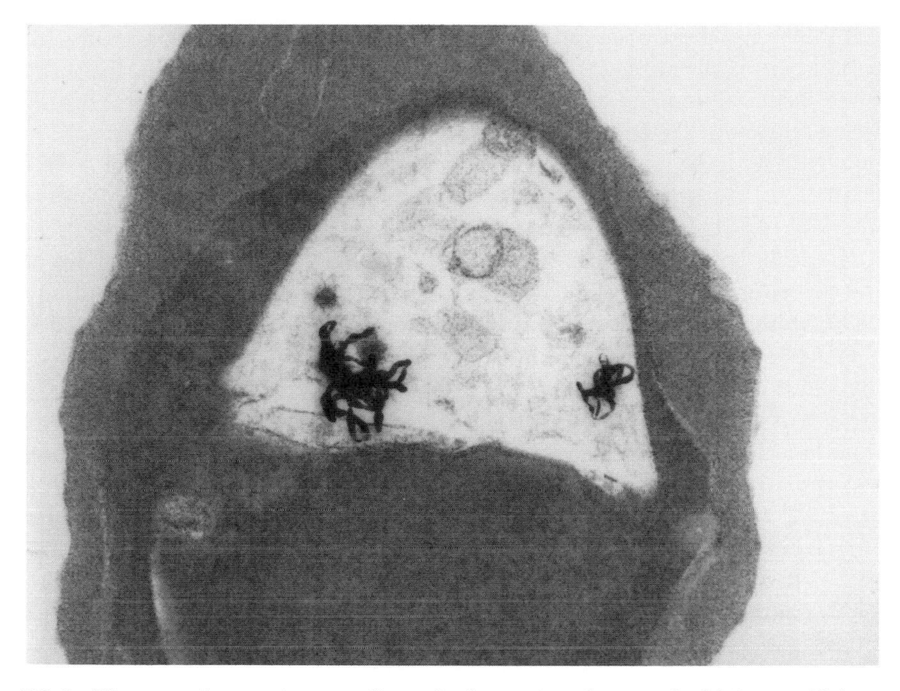

FIG. 6 Electron-microscopic autoradiograph of a section of mature *P. falciparum* (Colombian strain FCB_k+, chloroquine resistant) that was treated with a derivative of cyclosporine A (CsA). The concentration sites of CsA are present within a swollen food vacuole (Scheibel *et al.*, 1987, 1989). (Courtesy of M. Aikawa.)

Hidaka, 1984; Roufogalis, 1985). The rank order of activity of the clinically useful antimalarials within their class reflected their reported anti-CaM effect. In comparison to the other anti-CaM agents, these drugs were much more potent inhibitors of *in vitro* growth of *P. falciparum*, however. The exquisite sensitivity of plasmodia to these anti-CaM/antimalarial agents may result from those features clearly outlined by Hait and Lazo (1986) to explain vulnerability of tumor cells to manipulation of CaM function. In fact, tumor cells share many metabolic similarities to plasmodia (Scheibel *et al.*, 1989; Scheibel, 1988; Scheibel and Sherman, 1988).

It would therefore appear that the growth inhibition seen in the *in vitro* culture of *P. falciparum* is related to CaM antagonism because the biological response parallels anti-CaM potency. One criterion as to whether or not CaM is involved in a particular cell reaction is the existence of a direct relationship between the order of antagonistic potencies of a series of structurally related compounds and their anti-CaM activity (Roufogalis, 1982, 1985). Prozialeck and Weiss (1982) and Weiss *et al.* (1982) pointed out that while these drugs are of different chemical classes and are used for a number of purposes, they nevertheless share a number of structural features in common. Other criteria in support of these experimental results being related to CaM-mediated processes are (1) the antagonism of growth occurred with the use of pharmacological doses of drug, and (2) the *in vitro* growth of the parasite is Ca^{2+} dependent. Other workers (see Scheibel *et al.,* 1989, for references) found the inhibition of the *in vitro* growth of *P. falciparum* by these agents was nonstereospecific. Therefore, four out of the six criteria to support that the antiplasmodial effects of these drugs may be attributed to CaM antagonism have been met (Roufogalis, 1982, 1985).

Geary *et al.* (1986) questioned whether several classical CaM antagonists (i.e., chlorpromazine, calmidazolium, trifluoperazine) inhibit the growth of malaria by interacting with CaM or by inhibiting mitocchondrial function. Surprisingly, the parasites used by these workers appeared much less sensitive to these drugs than those in the study reported by Scheibel *et al.* (1987, 1989). In addition, one might question whether or not rhodamine-123 retention if a true reflection of mitochondrial function (Geary *et al.,* 1986). It may only reflect inside-negative membrane potential (Tanabe, 1983; Kato *et al.,* 1990). One should wait for detailed electron transport studies on isolated organelles before making definitive interpretations (Scheibel *et al.,* 1989). Others have reported on the antimalarial activity of agents that interfere with Ca^{2+}- or CaM-mediated processes. One such study by Tanabe *et al.* (1989) described inhibition of the growth and development of the FCR-3 strain of *P. falciparum* by Ca^{2+} antagonists (verapamil, nicardipine, diltiazem) and CaM antagonists (trifluoperazine, calmidazolium, W-7, W-5) but, as in the studies by Geary *et al.* (1986) and by Panijpan and Kantakanit (1983), the doses of some drugs required to inhibit growth were higher than those described by Scheibel *et al.* (1987, 1989) or Kristiansen and Jepsen (1985). This may be due to the differing strains of parasite being used or, more likely, to differences in experimental techniques. In this regard *B. bovis* (which also contains similar levels of CaM and depends on Ca^{2+}, as does *P. falciparum*) is likewise sensitive to calmidazolium (R24571) and trifluoperazine but to a lesser degree (Ray *et al.,* 1990). Tanabe *et al.* (1989) also observed a difference in sensitivity to Ca^{2+}/CaM inhibitors with increasing parasite maturity. In D-sorbitol-synchronized plasmodia the Ca^{2+} blockrs and CaM inhibitors used in their study exhibited a greater effect at the 24- to 48-hr period than they did the first 24 hr. The findings of Tanabe *et al.* (1989) differ somewhat from those of Matsumoto *et al.* (1987) and Scheibel *et al.* (1989), who observed the CaM inhibitors blocked invasion of red cells by merozoites at lower concentrations than required to inhibit schizont maturation. Recently, Ohnishi *et al.* (1989a,b) found prostaglandin oligomers MR-356 and MR-256 suppressed the growth of malaria parasites *in vivo* and *in vitro*. They attribute the mechanism to chelation of intracellular Ca^{2+} or inhibition of phospholipase and protease.

Because drugs of this type appear to inhibit the growth of plasmodia as a result of their anti-CaM potency, it was thought that other drugs known to affect Ca^{2+}CaM processes might interact with these agents and alter the *in vitro* antimalarial potency of each other. Using a modified semiautomated microdilution technique, concentrations of these drugs were used alone and with each other in studies of multidrug-resistant and drug-sensitive clones of *P. falciparum*. Fractional inhibitory concentration (FIC) indices reflected rather marked antagonism between pairs of drugs that bind to and presumably inhibit CaM (R24571, W-7, CsA, chlorpromazine). The antimalarials [mepacrine (quinacrine), quinine, chloroquine] are also known to inhibit CaM (Asano and Hidaka, 1984; Roufogalis, 1982, 1985; Loffler *et al.*, 1985; Nagai *et al.*, 1987). These were studied by Scheibel *et al.* (1987, 1989) in combination with R24571, W-7, and CsA. Significant antagonism resulted between many of the drug pairs, most notably in the multidrug-resistant clone of *P. falciparum*. Surprisingly, even the Ca^{2+} antagonists verapamil and diltiazam exhibited rather marked levels of antagonism, especially against R24571 and W-7, in the multidrug-resistant clone. This antagonism suggested that the drugs had the same site of action (i.e., competed for the same receptor-binding site).

Weiss *et al.*, (1982) and Prozialeck and Weiss (1982) pointed out that while various compounds of many diverse chemical classes modify the activity of CaM, these compounds have numerous architectural similarities. They go on to state that interactions of these drugs with CaM appear to involve two kinds of attachments. One attachment is a hydrophobic interaction between a very large lipophilic (ring) portion of the drug and nonpolar regions of CaM. The other point of attachment is a result of the electrostatic interaction between a positively charged amino group on the drug and a negatively charged acidic residue on CaM. Optimally the hydrophobic region of the drug consists of two aromatic rings and the side chain amino group is at least four atoms removed from the aromatic ring region. The binding of drug to CaM is influenced by the presence of Ca^{2+}. Other factors that also have a marked effect on anti-CaM activity are the introduction of substituents on the aromatic nucleus and modifications to the side chain amino group. These include altering the distance between the ring and amino group. Weiss *et al.* (1982) concluded that these structural prerequisites account for specific drug–receptor interactions between the drug and CaM. This interaction of these drugs with the Ca^{2+}/CaM complex would in turn inhibit the binding and subsequent activation by CaM of effector enzymes. Roufogalis (1982, 1985) stated that various classes of CaM anatgonists bind to distinct binding sites on CaM and allosteric interactions can occur between these sites. Roufogalis goes on to point out that some drugs interact directly with the effector enzyme and in so doing also prevent the subsequent activation by CaM. Furthermore, other receptor sites with structural features in common with the proposed drug-binding sites on CaM are also affected by these drugs even though they are not CaM-dependent enzyme systems. For example, trifluoperazine inhibits CaM-sensitive Ca^{2+}-ATPase as well as CaM-insensitive Na^{+},K^{+}- and Mg^{2+}-ATPase activities in human and rat red cells (Luthra, 1982). The CaM antagonists also inhibit a Ca^{2+}-dependent but CaM-independent protein kinase that in turn regulates a variety of cellular processes (Roufogalis, 1982, 1985). In addition, they may exert an even greater antagonistic effect on dopaminergic, α-adrenergic, serotonergic, muscarinic, and cholinergic receptors as well as prevent the release of neurotransmitters from nerve endings (Roufogalis, 1982; Snyder and Reynolds, 1985; Krueger, 1989; Zernig, 1990).

Protein kinase appears to be important to the plasmodium. Read and Mikkelsen (1990) demonstrated cAMP- and Ca^{2+}-dependent protein kinase activities in erythrocytic stages of *P. falciparum*. Activity of the plasmodial kinase is independent of CaM, unlike the Ca^{2+}-dependent kinases of mammalian tissues. In addition these workers identified an adenylate cyclase in the human malaria parasite that is distinct in several ways from that of the mammalian host cell by requiring Mn^{2+} and differing in pH sensitivity (Read and Mikkelsen, 1991a). The parasite is capable of taking up erythrocyte-derived ATP by an ATP/ADP transport system (Choi and Mikkelsen, 1990). When activities of adenylate cyclase and cAMP-dependent protein kinase in a gametocyte-producing strain were compared to those in a non-gametocyte-producing strain, there was a significant decrease in cAMP-dependent protein kinase in the non-gametocyte-producing strain (Read and Mikkelsen, 1991b). This suggested to the authors that the decrease in protein kinase deprives the parasite of the ability to respond to increases in intracellular levels of cAMP. While one might expect cAMP to play a role in sexual differentiation of the parasite, adenylate cyclase was unexpectedly the same in both strains.

Verapamil, diltiazem, and nifedipine bind to and inhibit CaM; this is in addition to their well-known role of inhibiting Ca^{2+} flux (Roufogalis, 1985; Johnson and Wittenauer, 1983; Epstein *et al.*, 1982). Chlorpromazine binds to CaM and inhibits CaM-mediated functions, but like other phenothiazines it can influence Ca^{2+} transport processes and function as a Ca^{2+} antagonist, too (Karaki *et al.*, 1982; Roufogalis, 1985). Ca^{2+} flux is also regulated in some cases by a cyclic AMP-dependent protein kinase (Krueger, 1989). Consequently, there is considerable overlap in action of the Ca^{2+} antagonists and the CaM antagonists.

The clinically effective antimalarials, i.e., quinacrine, quinine, and chloroquine, are known to interact with Ca^{2+}/CaM-mediated processes in the mammalian host (Asano and Hidaka, 1984; Loffler *et al.*, 1985; Weiss *et al.*, 1982; Prozialeck and Weiss, 1982; Volpi *et al.*, 1981a,b; Nagai *et al.*, 1987). Chloroquine poisoning has been successfully treated in 51 cases with diazepam (Riou *et al.*, 1988), a minor tranquilizer known to antagonize CaM (Asano and Hidaka, 1984). Treatment of the patient with chloroquine also appears to have a modulatory effect on the host immune system quite distinct from rather profound immunosuppression seen in *P. falciparum* infections in untreated nonimmune patients (Kremsner *et al.*, 1990).

Betticher *et al.* (1990) pointed out that the modulatory effect that chloroquine appears to have on the host immune system is presumably due to the inhibition of interleukin-1 production by monocytes. This is turn inhibits the generation of immunoglobulin-secreting cells (Salmeron and Lipsky, 1983). Chloroquine taken at prophylactic antimalarial doses is also known to reduce the T-cell-dependent antibody response to primary immunization with intradermal human diploid rabies vaccine (Pappaioanou *et al.*, 1986; Pappaioanou and Fishbein, 1986). The antimalarial quinidine is also capable of entering into immune-mediated reactions with the host (Cohen *et al.*, 1988; Reid and Shulman, 1988).

The immunosuppressive drug cyclosporin A (CsA) is antimalarial *in vitro* and *in vivo* both in the rodent malaria model and the *P. falciparum*–primate model (see Scheibel *et al.*, 1989, for references). Autoradiographic evidence suggests the uptake and deposition of CsA (Scheibel *et al.*, 1987, 1989) in the parasite is similar to that of the clinically useful antimalarials chloroquine, mefloquine, and mepacrine (Aikawa, 1972; Warhurst and Thomas, 1975; Jacobs *et al.*, 1987). As discussed earlier, flow cytometric analysis suggests competition between dansylated CsA and either the classical antimalarial drug chloroquine or the CaM inhibitor W-7 for binding sites in plasmodia. While the detailed mechanism of host immunosuppression by CsA as well as the inhibition of growth of *P. falciparum* is not completely understood it is known that both T cells and plasmodia rapidly take up the drug at micromolar concentrations. Also, CsA binds to a number of receptors, including CaM, suggesting Ca^{2+}/CaM-mediated reactions may be involved in the immunomodulating actions as well as antiparasitic activity of these drugs.

It would appear that the inhibition of the biological response seen in these studies may be related in some way to antagonism of CaM action, in terms of the criteria proposed by Roufogalis (1982, 1985). In addition, drug resistance would appear to be a result of drug binding to a receptor site that shares architectural features in common with CaM or CaM-mediated effector proteins. The single major difference that emerges when comparing the response to combinations of clinically effective antimalarials and CaM inhibitors in the drug-resistant clone versus the drug-sensitive clone is the marked antagonism that results in clones of resistant plasmodia. This suggests that the drugs are competing for the same receptor. As was clearly discussed by Weiss *et al.* (1982) and by Prozialeck and Weiss (1982), all these drugs share a very large hydrophobic region and a side chain amino group at least four atoms removed from the hydrophobic aromatic ring structure that bind to compatible features of the receptor. This could account for the cross-resistance that exists between seemingly dissimilar classes of antimalarials (Black *et al.,* 1986; Gustafsson *et al.,* 1987; Scheibel *et al.,* 1987, 1989).

Martin *et al.* (1987) reported *in vitro* synergy between the Ca^{2+} antagonist verapamil and the antimalarial anti-CaM drug chloroquine in chloroquine-resistant but not chloroquine-sensitive clones of *P. falciparum* parasites. This could be due simply to the inhibition of sequential reactions (i.e., Ca^{2+} flux and CaM functions). It is more likely a result of effects on Ca^{2+} metabolism other than simple Ca^{2+} channel blockade (Krishna and Squire-Pollard, 1990). Electron microscope studies showed that this combination of drugs in resistant parasites resulted in food vacuolar changes similar to those that occurred with CsA, chloroquine, or mefloquine in drug-sensitive parasites (Jacobs *et al.,* 1988; Scheibel *et al.,* 1987, 1989). Kyle *et al.* (1987) enlarged the group of Ca^{2+} antagonists that modulated chloroquine resistance. These are chlorpromazine (which functions both as a CaM antagonist and Ca^{2+} antagonist), verapamil, methoxyverapamil, diltiazem, and nifedipine. Similar results were observed in the *Plasmodium chabaudi* system *in vivo,* but verapamil increased the susceptibility to chloroquine of both susceptible and resistant strains (Tanabe *et al.,* 1990), in contrast to what is seen in *P. falciparum* (Martin *et al.,* 1987; Krogstad *et al.,* 1987). Gallopamil and devapamil (close structural analogs of the Ca^{2+} channel blocker and CaM antagonist, verapamil) also completely reverse chloroquine resistance of the human parasite *in vitro.* However, these stereoisomers do not bind to Ca^{2+} channels in cardiovascular tissue (Ye and VanDyke, 1988). Further work by Deloron and coworkers (P. Deloron, L. K. Basco, B. Dubois, C. Gaudin, F. Clavier, J. LeBras, and F. Verdier, unpublished observations) showed the dextrogyre enantiomer of the channel blocker amlodipine reverses chloroquine resistance *in vitro* in *P. falciparum* and *in vivo* in mice with *Plasmodium yoelii nigeriensis.* The dextrogyre enantiomer of amlodipine has almost no Ca^{2+} channel blockade potency. These studies as well as evidence presented by Krishna and Squire-Pollard (1990) would suggest that reversal of drug resistance is not dependent on Ca^{2+} channel blockade.

Bitonti *et al.* (1988) reported the CaM antagonists desipramine and imipramine (Asano and Hidaka, 1984), as well as several other tricyclic psychotropic drugs, reverse chloroquine resistance *in vitro* and *in vivo* in *P. falciparum.* Peters *et al.* (1989, 1990) added ketotifen, cyproheptadine, pizotifen, and others to the list of tricyclic antihistamines that reverse chloroquine resistance in the *in vitro* human and *in vivo* rodent malaria system. Desipramine appears to be more active than cyproheptadine, however (Bitonti and McCann, 1989), but this may be limited to certain strains of parasites (Basco and LeBras, 1990). It is not at all clear in the *in vivo* studies reported whether parasites were completely eradicated or if the animals recrudesce and die after a given dosage of drugs.

Krogstad *et al.* (1987) reported that drug-resistant strains of *P. falciparum* parasitizing red cells release chloroquine at a rate 40- to 50-fold greater than that of sensitive plasmodia. This enhanced release in resistant parasites is sensitive to inhibition by verapamil, diltiazem, and vinblastine, all known to be CaM antagonists. They therefore proposed that one mechanism of resistance to chloroquine results from enhanced efflux of antimalarials through a drug-sensitive efflux pump in chloroquine-resistant malaria-parasitized erythrocytes. It had been known for many years that red cells infected by chloroquine-resistant malaria parasites accumulate only one-half the amount of chloroquine found in chloroquine-sensitive cells (Macomber *et al.*, 1966, 1967). These workers proposed that the antimalarial activity of the drug may be a result of the concentrating mechanism within the parasitized cell and resistance may be a result of some impairment to this mechanism. This is compatible with the mechanism proposed by Martin *et al.* (1987), where verapamil inhibits the active efflux of drugs in resistant cells. The findings of Macomber *et al.* (1966, 1967) also agree with those of Krogstad *et al.* (1987). Martin *et al.* (1987) as well as Krogstad *et al.* (1987) pointed to the similarity that appears to exist between cultures of drug-resistant neoplastic cells and those of *P. falciparum*-infected erythrocytes. This is that verapamil inhibits the active efflux of drugs from both types of resistant cells.

Multidrug resistance (MDR) appears to result from an altered ability to intracellularly accumulate drugs in a variety of cell types (Higgins, 1989a,b; Pastan and Gottesman, 1987; Ames, 1986; Gros *et al.*, 1986; Chen *et al.*, 1986; Gerlach, 1986; Juranka *et al.*, 1989). In these cells the drug is actively pumped out against a concentration gradient and toxicity to the cell is reduced. Resistance varies inversely to the amount of drug that accumulates and as resistance increases so does drug efflux. A number of characteristics are seemingly shared between multidrug-resistant tumor cells and chloroquine resistance in *P. falciparum*. A portion of DNA is amplified and this gene amplification appears to correlate quantitatively with the level of resistance at least in tumor cells. Multidrug-resistant cells appear to consistently reflect an increase in a 170-kDa plasma membrane glycoprotein known as the P-glycoprotein or MDR protein. This protein has a high degree of homology to other proteins in various cells that are involved in energy coupling and the transport of ions and other molecules.

This P-glycoprotein is absent from drug-sensitive cells, however. The MDR protein consists of two parts, each containing a hydrophobic domain of transmembrane loops and a hydrophilic ATP-binding domain on the intracellular side of the cell membrane. The hydrolysis of ATP provides the energy for transport or efflux across the membrane. In addition, Pastan and Gottesman (1987) pointed out that many of the substrates for the multidrug-resistant gene are compounds with complex hydrophobic ring structures and carry a positive charge. This may account for the cross-resistance that occurs in tumor cells to a variety of seemingly unrelated compounds. Is it possible that structural features that contribute to the binding of the various CaM antagonists to CaM also contribute to their binding to the P-glycoprotein? A variety of lipophilic compounds, e.g., nonionic detergents, local anesthetics, Ca^{2+} channel blockers, CaM inhibitors, steroids, nontoxic drug analogs, and cyclosporines, overcome resistance (Slater *et al.,* 1986; Juranka *et al.,* 1989) and apparently compete with the export of antitumor agents (Zernig, 1990). Many of these chemosensitizing agents bind to and antagonize CaM (Hait and Lazo, 1986), and there is now evidence that their ability to reverse resistance is not necessarily related to Ca^{2+} channel blockade but may instead be due to similar binding sites that exist on CaM and the P-glycoprotein as well as other pump proteins. Trifluoperazine but not verapamil synergize CsA in the inhibition of lymphocyte proliferation (Tesi *et al.,* 1985). In addition, various enantiomers of Ca^{2+} antanonists that lack channel blockade activity also reverse resistance. The mechanism by which a drug such as verapamil inhibits the efflux of antitumor agents in cancer cells has not been clearly explained, but Martin *et al.* (1987) pointed out that verapamil binds to the inner surface of the plasmalemma in cardiac tissues and inhibits a Ca^{2+}-activated ATPase. They go on to state that while there is no evidence that this pump is related to the P-glycoprotein-associated efflux pump in tumor cells such a pump may explain chloroquine resistance in *P. falciparum.*

Plasmodium falciparum with multiple drug resistance contains at least two genes similar in sequence to the MDR genes in human multidrug-resistant tumor cells (Wilson *et al.,* 1989). In addition, at least one of these genes is expressed at a higher level and is present in higher copy number in the resistant malaria parasite than is seen in the sensitive strain. These workers concede, however, that other mechanisms of drug resistance may also take place in *P. falciparum.* Similar findings were independently reported in two chloroquine-resistant isolates by Foote *et al.* (1989, 1990). These workers state that while none of the three drug-sensitive isolates of *P. falciparum* show any amplification of the multidrug-resistance MDR gene, not all the chloroquine-resistant parasites show changes in copy number of the MDR gene either. Three chloroquine-resistant isolates have no amplification of the MDR gene encoding the protein capable of transporting drug from the cell. In addition, further work by Wellems *et al.* (1990) indicates no relationship between rapid efflux of drug, the resistance of chloroquine phenotype, multidrug-resistant MDR genes (*pfmdr 1* or *pfmdr 2*) or amplification of these genes. Mefloquine resistance is also not linked to the *pfmdr 1* gene. Even though the structure of the P-glycoprotein is conserved in species phylogenetically distinct from one another and it is a member of the same transport protein family in both eukaryotes and prokaryotes (Juranka *et al.*, 1989), the degree of amino acid sequence conservation between one of the *Plasmodium pfmdr* genes to the human P-glycoprotein class I gene is only 57%.

Watt *et al.* (1990) used Western blots of purified plasma membranes and probed with mouse monoclonal antibody directed against the cytoplasmic domain of the P-glycoprotein of various mammalian cells and *P. falciparum*-parasitized red cells. They found only those cells in which reversing agents enhanced drug accumulation reacted positively for P-glycoprotein, but these agents may block export from normal tissues as well. This could result in undesirable toxicity complicating the use of these agents.

Therefore, it would appear that reversal of resistance in malarial parasites as in tumor cells goes beyond the presence of a surface glycoprotein or efflux pump (Nakagawa *et al.*, 1986; Hait and Lazo, 1986; Scheibel *et al.*, 1989; Newbold, 1990). Further evidence of this follows. There is a well-described CaM-regulated Ca^{2+} efflux pump present in erythrocyte membranes (Penniston *et al.*, 1980). R24571 inhibits the CaM-mediated Ca^{2+}-ATPase transport pump in cell membranes rather profoundly, with some degree of specificity (Van Belle, 1981; Gietzen *et al.*, 1981). One would expect marked potentiation between R24571 and the antimalarials if this efflux pump is essential for drug resistance, as had been suggested in some tumor cells (Hait and Lazo, 1986). In contrast, data published by Scheibel *et al.* (1987, 1989) showed a high degree of anatagonism between chloroquine and R24571 in the resistant clone of *P. falciparum*. Therefore, their data would indicate drug resistance depends on other factors besides enhanced efflux of antimalarials by this pump. There may be commonality between the drug-binding site of various pump proteins that are present in cells as well as a similar site on CaM, however. This may be the reason why many CaM antagonists reverse drug resistance and apparently bind to *P. glycoprotein*. In any event, much remains to be learned about the structure of Ca^{2+}-ATPases. Certainly there are a variety of cation-transporting membrane ATPases in eukaryotes. There is an inward movement of Ca^{2+} and other components in plasmodia, as discussed previously here and elsewhere (Scheibel, 1988; Scheibel and Sherman, 1988; Izumo *et al.*, 1988, 1989; Mikkelsen *et al.*, 1982, 1986; Tanabe *et al.*, 1982a; Tanabe,

1983, 1990). In eukaryotic cells this Ca^{2+} is sequestered into cell organelles. Tanabe *et al.* (1982a) showed Ca^{2+} transport processes in malaria parasites are less sensitive to the inhibitors of mitochondrial electron transport (1 m*M* KCN and 10 n*M* NaN_3). They interpreted this to mean that plasmodial mitochondria are not actively involved in the regulation of cytoplasmic Ca^{2+} levels. They therefore considered the Ca^{2+}-ATPase to be of vital importance to regulation of malarial cytoplasmic Ca^{2+} levels. In this regard, Murakami *et al.* (1991) reported *Plasmodium yoelii* has a cation-transporting ATPase expressed during the erythrocytic stage of development. There is homology between this protein and a similar one in *P. falciparum*. In addition, they reported the amino acid sequence of the functionally important domains appears to be conserved even in evolutionarily distant organisms. Overall homology, however, may be considerably lower. Differences such as this may be reflected in a variety of responses to pharmacological agents.

In the free and intraerythrocytic merozoite (Scheibel *et al.*, 1987, 1989), immunoelectron microscopy revealed CaM to be concentrated at the apical complex (see Fig. 5), which is believed to play a role in the penetration of the host red cells. Evidence published by Wasserman *et al.* (1982) suggests this is a Ca^{2+}-dependent processes. These data would indicate that CaM is also involved in this process. When the CaM inhibitors W-7, R24571, or trifluoperazine (TFP) were added to synchronized cultures of *P. falciparum,* invasion of red cells by merozoites was inhibited. In addition, maturation of the parasite to the schizont stage was also inhibited (Matsumoto *et al.*, 1987; Scheibel *et al.*, 1989). Lower doses of these agents were required to inhibit the appearance of ring stages than what was required for the inhibition of schizont maturation. Cyclosporin (CsA) was equally effective in the inhibition of both processes at low concentrations. W-5, a chlorine-deficient analog of W-7 and known to have less affinity for CaM, is a less effective inhibitor of these processes, supporting the view that these processes are CaM mediated. When merozoites were grown in the presence of 10 μM trifluoperazine but not W-5, W-7, R24571, or EGTA no immunoreactivity was observed at the apical end by immunoelectron microscopy. Because rhopty components located at the apical end of the merozoite appear to participate in penetration of the red cell, the above data suggest CaM plays a role in the release of rhoptry components. Alternatively, CaM may regulate microtubule assembly and disassembly underlying the plasma membrane at the apical end and play a role in mobility of the merozoite (Matsumoto *et al.*, 1987). Vinblastine (5×10^{-5} *M*) and colchicine (5×10^{-4} *M*), agents inhibitory to microtubular processes, are known to exert a profound effect on *P. knowlesi* invasion of erythrocytes *in vitro*. High doses of cAMP (10^{-5} *M*) partially antagonized this effect (McColm *et al.*, 1980).

The *in vitro* induction of gametocytogenesis is enhanced in a number of strains of *P. falciparum* in RPMI-FSC reactive medium by caffeine (2 m*M*/ml) or calmodulin (5 μg/ml; source of CaM unknown) (Ono and Nakabayashi, 1989). These gametocytes continue to deveop to ookinetes in the candle jar in Waymouth's medium. The mechanism by which this takes place is not completely understood, however. Kaushal *et al.* (1980) reported that 1 m*M* cAMP stimulates gametocytogenesis in culture. CaM would be expected to stimulate adenylate cyclase in mammalian cells and facilitate production of cAMP. In addition, caffeine would be expected to inhibit PDE, the enzyme that degrades cAMP. Ono and Nakabayashi (1989) and others, however, were unable to confirm any stimulation of gametocytogenesis by cAMP. Several years ago Brockelman (Ono and Nakabayashi, 1989) reported caffeine improves conditions for gametocytogenesis. Kaushal *et al.* (1980) conceded that prolonged exposure to 1 m*M* cAMP is lethal to parasites, so the timing of this procedure as well as other conditions may be critical. It is known that a number of factors do play a role in the process of gametocytogenesis (Carter and Graves, 1989). It would be interesting to compare these cultures to those of Read and Mikkelsen (1990, 1991a,b). They reported that levels of a Ca^{2+}-dependent, CaM-independent protein kinase, not adenylate cyclase, reflect the gametocyte-producing potential of a strain of *P. falciparum*.

VII. Concluding Remarks

An enormous amount of painstaking work has been devoted to the study of Ca^{2+}/CaM-mediated processes in protozoa. This includes the sequencing of CaM, fractionation of cilia and flagella, purifying subcellular fractions, etc. Nevertheless, a number of important questions remain to be answered. It is generally agreed that CaM is present in all eukaryotic cells occupying a critical role in cell survival. We have reviewed a number of reports where Ca^{2+}-dependent processes are stimulated by CaM in mammalian tissue but some of the same processes in protozoa appear to be CaM independent. Most notable are the Ca^{2+}-stimulated ATPases described before CaM was recognized. More recently, however, there have been reports of adenylate cyclase, phosphodiesterase, Ca^{2+}-ATPase in *Trypanosoma* and *Tetrahymena,* as well as a Ca^{2+}-dependent protein kinase in *Plasmodium,* all of which are CaM independent. Dolichol kinase from *Tetrahymena* and rat brain has a Ca^{2+}/CaM requirement, while that from rat liver does not. In the light of evolving technology would it be possible that if investigators had reagent-grade CaM available to them, could they have demonstrated CaM dependence in these reactions? Of greater importance, if autologous CaM had been available to these investigators would greater CaM dependence have been recognized, based on possible selective interaction with novel regions of protozoal CaM? For example, the cAMP phosphodiesterase from *T. cruzi* is activated to a greater degree in the presence of *T. cruzi* CaM. *Tetrahymena* guanylate cyclase has a specific requirement for *Tetrahymena* CaM. The trimethyllated lysine at position 115 results in different functional activity than its unmethyllated counterpart. This specificity extends to other conserved proteins. *Naegleria* actin is unusual in several ways compared to actin from the ameboid forms of *Physarum, Dictyostelium,* or *Acanthamoeba.* The amino acid sequence of β-tubulin from *Trypanosoma, Leishmania, Euglena,* and *Chlamydomonas* are very similar to one another, but the human form is different. If there are differences in these conserved proteins could there be differences in another conserved protein, the P-glycoprotein?

It is generally accepted that specific cellular responses can be elicited by changes in Ca^{2+}/CaM (Hait and Lazo, 1986; Ruben *et al.,* 1990). A few of these are differences in:

1. Distribution of CaM in the cell
2. Concentration of CaM in the cell
3. Structural features of CaM in the cell
4. Ca^{2+}- or CaM-binding proteins
5. Intracellular free Ca^{2+} or Ca^{2+} flux

Further knowledge about the subtle differences between these processes in various cells may not only increase our understanding of comparative biochemistry/pharmacology but also improve our use of antiprotozoal agents.

Acknowledgments

I am deeply indebted to Dr. Mark Adelman, Uniformed Services University of the Health Sciences, Bethesda, Maryland, and Dr. Larry Ruben, Southern Methodist University, Dallas, Texas, for reviewing the manuscript and for their many helpful suggestions. I also thank Dr. W. Y. Cheung and the publishers of "The Cell" (Garland Publishing Co., New York) for permission to use adaptations from previously published figures and to Drs. L. Ruben, C. L. Patton, and Matsumishi Aikawa, Case Western Reserve University, Cleveland, Ohio, for graciously supplying figures and photos for the manuscript. Last but not least I am most grateful to Dr. K. W. Jeon, Editor of the *International Review of Cytology,* for his patience in waiting for this lengthy undertaking to be completed, as well as to Dr. Martin Watterson for his encouragement during the writing of this manuscript and for supplying me with helpful references.

References

Adrian, G. S., and Keenan, R. W. (1979). *Biochim. Biophys. Acta* **575,** 431–438.
Aikawa, M. (1972). *Am J. Pathol.* **67,** 277–280.
Albanesi, J. P., Hammer, J. A., III, and Korn, E. D. (1983). *J. Biol. Chem.* **258,** 10176–10181.
Albanesi, J. P., Fujisaki, H., Hammer, J. A., III, Korn, E. D., Jones, R., and Sheetz, M. P. (1985a). *J. Biol. Chem.* **260,** 8649–8652.
Albanesi, J. P., Fujisaki, H., and Korn, E. D. (1985b). *J. Biol. Chem.* **260,** 11174–11179.
Albanesi, J. P., Coue, M., Fujisaki, H., and Korn, E. D. (1985c). *J. Biol. Chem.* **260,** 13276–13280.
Alberts, B., Bray, D., Lewis, J., Raff, M., Roberts, K., and Watson, J. D. (1989). "Molecular Biology of The Cell," 2nd Ed., pp. 613–680. Garland, New York.
Allen, R. D. (1968). *J. Cell Biol.* **37,** 825–831.
Ames, G. F.-L. (1986). *Cell* **47,** 323–324.
Amos, W. B., Routledge, L. M., and Yew, F. F. (1975). *J. Cell Sci.* **19,** 203–213.
Anderson, A., Drakenberg, T., and Forsen, S. (1985). *In* "Calmodulin Antagonists and Cellular Physiology" (H. Hidaka and D. J. Hartshorne, eds.), pp. 27–44. Academic Press, Orlando, Florida.
Anderson, B., Osborn, M., and Weber, K. (1978). *Cytobiologie* **17,** 354–364.
Andrivon, C., Brugerolle, G., and Delachambre, D. (1983). *Biol. Cell.* **47,** 351–364.
Appleman, M. M., Thompson, W. J., and Russell, T. R. (1973). *Adv. Cyclic Nucleotide Res.* **3,** 65–98.
Asano, M., and Hidaka, H. (1984). *In* "Calcium and Cell Function" (W. Y. Cheung, ed.), Vol. 5, pp. 123–164. Academic Press, San Diego.
Bababunmi, E. A., Emeh, J. K., and Bolaj, D. M. (1982). *Ann. N.Y. Acad. Sci.* **402,** 435–437.
Baker, J. R. (1977). *Parasit. Protozoa* **1,** 35–56.
Barnwell, J. W. (1989). *Exp. Parasitol.* **69,** 407–412.
Barnwell, J. W., Asch, A. S., Nachman, R. L., Yamaya, M., Aikawa, M., and Igravallo, P. (1989). *J. Clin. Invest.* **84,** 765–772.
Basco, L. K., and LeBras, J. (1990). *Lancet* **1,** 422.
Baugh, L. C., Satir, P., and Satir, B. (1976). *J. Cell Biol.* **70,** 66a (abstr.).
Benaim, G., and Romero, P. J. (1990). *Biochim. Biophys. Acta* **1027,** 79–84.
Benaim, G., Szabo, V., and Cornivelli, L. (1987). *Acta Cient. Venez.* **38,** 289–291.
Bergquist, B. L. (1989). *Trans. Am. Microsc. Soc.* **108,** 369–379.
Bergquist, B. L., and Bovee, E. C. (1976). *Acta Protozool.* **15,** 471–483.
Bergquist, B. L., Wyttenbach, E., and Bovee, E. C. (1986). *Biol. Trace Elem. Res.* **9,** 125–138.
Betticher, D. C., Nicole, A., Pugin, P., and Regamey, C. (1990). *J. Infect. Dis.* **161,** 157–159.
Bilinski, M., Plattner, H., and Tiggemann, R. (1981). *Eur. J. Cell Biol.* **24,** 108–115.
Bitonti, A. J., and McCann, P. P. (1989). *Lancet* **2,** 1282–1283.
Bitonti, A. J., Sjoerdsma, A., McCann, P. P., Kyle, D. E., Oduola, A. M. J., Rossan, R. N., Milhous, W. K., and Davidson, D. E., Jr. (1988). *Science* **242,** 1301–1303.
Black, R. H., Canfield, C. I., Clyde, D. F., Peters, W., and Wernsdorfer, W. H. (1986). *In* "Chemotherapy of Malaria" (L. J. Bruce-Chwatt, ed.), 2nd Rev. Ed., pp. 65–66. World Health Organ., Geneva, Switzerland.
Blum, J. J. (1967). *Proc. Natl. Acad. Sci. U.S.A.* **58,** 81–88.
Blum, J. J. (1970). *Arch. Biochem. Biophys.* **137,** 65–74.
Blum, J. J., and Hayes, A. (1984). *J. Cell. Biochem.* **24,** 373–384.
Blum, J. J., Hayes, A., Jamieson, G. A., Jr., and Vanaman, T. C. (1980). *J. Cell Biol.* **87,** 386–397.

Blum, J. J., Hayes, A., Jamieson, G. A., Jr., and Vanaman, T. C. (1981). *Arch. Biochem. Biophys.* **210,** 363–371.

Blum, J. J., Hayes, A., Vanaman, T., and Schachat, F. H. (1982). *J. Cell. Biochem.* **19,** 45–57.

Bookchin, R. M., Lew, V. L., Nagel, R. L., and Raventos, C. (1981). *J. Physiol.* (*London*) **312,** 65P.

Boothroyd, J. C. (1985). *Annu. Rev. Microbiol.* **39,** 475–502.

Browning, J. L., and Nelson, D. L. (1976). *Biochim. Biophys. Acta* **448,** 338–351.

Carboni, J. M., Conzelman, K. A., Adams, R. A., Kaiser, D. A., Pollard, T. D., and Mooseker, M. S. (1988). *J. Cell Biol.* **107,** 1749–1757.

Carter, R., and Graves, P. M. (1988). *In* "Malaria Principles and Practice of Malariology" (W. H. Wernsdorfer and I. McGregor, eds.), Vol. I, pp. 253–305. Churchill Livingstone, Edinburgh, Scotland.

Chafouleas, J. G., Dedman, J. R., Munjaal, R. P., and Means, A. R. (1979). *J. Biol. Chem.* **254,** 10262–10267.

Chan, M. M.-Y., and Fong, D. (1990). *Science* **249,** 924–926.

Chen, C., Chin, J. E., Ueda, K., Clark, D. P., Pastan, I., Gottesman, M. M., and Roninson, I. B. (1986). *Cell* **47,** 381–389.

Cheung, W. Y. (1970). *Biochem. Biophys. Res. Commun.* **38,** 533–538.

Cheung, W. Y. (1971). *J. Biol. Chem.* **246,** 2859–2869.

Cheung, W. Y. (1980). *Science* **207,** 19–27.

Cheung, W. Y. (1988). *In* "Calcium Antagonists, Pharmacology and Clinical Research" (P. M. Vanhoutte, R. Paoletti, and S. Govoni, eds.), pp. 74–87. N.Y. Acad. Sci., New York.

Choi, I., and Mikkelsen, R. B. (1990). *Exp. Parasitol.* **71,** 452–462.

Chua, B., Elson, C., and Shrago, E. (1977). *J. Biol. Chem.* **252,** 7548–7554.

Clarke, M., and Spudich, J. A. (1974). *J. Mol. Biol.* **86,** 209–222.

Clarkson, A. B., Jr., and Amole, B. O. (1982). *Science* **216,** 1321–1323.

Cohen, M. G., Kevat, S., Prowse, M. V., and Ahern, M. J. (1988). *Ann. Intern. Med.* **108,** 369–371.

Collins, J. H., and Korn, E. D. (1980). *J. Biol. Chem.* **255,** 8011–8014.

Collins, J. H., and Korn, E. D. (1981). *J. Biol. Chem.* **256,** 2586–2595.

Collins, J. H., Cote, G. P., and Korn, E. D. (1982a). *J. Biol. Chem.* **257,** 4529–4534.

Collins, J. H., Kuznicki, J., Bowers, B., and Korn, E. D. (1982b). *Biochemistry* **21,** 6910–6915.

Colombani, P. M., Robb, A., and Hess, A. D. (1985). *Science* **228,** 337–339.

Colombani, P. M., Bright, E. C., and Hess, A. D. (1986). *Transplant. Proc.* **18,** 866–869.

Conner, R. L., Chook, S. A., and Ray, E. (1963). *J. Cell Biol.* **19,** 15a–16a (abstr.).

Cote, G. P., Collins, J. H., and Korn, E. D. (1981). *J. Biol. Chem.* **256,** 12811–12816.

Cote, G. P., Albanesi, J. P., Ueno, T., Hammer, J. A., III, and Korn, E. D. (1985). *J. Biol. Chem.* **260,** 4543–4546.

deAlmeida, M. L. C., and Turner, M. J. (1983). *Nature* (*London*) **302,** 349–352.

D'hondt, J., Van Meirvenne, N., Moens, L., and Kondo, M. (1979). *Nature* (*London*) **282,** 613–615.

Dickens, M. S., Lucas-Lenard, J., and Roth, J. S. (1975). *Biochem. Biophys. Res. Commun.,* **67,** 1319–1325.

Dickinson, J. R., Graves, M. G., and Swoboda, B. E. P. (1976). *FEBS Lett.* **65,** 152–154.

Dolan, M. T., Reid, C. G., and Voorheis, H. P. (1986). *J. Cell Sci.* **80,** 123–140.

Doughty, M. J. (1978a). *Comp. Biochem. Physiol.* **60B,** 339–345.

Doughty, M. J. (1978b). *Comp. Biochem. Physiol.* **61C,** 369–373.

Doughty, M. J. (1978c). *Comp. Biochem. Physiol.* **61C,** 375–384.

Doughty, M. (1979). *Comp. Biochem. Physiol.* **64B,** 255–266.
Doughty, M. J., and Kaneshiro, E. S. (1983). *J. Protozool.* **30,** 565–573.
Dunlap, K. (1977). *J. Physiol.* (*London*) **271,** 119–133.
Ebashi, S. (1988). *Ann. N.Y. Acad. Sci.* **522,** 51–59.
Ebashi, S., and Endo, M. (1968). *Prog. Biophys. Mol. Biol.* **18,** 123–183.
Eckert, R. (1972). *Science* **176,** 473–481.
Eckert, R., and Brehm, P. (1979). *Ann. Rev. Biophys. Bioeng.* **8,** 353–383.
Edmunds, L. N., Jr., Laval-Martin, D. L., and Goto, K. (1987). *Ann. N.Y. Acad. Sci.* **503,** 459–475.
Epstein, P. M., Fiss, K., Hachisu, R., and Andrenyak, D. M. (1982). *Biochem. Biophys. Res. Commun.* **105,** 1142–1149.
Ettienne, E. M. (1970). *J. Gen. Physiol.* **56,** 168–179.
Fisher, G., Kaneshiro, E. S., and Peters, P. D. (1976). *J. Cell Biol.* **69,** 429–442.
Flicker, P. F., Peltz, G., Sheetz, M. P., Parham, P., and Spudich, J. A. (1985). *J. Cell Biol.* **100,** 1024–1030.
Fok, A. K., Allen, R. D., and Kaneshiro, E. S. (1981). *Eur. J. Cell Biol.* **25,** 193–201.
Fong, D., and Lee, B. (1988). *Mol. Biochem. Parasitol.* **31,** 97–106.
Foote, S. J., Thompson, J. K., Cowman, A. F., and Kemp, D. J. (1989). *Cell* **57,** 921–930.
Foote, S. J., Kyle, D. E., Martin, R. K., Oduola, A. M. J., Forsyth, K., Kemp, D. J., and Cowman, A. F. (1990). *Nature* (*London*) **345,** 255–258.
Frash, A. C. C., Segura, E. L., Cazzulo, J. J., and Stoppani, A. O. M. (1978a). *Comp. Biochem. Physiol. B* **60B,** 271–275.
Frash, A. C. C., Cazzulo, J. J., and Stoppani, A. O. M. (1978b). *Comp. Biochem. Physiol. B* **61B,** 207–212.
Fujisaki, H., Albanesi, J. P., and Korn, E. D. (1985). *J. Biol. Chem.* **260,** 11183–11189.
Fulton, C. (1977a). *Annu. Rev. Microbiol.* **31,** 597–629.
Fulton, C. (1977b). *J. Supramol. Struct.* **6,** 13–43.
Fulton, C. (1983). *J. Protozool.* **30,** 192–198.
Fulton, C., and Lai, E. Y. (1980). *J. Cell Biol.* **87,** 282a (abstr.).
Fulton, C., and Walsh, C. (1980). *J. Cell Biol.* **85,** 346–360.
Fulton, C., Cheng, K.-L., and Lai, E. Y. (1986). *J. Cell Biol.* **102,** 1671–1678.
Gadasi, H., and Korn, E. D. (1979). *J. Biol. Chem.* **254,** 8095–8098.
Gadasi, H., and Korn, E. D. (1980). *Nature* (*London*) **286,** 452–456.
Gadasi, H., Maruta, H., Collins, J. H., and Korn, E. D. (1979). *J. Biol. Chem.* **254,** 3631–3636.
Gandhi, C. R., and Keenan, R. W. (1983a). *J. Biol. Chem.* **258,** 7639–7643.
Gandhi, C. R., and Keenan, R. W. (1983b). *Biochem. J.* **216,** 317–323.
Gbenle, G. O. (1985). *Mol. Biochem. Parasitol.* **15,** 37–47.
Gbenle, G. O., and Akinrimisi, E. O. (1982). *Mol. Biochem. Parasitol.* **5,** 213–220.
Gbenle, G. O., Opperdoes, F. R., and Van Roy, J. (1986). *Acta Trop.* **43,** 295–305.
Geary, T. G., Divo, A. A., and Jensen, J. B. (1986). *Antimicrob. Agents Chemother.* **30,** 785–788.
Gerlach, J. H., Endicott, J. A., Juranka, P. F., Henderson, G., Sarangi, F., Deuchars, K. L., and Ling, V. (1986). *Nature* (*London*) **324,** 485–489.
Gibbons, I. R. (1966). *J. Biol. Chem.* **241,** 5590–5596.
Gibbons, I. R. (1982). *Prog. Clin. Biol. Res.* **80,** 87–93.
Gibbons, I. R., and Grimstone, A. V. (1960). *J. Biophys. Biochem. Cytol.* **7,** 697–715.
Gietzen, K., Wüthrich, A., and Bader, H. (1981). *Biochem. Biophys. Res. Commun.* **101,** 418–425.
Gitelman, S. E., and Witman, G. B. (1980). *J. Cell Biol.* **87,** 764–770.
Goldberg, N. D., and Haddox, M. K. (1977). *Annu. Rev. Biochem.* **46,** 823–896.

Goncalves, M. F., Zingales, B., and Colli, W. (1980). *Mol. Biochem. Parasitol.* **1,** 107–118.
Gonzales-Perdomo, M., Romero, P., and Goldenberg, S. (1988). *Exp. Parasitol.* **66,** 205–212.
Goto, K., Laval-Martin, D. L., and Edmunds, L. N., Jr. (1985). *Science* **228,** 1284–1288.
Graves, M. G., Dickinson, J. R., and Swoboda, B. E. P. (1976). *FEBS Lett.* **69,** 165–166.
Gray, N. C. C., Dickinson, J. R., and Swoboda, B. E. P. (1977). *FEBS Lett.* **81,** 311–314.
Gros, P., Croop, J., and Housman, D. (1986). *Cell* **47,** 371–380.
Gustafsson, L. L., Beermann, B., and Abdi, Y. A. (1987). "Handbook of Drugs for Tropical Parasitic Infections," pp. 70–72. Taylor & Francis, Philadelphia, Pennsylvania.
Hait, W. N., and Lazo, J. S. (1986). *J. Clin. Oncol.* **4,** 994–1012.
Hammer, J. A., III, Albanesi, J. P., and Korn, E. D. (1983). *J. Biol. Chem.* **258,** 10168–10175.
Hammer, J. A., III, Korn, E. D., and Paterson, B. M. (1984). *J. Biol. Chem.* **259,** 11157–11159.
Hartshorne, D. J. (1985). *In* "Calmodulin Antagonists and Cellular Physiology" (H. Hidaka and D. J. Hartshorne, eds.), pp. 3–12. Academic Press, Orlando, Florida.
Heilbrunn, L. V. (1940). *Physiol. Zool.* **13,** 88–94.
Heilbrunn, L. V., and Wiercinski, F. J. (1947). *J. Cell. Comp. Physiol.* **29,** 15–32.
Hemming, F. W. (1974). *In* "Biochemistry of the Lipids" (T. W. Goodwin, ed.), pp. 39–97. Univ. Park Press, Baltimore, Maryland.
Hess, A. D., Colombani, P. M., Donnenberg, A. D., Fischer, A. C., and Ryffel, B. (1985). *Transplant. Proc.* **17,** 1419–1427.
Hess, A. D., Tuszynski, T., Engel, P., Colombani, P. M., Farrington, J., Wenger, R., and Ryffel, B. (1986). *Transplant. Proc.* **18,** 861–865.
Hidaka, H., and Tanaka, T. (1985). *In* "Calmodulin Antagonists and Cellular Physiology" (H. Hidaka and D. J. Hartshorne, eds.), pp. 13–23. Academic Press, Orlando, Florida.
Higgins, C. (1989a). *Nature* (*London*) **340,** 342.
Higgins, C. (1989b). *Nature* (*London*) **341,** 103.
Hirano, J., and Watanabe, Y. (1985). *Exp. Cell Res.* **157,** 441–450.
Hirano-Ohnishi, J., and Watanabe, Y. (1988). *Exp. Cell Res.* **178,** 18–24.
Hirano-Ohnishi, J., and Watanabe, Y. (1989). *J. Biochem.* (*Tokyo*) **105,** 858–860.
Holmes, R. P., and Stewart, P. R. (1979). *Biochim. Biophys. Acta* **545,** 94–105.
Inagaki, M., Naka, M., Nozawa, Y., and Hidaka, H. (1983). *FEBS Lett.* **151,** 67–70.
Iwasa, Y., Yonemitsu, K., Matsui, K., Fukunaga, K., and Miyamoto, E. (1981). *FEBS Lett.* **128,** 311–314.
Izumo, A., Tanabe, K., and Kato, M. (1988). *Comp. Biochem. Physiol. B* **91B,** 735–739.
Izumo, A., Tanabe, K., Kato, M., Doi, S., Maekawa, K., and Takada, S. (1989). *Parasitology* **98,** 371–379.
Jacobs, G. H., Aikawa, M., Milhous, W. K., and Rabbege, J. R. (1987). *Am. J. Trop. Med. Hyg.* **36,** 9–14.
Jacobs, G. H., Oduola, A. M. J., Kyle, D. E., Milhous, W. K., Martin, S. K., and Aikawa, M. (1988). *Am. J Trop. Med. Hyg.* **39,** 15–20.
Jamieson, G. A., Jr., and Vanaman, T. C. (1979). *Biochem. Biophys. Res. Commun.* **90,** 1048–1056.
Jamieson, G. A., Jr., Vanaman, T. C., and Blum, J. J. (1979). *Proc. Natl. Acad. Sci. U.S.A.* **76,** 6471–6475.
Jamieson, G. A., Jr., Vanaman, T. C., Hayes, A., and Blum, J. J. (1980). *Ann N.Y. Acad. Sci.* **356,** 391–392.
Janakidevi, K., Dewey, V. C., and Kidder, G. W. (1966a). *Arch. Biochem. Biophys.* **113,** 758–759.
Janakidevi, K., Dewey, V. C., and Kidder, G. W. (1966b). *J. Biol. Chem.* **241,** 2576–2578.
Johnson, J. D., and Wittenauer, L. A. (1983). *Biochem. J.* **211,** 473–479.
Johnson, J. G., Epstein, N., Shiroishi, T., and Miller, L. H. (1980). *Parasitology* **80,** 539–550.
Jones, H. P., Mathews, J. C., and Cormier, M. J. (1979). *Biochemistry* **18,** 55–60.

Juranka, P. E., Zastawny, R. L., and Ling, V. (1989). *FASEB J.* **3,** 2583–2592.
Kakiuchi, S., and Yamazaki, R. (1970a). *Biochem. Biophys. Res. Commun.* **41,** 1104–1110.
Kakiuchi, S., and Yamazaki, R. (1970b). *Proc. Jpn. Acad.* **46,** 387–392.
Kakiuchi, S., Yamazaki, R., and Nakajima, H. (1969). *Bull. Jpn. Neurochem. Soc.* **8,** 17–20.
Kakiuchi, S., Yamazaki, R., and Nakajima, H. (1970). *Proc. Jpn. Acad.* **46,** 587–592.
Kakiuchi, S., Yamazaki, R., and Teshima, Y. (1972). *Adv. Cyclic Nucleotide Res.* **1,** 455–477.
Kakiuchi, S., Sobue, K., Yamazaki, R., Nagao, S., Umeki, S., Nozawa, Y., Yazawa, M., and Yagi, K. (1981). *J. Biol. Chem.* **256,** 19–22.
Kamada, T., and Kinosita, H. (1943). *Jpn. J. Zool.* **10,** 469–493.
Kameyama, Y., Kudo, S., Ohki, K., and Nozawa, Y. (1982). *Jpn. J. Exp. Med.* **52,** 183–192.
Karaki, H., Murakami, K., Nakagawa, H., Ozaki, H., and Urakawa, N. (1982). *Br. J. Pharmacol.* **77,** 661–666.
Kariya, K., Saito, K., and Iwata, H. (1974). *Jpn. J. Pharmacol.* **24,** 129–134.
Kassis, S., and Kindler, S. H. (1975). *Biochim. Biophys. Acta* **391,** 513–516.
Kato, M., Tanabe, K., Miki, A., Ichimori, K., and Waki, S. (1990). *FEMS Microbiol. Lett.* **69,** 283–288.
Kaushal, D. C., Carter, R., Miller, L. H., and Krishna, G. (1980). *Nature (London)* **286,** 490–492.
Keith, K., Hide, G., and Tait, A. (1990). *Mol. Biochem. Parasitol* **43,** 107–116.
Keogh, P. P., and Shaw, P. M. (1944). *Aust. J. Exp. Biol. Med. Sci.* **22,** 139–147.
Kim, J. V., Kudzina, L. J., Zinchenko, V. P., and Evtodienko, J. V. (1984). *Cell Calcium* **5,** 29–41.
Kim, Y. V., Kudzina, L. Y., Zinchenko, V. P., and Evtodienko, Y. V. (1985). *J. Cell Sci.* **77,** 47–56.
Kippert, F. (1987). *Ann. N.Y. Acad. Sci.* **503,** 476–495.
Klee, C. B., Crouch, T. H., and Richman, P. G. (1980). *Annu. Rev. Biochem.* **49,** 489–515.
Kobayashi, R., Tawata, M., and Hidaka, H. (1979). *Biochem. Biophys. Res. Commun.* **88,** 1037–1045.
Kohama, K. (1981). *J. Biochem. (Tokyo)* **90,** 1829–1832.
Kohama, K., and Ebashi, S. (1986). *In* "The Molecular Biology of *Physarum polycephalum*" (W. F. Dove, J. Dee, S. Hatano, F. B. Haugli, and K.-E. Wohlfarth-Bottermann, eds.), pp. 175–190. Plenum, New York.
Kohama, K., and Kendrick-Jones, J. (1986). *J. Biochem. (Tokyo)* **99,** 1433–1446.
Kohama, K., and Kohama, T. (1984). *Proc. Jpn. Acad.* **60B,** 435–439.
Kohama, K., and Shimmen, T. (1985). *Protoplasma* **129,** 88–91.
Kohama, K., and Takano-Ohmuro, H. (1984). *Proc. Jpn. Acad.* **60B,** 431–434.
Kohama, K, Kobayashi, K., and Mitani, S. (1980). *Proc. Jpn. Acad.* **56B,** 591–596.
Kohama, K., Uyeda, T. Q. P., Takano-Ohmuro, H., Tanaka, T., Yamaguchi, T., Maruyama, K., and Kohama, T. (1985). *Proc. Jpn. Acad.* **61B,** 501–505.
Kohama, K., Takano-Ohmuro, H., Tanaka, T., Yameguchi, Y., and Kohama, T. (1986). *J. Biol. Chem.* **261,** 8022–8027.
Korn, E. D. (1982). *Physiol. Rev.* **62,** 672–737.
Korn, E. D., and Hammer, J. A., III (1988). *Annu. Rev. Biophys. Biophys. Chem.* **17,** 23–45.
Kremsner, P. G., Zotter, G. M., Feldmeier, H., Graninger, W., Rocha, R. M., Jansen-Rosseck, R., and Bienzle, U. (1990). *J. Infect. Dis.* **161,** 1025–1028.
Krishna, S., and Ng, L. L. (1989). *Exp. Parasitol.* **69,** 402–406.
Krishna, S., and Squire-Pollard, L. (1990). *Parasitol. Today* **6,** 196–198.
Kristiansen, J. E., and Jepsen, S. (1985). *Acta Pathol. Microbiol. Immunol. Scand. Sect. B* **93,** 249–251.
Krogstad, D. J., Gluzman, I. Y., Kyle, D. E., Oduola, A. M. J., Martin, S. K., Milhous, W. K., and Schlesinger, P. H. (1987). *Science* **238,** 1283–1285.

Krueger, B. K. (1989). *FASEB J.* **3,** 1906–1914.
Krungkrai, J., and Yuthavong, Y. (1983). *Mol. Biochem. Parasitol,* **7,** 227–235.
Kuczmarski, E. R., and Spudich, J. A. (1980). *Proc. Natl. Acad. Sci. U.S.A.* **77,** 7292–7296.
Kudo, S., and Nozawa, Y. (1983). *J. Protozool.* **30,** 30–36.
Kudo, S., Nakazawa, K., and Nozawa, Y. (1980). *J. Prontozool.* **17,** 342–345.
Kudo, S., Nagao, S., Kasai, R., and Nozawa, Y. (1981a). *J. Protozool.* **28,** 165–167.
Kudo, S., Nagao, S., Kameyama, Y., and Nozawa, Y. (1981b). *Cell Differ.* **10,** 237–242.
Kudo, S., Nakazawa, K., Nagao, S., and Nozawa, Y. (1982a). *Jpn. J. Exp. Med.* **52,** 193–200.
Kudo, S., Muto, Y., Nagao, S., Naka, M., Hidaka, H., Sano, M., and Nozawa, Y. (1982b). *FEBS Lett.* **149,** 271–276.
Kudo, S., Muto, Y., Nagao, S., and Nozawa. Y. (1983). *Biochem. Int.* **7,** 361–367.
Kudo, S., Muto, Y., Inagaki, M., Hidaka, H., and Nozawa, Y. (1985a). *Comp. Biochem. Physiol. B* **80B,** 455–498.
Kudo, S., Muto, Y., and Nozawa, Y. (1985b). *Comp. Biochem. Physiol. B* **80B,** 813–816.
Kumagai, H., Nishida, E., Ishiguro, K., and Murofushi, H. (1980). *J. Biochem. (Tokyo)* **87,** 667–670.
Kung, C., Chang, S.-Y., Satow, Y., VanHouten, J., and Hansma, H. (1975). *Science* **188,** 898–904.
Kusamran, K., Mattox, S. M., and Thompson, G. A., Jr. (1980). *Biochim. Biophys. Acta* **598,** 16–26.
Kuznicki, J., Kuznicki, L., and Drabikowski, W. (1979). *Cell Biol. Int. Rep.* **3,** 17–23.
Kyle, D. E., Oduola, A. M. J., Martin, S. K., and Milhous, W. K. (1987). *Proc. Int. Congr. Malaria Babesiosis, 3rd* p. 257 (abstr. 113 VII).
Lai, E. Y., Walsh, C., Wardell, D., and Fulton, C. (1979). *Cell* **17,** 867–878.
LeGrue, S. J., Turner, R., Weisbrodt, N., and Dedman, J. R. (1986). *Science* **234,** 68–71.
Leida, M. N., Mahoney, J. R., and Eaton, J. W. (1981). *Biochem. Biophys. Res. Commun.* **103,** 402–406.
Loffler, B.-M., Bohn, E., Hesse, B., and Kunze, H. (1985). *Biochim. Biophys. Acta* **835,** 448–455.
London, J. F., Charp, P. A., and Whitson, G. L. (1979). *J. Cell Biol.* **83,** 9a (abstr.).
Lonergan, T. A. (1984). *J. Cell Sci.* **71,** 37–50.
Lonergan, T. A. (1985). *J. Cell. Sci.* **77,** 197–208.
Lukas, T. J., Haiech, J., Lau, W., Craig, T. A., Zimmer, W. E., Shattuck, R. L., Shoemaker, M. D., and Watterson, D. M. (1988). *In* "Cold Spring Harbor Symnposia on Quantitative Biology, Molecular Biology on Signal Transduction," pp. 185–193. Cold Spring Harbor Laboratory, Cold Spring Harbor, New York.
Lukas, T. J., Marshak, D. R., and Watterson, D. M. (1985). *Biochemistry* **24,** 151–157.
Luthra, M. C. (1982). *Biochim. Biophys. Acta* **692,** 271–277.
Macomber, P. B., O'Brien, R. L., and Hahn, F. E. (1966). *Science* **152,** 1374–1375.
Macomber, P. B., Sprinz, H., and Tousimis, A. J. (1967). *Nature (London)* **214,** 937–939.
Maihle, N. J., and Satir, B. H. (1979). *J. Protozool.* **26,** 10A (abstr. 18).
Maihle, N. J., and Satir, B. H. (1980). *Ann. N.Y. Acad. Sci.* **356,** 408–409.
Maihle, N. J., Garofalo, R. S., and Satir, B. H. (1979). *J. Cell. Biol.* **83,** 476a (abstr. Z2919).
Majumder, G. C., Shrago, E., and Elson, C. E. (1975). *Biochim. Biophys. Acta* **384,** 399–412.
Mancini, P. E., and Patton, C. L. (1983). *J. Cell. Biochem., Suppl.* **7A,** 14.
Mar, J., Lee, J. H., Shea, D., and Walsh, C. J. (1986). *J. Cell Biol.* **102,** 353–361.
Marshak, D. R., Lukas, T. J., and Watterson, D. M. (1985a). *In* "Calmodulin antagonists and Cellular Physiology" (H. Hidaka and D. J. Harshorne, eds.), pp. 495–510. Academic Press, Orlando, Florida.
Marshak, D. R., Lukas, T. J., and Watterson, D. M. (1985b). *Biochemistry* **24,** 144–150.
Martin, S. K., Oduola, A. M., and Milhous, W. K. (1987). *Science* **235,** 899–901.

Maruta, H., and Isenberg, G. (1983). *J. Biol. Chem.* **258,** 10151–10158.
Maruta, H., and Korn, E. D. (1977a). *J. Biol. Chem.* **252,** 8329–8332.
Maruta, H., and Korn, E. D. (1977b). *J. Biol. Chem.* **252,** 399–402.
Maruta, H., and Korn, E. D. (1977c). *J. Biol. Chem.* **252,** 6501–6509.
Maruta, H., and Korn, E. D. (1981a). *J. Biol. Chem.* **256,** 499–502.
Maruta, H., and Korn, E. D. (1981b). *J. Biol. Chem.* **256,** 503–506.
Maruta, H., Gadasi, H., Collins, J. H., and Korn, E. D. (1978). *J. Biol. Chem.* **253,** 6297–6300.
Maruta, H., Gadasi, H., Collins, J. H., and Korn, E. D. (1979). *J. Biol. Chem.* **254,** 3624–3630.
Maruta, H., Isenberg, G., Schreckenbach, T., Hallmann, R., Risse, G., Shibayama, T., and Hesse, J. (1983). *J. Biol. Chem.* **258,** 10144–10150.
Matsumoto, Y., Perry, G., Scheibel, L. W., and Aikawa, M. (1987). *Eur. J. Cell Biol.* **45,** 36–43.
Matt, H., Bilinski, M., and Plattner, H. (1978). *J. Cell. Sci.* **32,** 67–86.
McCartney, J. E., Blum, J. J., and Vanaman, T. C. (1984). *Biochemistry* **23,** 5956–5963.
McColm, A. A., Hommel, M., and Trigg, P. I. (1980). *Mol. Biochem. Parasitol.* **1,** 119–127.
McLaughlin, J. (1985). *Mol. Biochem. Parasitol.* **15,** 189–201.
McLaughlin, J., and Muller, M. (1979). *J. Protozool.* **26,** 10A–11A.
McLaughlin, J., and Muller, M. (1981). *Mol. Biochem. Parasitol.* **3,** 369–379.
Mikkelsen, R. B., Tanabe, K., and Wallach, D. F. H. (1982). *J. Cell Biol.* **93,** 685–689.
Mikkelsen, R. B., Geller, E., VanDoren, E., and Asher, C. R. (1984). *Prog. Clin. Biol. Res.* **155,** 25–34.
Mikkelsen, R. B., Wallach, D. F. H., VanDoren, E., and Nillni, E. A. (1986). *Mol. Biochem. Parasitol.* **21,** 83–92.
Molla, A., Kihoffer, M.-C., Ferraz, C., Audemard, E., Walsh, M. P., and Demaille, J. G. (1981). *J. Biol. Chem.* **256,** 15–18.
Morcos, N. C., and Drummond, G. I. (1980). *Biochim. Biophys. Acta* **598,** 27–39.
Munk, N., and Rosenberg, H. (1969). *Biochim. Biophys. Acta* **177,** 629–640.
Munoz, M. L., Weinbach, E. C., Wieder, S. C., Clagget, C. E., and Levenbook, L. (1987). *Exp. Parasitol.* **63,** 42–48.
Munoz, M. L., Claggett, C. E., and Weinbach, E. C. (1988). *Comp. Biochem. Physiol.* **91B,** 137–142.
Murad, F., Arnold, W. P., Mittal, C. K., and Braughler, J. M. (1979). *Adv. Cyclic Nucleotide Res.* **11,** 175–204.
Murakami, K., Tanabe, K., and Takada, S. (1991). *J. Cell Sci.* (in press).
Murofushi, H. (1973). *Biochim. Biophys. Acta* **327,** 354–364.
Murofushi, H. (1974). *Biochim. Biophys. Acta* **370,** 130–139.
Muto, Y., and Nozawa, Y. (1984). *Biochim. Biophys. Acta* **777,** 67–74.
Muto, Y., and Kudo, S., and Nozawa, Y. (1983). *Biochem. Pharmacol.* **32,** 3559–3563.
Muto, Y., and Kudo, S., and Nozawa, Y. (1984). *J. Protozool.* **31,** 164–166.
Muto, Y., and Kudo, S., Nagao, S., and Nozawa, Y. (1985). *Exp. Cell Res.* **159,** 267–271.
Nagai, F., Ushiyama, K., Kano, I., Nakagawa, A., Nakao, T., and Nakajima, A. (1987). *Biochem. Pharmacol.* **36,** 3433–3437.
Nagao, S., and Nozawa, Y. (1985). *Comp. Biochem. Physiol.* **82B,**689–693.
Nagao, S., and Nozawa, Y. (1987). *Arch. Biochem. Biophys.* **252,** 179–187.
Nagao, S., Suzuki, Y., Watanabe, Y., and Nozawa, Y. (1979). *Biochem. Biophys. Res. Commun.* **90,** 261–268.
Nagao, S., Kudo, S., and Nozawa, Y. (1981a). *Biochem. Parmacol.* **30,** 2709–2712.
Nagao, S., Banno, Y., Nozawa, Y., Sobue, K., Yamazaki, R., and Kakiuchi, S. (1981b). *J. Biochem. (Tokyo)* **90,** 897–899.
Nagao, S., Kudo, S., and Nozawa, Y. (1983). *Biochem. Pharmacol.* **32,** 2501–2504.

Naitoh, Y., and Kaneko, H. (1972). *Science* **176,** 523–524.
Naitoh, Y., and Kaneko, H. (1973). *J. Exp. Biol.* **58,** 657–676.
Nakagawa, M., Akiyama, S., Yamaguchi, T., Shiraishi, N., Ogata, J., and Kuwano, M. (1986). *Cancer Res.* **46,** 4453–4457.
Nakazawa, K., Shimonaka, H., Nagao, S., Kudo, S., and Nozawa, Y. (1979). *J. Biochem. (Tokyo)* **86,** 321–324.
Nandini-Kishore, S. G., and Thompson, G. A., Jr. (1979). *Proc. Natl. Acad. Sci. U.S.A.* **76,** 2708–2711.
Newbold, C. (1990). *Nature (London)* **345,** 202–203.
Nilsson, J. R., and Coleman, J. R. (1977). J. Cell Sci. **24,** 311–325.
Noguchi, M., Inoue, H., and Kubo, K. (1979). *J. Biochem. (Tokyo)* **85,** 367–373.
Nozawa, Y., and Nagao, S. (1986). *Insect Sci. Appl.* **7,** 267–277.
Nozawa, Y., and Thompson, G. A., Jr. (1971a). *J. Cell Biol.* **49,** 712–721.
Nozawa, Y., and Thompson, G. A., Jr. (1971b). *J. Cell Biol.* **49,** 722–730.
Ochiai, T., Kato, M., Ogawa, T., and Asai, H. (1988). *Experientia* **44,** 768–771.
Ogihara, S., Ikebe, M., Takahashi, K., and Tonomura, Y. (1983). *J. Biochem. (Tokyo)* **93,** 205–223.
Ogura, A., and Takahashi, K. (1976). *Nature (London)* **264,** 170–172.
Ohnishi, K., and Watanabe, Y. (1983). *J. Biol. Chem.* **258,** 13978–13985.
Ohnishi, K., Suzuki, Y., and Watanabe, Y. (1982). *Exp. Cell Res.* **137,** 217–227.
Ohnishi, S. T., Ohnishi, N., Oda, Y., and Katsuoka, M. (1989a). *Cell Biochem. Funct.* **7,** 105–109.
Ohnishi, S. T., Sadanaga, K., Katsuoka, M., and Weidanz, W. P. (1989b). *Mol. Cell. Biochem.* **91,** 159–165.
Oliveira, M. M., Antunes, A., and DeMello, F. G. (1984). *Mol. Biochem. Parasitol.* **11,** 283–292.
Ono, T., and Nakabayashi, T. (1989). *Parasitol. Res.* **75,** 348–352.
Panijpan, B., and Kantakanit, N. (1983). *J. Pharm. Pharmacol.* **35,** 473–475.
Pantaloni, D. (1985). *J. Biol. Chem.* **260,** 11180–11182.
Pappaioanou, M., and Fishbein, D. B. (1986). *N. Eng. J. Med.* **315,** 712–713.
Pappaioanou, M., Fishbein, D. B., Dreesen, D. W., Schwartz, I. K., Campbell, G. H., Sumner, J. W., Patchen, L. C., and Brown, W. J. (1986). *N. Engl. J. Med.* **314,** 280–284.
Pastan, I., and Gottesman, M. (1987). *N. Engl. J. Med.* **316,** 1388–1393.
Pearson, R. D., Manian, A. A., Harcus, J. L., Hall, D., and Hewlett, E. L. (1982). *Science* **217,** 369–371.
Peltz, G., Kuczmarski, E. R., and Spudich, J. A. (1981). *J. Cell Biol.* **89,** 104–108.
Penniston, J. T., Graf, E., and Itano, T. (1980). *Ann. N.Y. Acad. Sci.* **356,** 245–257.
Peters, W., Ekong, R., Robinson, B. L., Warhurst, D. C., and Pan, X.-Q. (1989). *Lancet* **2,** 334–335.
Peters, W., Ekong, R., Robinson, B. L., Warhurst, D. C., and Pan, X.-Q. (1990). *Ann. Trop. Med. Parasitol.* **84,** 541–551.
Piperno, G., and Luck, D. J. L. (1982). *Prog. Clin. Biol. Res.* **80,** 95–99.
Pitelka, D. R. (1974), *In* ''Cilia and Flagella'' (M. A. Sleigh, ed.), pp. 437–470. Academic Press, New York.
Plattner, H. (1975). *J. Cell Sci.* **18,** 257–269.
Plattner, H., Reichel, K., and Matt, H. (1977). *Nature (London)* **267,** 702–704.
Pleshinger, J., and Weidner, E. (1985). *J. Cell Biol.***100,** 1834–1838.
Pollard, T. D., and Korn, E. D. (1973a). *J. Biol. Chem.* **248,** 4682–4690.
Pollard, T. D., and Korn, E. D. (1973b). *J. Biol. Chem.* **248,** 4691–4697.
Pollard, T. D., Stafford, W. F., III, and Porter, M. E. (1978). *J. Biol. Chem.* **253,** 4798–4808.
Preston, R. R. (1990). *Science* **250,** 285–288.

Prozialeck, W. C., and Weiss, B. (1982). *J. Pharmacol. Exp. Ther.* **222,** 509–516.
Quinones, J. A., and Renaud, F. L. (1975). *J. Cell Biol.* **67,** 347a (abstr.).
Rahmsdorf, H. J., Malchow, D., and Gerisch, G. (1978). *FEBS Lett.* **88,** 322–326.
Ramanathan, S., and Chou, S. C. (1973a). *Experientia* **29,** 814.
Ramanathan, S., and Chou, S. C. (1973b). *Comp. Biochem. Physiol.* **46B,** 93–97.
Rangel-Aldao, R., Allende, O., Triana, F., Piras, R., Henriquez, D., and Piras, M. (1987). *Mol. Biochem. Parasitol.* **22,** 39–43.
Rauh, J. J., and Nelson, D. L. (1981). *J. Cell Biol.* **91,** 860–865.
Rauh, J., Levin, A. E., and Nelson, D. L. (1980). *In* "Calcium-Binding Proteins: Structure and Function" (F. L. Siegel, E. Carafoli, R. H. Kretsinger, D. H. MacLennan, and R. H. Wasserman, eds.), pp. 231–232. Elsevier/North-Holland, New York.
Ray, A., Quade, J., Carson, A. C., and Ray, B. K. (1990). *J. Parasitol.* **76,** 153–161.
Read, L. K., and Mikkelsen, R. B. (1990). *Exp. Parasitol.* **71,** 39–48.
Read, L. K., and Mikkelsen, R. B. (1991a). *Mol. Biochem. Parasitol.* **45,** 109–120.
Read, L. K., and Mikkelsen, R. B. (1991b). *J. Parasitol.* **77,** 346–352.
Read, W., and Satir, P. (1980). *Ann. N.Y. Acad. Sci.* **356,** 423–426.
Reid, D. M., and Shulman, N. R. (1988). *Ann. Intern. Med.* **108,** 206–208.
Rice, H., Ruben, L., Gould, S., Njogu, R., and Patton, C. L. (1987). *Trans. R. Soc. Trop. Med. Hyg.* **81,** 932.
Riddle, L. M., Rauh, J. J., and Nelson, D. L. (1982). *Biochim. Biophys. Acta* **688,** 525–540.
Ringer, S. (1883). *J. Physiol.* (*London*) **4,** 29–42.
Ringo, D. L. (1967). *J. Ultrastruct. Res.* **17,** 266–277.
Riou, B., Barriot, P., Rimailho, A., and Baud, F. J. (1988). *N. Engl. J. Med.* **318,** 1–6.
Roberts, D. M., Burgess, W. H., and Watterson, D. M. (1984). *Plant Physiol.* **75,** 796–798.
Roberts, D. M., Crea, R., Malecha, M., Alvarado-Urbina, G., Chiarello, R. H., and Watterson, D. M. (1985). *Biochemistry* **24,** 5090–5098.
Roberts, D. M., Rowe, P. M., Siegel, F. L., Lukas, T. J., and Watterson, D. M. (1986). *J. Biol. Chem.* **261,** 1491–1494.
Robson, K. J. H., and Jennings, M. W. (1991). *Mol. Biochem. Parasitol.* **46,** 19–34.
Ross, E. M., and Gilman, A. G. (1980). *Annu. Rev. Biochem.* **49,** 533–564.
Roufogalis, B. D. (1982). *In* "Calcium and Cell Function" (W. Y. Cheung, ed.), Vol. 3, pp. 129–159. Academic Press, San Diego.
Roufogalis, B. D. (1985). *In* "Calcium and Cell Physiology" (D. Marmé, ed.), pp. 148–169. Springer-Verlag, New York.
Rozensweig, Z., and Kindler, S. H. (1972). *FEBS Lett.* **25,** 221–223.
Ruben, L. (1987). *J. Protozool.* **34,** 367–370.
Ruben, L., and Haghighat, N. (1990). *In* "Calcium as an Intracellular Messenger in Eucaryotic Microbes" (D. O'Day, ed.), pp. 317–342. Am. Soc. Microbiol., Washington, D.C.
Ruben, L., and Patton, C. L. (1985a). *Mol. Biochem. Parasitol.* **17,** 331–341.
Ruben, L., and Patton, C. L. (1985b). *Immunology* **56,** 227–233.
Ruben, L., and Patton, C. L. (1987). *In* "Methods in Enzymology" (A. R. Means and P. M. Conn, eds.), Vol. 139, pp. 262–276. Academic Press, Orlando, Florida.
Ruben, L., Egwagu, C., and Patton, C. L. (1983). *Biochim. Biophys. Acta* **758,** 104–113.
Ruben, L., Strickler, J. E., Egwuagu, C., and Patton, C. L. (1984). *UCLA Symp. Mol. Cell. Biol.* **13,** 267–278.
Ruben, L., Haghighat, N., and Campbell, A. (1990). *Exp. Parasitol.* **70,** 144–153.
Ruben, L., Ridgley, E. L., Haghighat, N. G., and Chan, E. (1991). *Mol. Biochem. Parasitol.* **46,** 123–136.
Saimi, Y., and Ling K.-Y. (1990). *Science* **249,** 1441–1444.
Sairam, M. R. (1989). *FASEB J.* **3,** 1915–1925.
Salmeron, G., and Lipsky, P. E. (1983). *Am. J. Med.* **75,** 19–24.

Sastre, M. B. R., and Stoppani, A. O. M. (1973). *FEBS Lett.* **31,** 137–142.
Satir, B. H., and Oberg, S. G. (1978). *Science* **199,** 536–538.
Satir, B. H., Garofalo, R. S., Gilligan, D. M., and Maihle, N. J. (1980). *Ann. N.Y. Acad. Sci.* **356,** 83–91.
Schanne, F. A. X., Kane, A. B., Young, E. E., and Farber, J. L. (1979). *Science* **206,** 700–702.
Scheibel, L. W. (1988). *In* "Malaria" (W. Wernsdorfer and I. McGregor, eds.), pp. 171–217. Churchill Livingstone, Edinburgh, Scotland.
Scheibel, L. W., and Sherman, I. W. (1988). *In* "Malaria" (W. Wernsdorfer and I. McGregor, eds.), pp. 219–252. Churchill Livingstone, Edinburgh, Scotland.
Scheibel, L. W., Colombani, P. M., Hess, A. D., Aikawa, M., Atkinson, C. T., and Milhous, W. K. (1987). *Proc. Natl. Acad. Sci. U.S.A.* **84,** 7310–7314.
Scheibel, L. W., Colombani, P. M., Hess, A. D., Aikawa, M., Atkinson, C. T., Igarashi, I., Matsumoto, Y., and Milhous, W. K. (1989). *Prog. Clin. Parasitol.* **1,** 21–56.
Schultz, J. E., and Klumpp, S. (1984). *Adv. Cyclic Nucleotide Protein Phosphorylation Res.* **17,** 275–283.
Schultz, J. E., Schonefeld, U., and Klumpp, S. (1983).*Eur. J. Biochem.* **137,** 89–94.
Schuster, F. L., and Mandel, N. (1984). *Antimicrob. Agents Chemother.* **25,** 109–112.
Seebeck, T., and Gehr, P. (1983). *Mol. Biochem. Parasitol.* **9,** 197–208.
Shalev, O., Leida, M. N., Hebbel, R. P., Jacob, H. S., and Eaton, J. W. (1981). *Blood* **58,** 1232–1235.
Shea, D. K., and Walsh, C. J. (1987). *J. Cell Biol.* **105,** 1303–1309.
Shimizu, T., and Hatano, M. (1983). *FEBS Lett.* **160,** 182–186.
Shimizu, T., Hatano, M., Nagao, S., and Nozawa, Y. (1982). *Biochem. Biophys. Res. Commun.* **106,** 1112–1118.
Shimizu, T., Hatano, M., Muto, Y., and Nozawa, Y. (1984). *FEBS Lett.* **166,** 373–377.
Shimonaka, H., and Nozawa, Y. (1977). *Cell Struct. Funct.* **2,** 81–89.
Sibley, L. D., Krahenbuhl, J. L., Adams, G. M. W., and Weidner, E. (1986). *J. Cell Biol.* **103,** 867–874.
Singh, J., and Chatterjee, S. (1988). *Cytobios* **55,** 95–103.
Slater, L. M., Sweet, P., Stupecky, M., and Gupta, S. (1986). *J. Clin Invest.* **77,** 1405–1408.
Synder, S. H., and Reynolds, I. J. (1985). *N. Engl. J. Med.* **313,** 995–1002.
Stekhoven, F. S., and Bonting, S. L. (1981). *Physiol. Rev.* **61,** 1–76.
Stiles, P. G. (1903). *Am. J. Physiol.* **5,** 338–357.
Sussman, D. J., Lai, E. Y., and Fulton, C. (1984). *J. Biol. Chem.* **259,** 7355–7360.
Suzuki, Y., Hirabayashi, T., and Watanabe, Y. (1979). *Biochem. Biophys. Res. Commun.* **90,** 253–260.
Suzuki, Y., Nagao, S., Abe, K., Hirabayashi, T., and Watanabe, Y. (1981). *J. Biochem. (Tokyo)* **89,** 333–336.
Suzuki, Y., Ohnishi, K., Hirabayashi, T., and Watanabe, Y. (1982). *Exp. Cell Res.* **137,** 1–14.
Swan, D. G., Hale, R. S., Dhillon, N., and Leadlay, P. F. (1987). *Nature (London)* **329,** 84–85.
Takagi, T., Nemoto, T., Konishi, K., Yazawa, M., and Yagi, K. (1980). *Biochem. Biophys. Res. Commun.* **96,** 377–381.
Takahashi, M., Onimaru, H., and Naitoh, Y. (1980). *Proc. Jpn. Acad.* **56,** 585–590.
Takahashi, K., Ogihara, S., Ikebe, M., and Tonomura, Y. (1983). *J. Biochem. (Tokyo)* **93,** 1175–1183.
Takeda, T., and Yamamoto, M. (1987). *Proc. Natl. Acad. Sci. U.S.A.* **84,** 3580–3584.
Takemasa, T., Ohnishi, K., Kobayashi, T., Takagi, T., Konishi, K., and Watanabe, Y. (1989). *J. Biol. Chem.* **264,** 19293–19301.
Tanabe, K. (1983). *J. Protozool.* **30,** 707–710.
Tanabe, K. (1990). *Parasitol. Today* **6,** 225–229.

Tanabe, K., Mikkelsen, R. B., and Wallach, D. F. H. (1982a). *J. Cell Biol.* **93,** 680–684.
Tanabe, K., Mikkelsen, R. B., and Wallach, D. F. H. (1982b). *In* "Parasites—Their World and Ours" (D. F. Mettrick and S. S. Desser, eds.), pp. 144–148. Elsevier, Amsterdam.
Tanabe, K., Mikkelsen, R. B., and Wallach, D. F. H. (1983). *Ciba Found. Symp.* **94,** 64–73.
Tanabe, K., Izumo, A., Kato, M., Miki, A., and Doi, S. (1989). *J. Protozool.* **36,** 139–143.
Tanabe, K., Kato, M., Izumo, A., Hagiwara, A., and Doi, S. (1990). *Exp. Parasitol.* **70,** 419–426.
Tanowitz, H. B., Morris, S. A., Weiss, L. M., Bilezikian, J. P., Factor, S. M., and Wittner, M. (1989). *Am. J. Trop. Med. Hyg.* **41,** 643–649.
Tash, J. S., and Means, A. R. (1982). *Biol. Reprod.* **26,** 745–763.
Tellez-Inon, M. T., Ulloa, R. M., Torruella, M., and Torres, H. N. (1985). *Mol. Biochem. Parasitol.* **17,** 143–153.
Teo, T. S., and Wang, J. H. (1973). *J. Biol. Chem.* **248,** 5950–5955.
Teo, T. S., Wang, T. H., and Wang, J. H. (1973). *J. Biol. Chem.* **248,** 588–595.
Tesi, R. J., Wait, R. B., Butt, K. M. H., Jaffe, B. M., and McMillen, M. A. (1985). *Surg. Forum* **36,** 339–340.
Tiggemann, R., and Plattner, H. (1982). *FEBS Lett.* **148,** 226–230.
Tomlinson, S., MacNeil, S., Walker, S. W., Ollis, C. A., Merritt, J. E., and Brown, B. L. (1984). *Clin Sci.* **66,** 497–508.
Tschudi, C., and Ullu, E. (1988). *EMBO J.* **7,** 455–463.
Tschudi, C., Young, A. S., Ruben, L., Patton, C. L., and Richards, F. F. (1985). *Proc. Natl. Acad. Sci. U.S.A.* **82,** 3998–4002.
Tsien, R. Y., Pozzan, T., and Rink, T. J. (1984). *Adv. Cyclic Nucleotide Protein Phosphorylation Res.* **17,** 535–541.
Tsuchiya, T. (1976). *Experientia* **32,** 1176–1177.
Ulloa, R. M., Mesri, E., Esteva, M., Torres, H. N., and Tellez-Inon, M. T. (1988). *Biochem. J.* **255,** 319–326.
Vanaman, T. C., Sharief, F., Awramik, J. L., Mendel, P. A., and Watterson, D. M. (1976). *In* "Contractile Systems in Non-Muscle Tissues" (S. V. Perry, A. Margreth, and R. S. Adelstein, eds.), pp. 165–176. Elsevier/North-Holland, Amsterdam.
Van Belle, H. (1981). *Cell Calcium* **2,** 483–494.
Van Eldik, L. J., and Watterson, D. M. (1981). *J. Biol. Chem.* **256,** 4205–4210.
Van Eldik, L., Grossman, A. R., Iverson, D. B., and Watterson, D. M. (1980a). *Proc. Natl. Acad. Sci. U.S.A.* **77,** 1912–1916.
Van Eldik, L. J., Piperno, G., and Watterson, D. M. (1980b). *Proc. Natl. Acad. Sci. U.S.A.* **77,** 4779–4783.
Van Eldik, L. J., Fok, K.-F., Erickson, B. W., and Watterson, D. M. (1983a). *Proc. Natl. Acad. Sci. U.S.A.* **80,** 6775–6779.
Van Eldik, L. J., Watterson, D. M., Fok, K.-F., and Erickson, B. W. (1983b). *Arch. Biochem. Biophys.* **227,** 522–533.
Vigues, B., and Groliere, C.-A. (1985). *Exp. Cell Res.* **159,** 366–376.
Voichick, J., Elson, C., Granner, D., and Shrago, E. (1973). *J. Bacteriol.* **115,** 68–72.
Volpi, M., Sha'afi, R. I., Epstein, P. M., Andrenyak, D. M., and Feinstein, M. B. (1981a). *Proc. Natl. Acad. Sci. U.S.A.* **78,** 795–799.
Volpi, M., Sha'afi, R. I., and Feinstein, M. B. (1981b). *Mol. Pharmacol.* **20,** 363–370.
Voorheis, H. P., and Martin, B. R. (1981). *Eur. J. Biochem.* **116,** 471–477.
Voorheis, H. P., Bowles, D. J., and Smith, G. A. (1982). *J. Biol. Chem.* **257,** 2300–2304.
Waechter, C. J., and Lennarz, W. J. (1976). *Annu. Rev. Biochem.* **45,** 95–112.
Walker, G. M., and Zeuthen, E. (1980). *Exp. Cell Res.* **127,** 487–490.
Wallace, R. W., and Cheung, W. Y. (1979). *J. Biol. Chem.* **254,** 6564–6571.
Walsh, M. P., and Hartshorne, D. J. (1983). *In* "Biochemistry of Smooth Muscle" (N. L. Stephens, ed.), Vol. II, pp. 1–84. CRC Press, Boca Raton, Florida.

Walter, R. D., and Opperdoes, F. R. (1982). *Mol. Biochem. Parasitol.* **6,** 287–295.
Warhurst, D. C., and Thomas, S. C. (1975). *Ann. Trop. Med. Parasitol.* **69,** 417–420.
Warner, F. (1970). *J. Cell Biol.* **47,** 159–182.
Warner, F. D. (1976). *Cold Spring Harbor Symp. Cell Proliferation* **3,** 891–914.
Warner, F. D., Mitchell, D. R., and Perkins, C. R. (1977). *J. Mol. Biol.* **114,** 367–384.
Warrell, D. A., Molyneux, M. E., and Beales, D. F. (1990). *Trans. R. Soc. Trop. Med. Hyg.* (*London*) **84** (Suppl. 2), 1–65.
Warrick, H. M., and Spudich, J. A. (1987). *Annu. Rev. Cell Biol.* **3,** 379–421.
Wasserman, M., Alarcon, C., and Mendoza, P. M. (1982). *Am. J. Trop. Med. Hyg.* **31,** 711–717.
Watanabe, Y., and Nozawa, Y. (1982). *In* "Calcium and Cell Function" (W. Y. Cheung, ed.), Vol. 2, pp. 297–323. Academic Press, San Diego.
Watt, G., Long, G. W., Grogl, M., and Martin, S. (1990). *Trans. R. Soc. Trop. Med. Hyg.* **84,** 187–190.
Watterson, D. M., and Vanaman, T. C. (1976). *Biochem. Biophys. Res. Commun.* **73,** 40–46.
Watterson, D. M., Sharief, F. S., and Vanaman, T. C. (1980). *J. Biol. Chem.* **255,** 962–975.
Watterson, D. M., Burgess, W. H., Lukas, T. J., Iverson, D., Marshak, D. R., Schleicher, M., Erickson, B. W., Fok, K.-F., and Van Eldik, L. J. (1984). *Adv. Cyclic Nucleotide Res.* **16,** 205–226.
Weiss, B., Prozialeck, W. C., and Wallace, T. L. (1982). *Biochem. Pharmacol.* **31,** 2217–2226.
Weiss, B., Sellinger-Barnette, M., Winker, J. D., Schechter, L. E., and Prozialeck, W. C. (1985). *In* "Calmodulin Antagonists and Cellular Physiology" (H. Hidaka and D. J. Hartshorne, eds.), pp. 45–62. Academic Press, Orlando, Florida.
Wellems, T. E., Panton, L. J., Gluzman, I. Y., Rosario, V. E., Gwadz, R. W., Walker-Jonah, A., and Krogstad, D. J. (1990). *Nature* (*London*) **345,** 253–255.
Willadsen, P. (1984). *Mol. Biochem. Parasitol.* **12,** 195–205.
Wilson, C. M., Serrano, A. E., Wasley, A., Bogenschutz, M. P., Shankar, A. H., and Wirth, D. F. (1989). *Science* **244,** 1184–1186.
Wiser, M. F., and Schweiger, H.-G. (1985). *Mol. Biochem. Parasitol.* **17,** 179–189.
Wiser, M. F., Wood, P. A., Eaton, J. W., and Sheppard, J. R. (1982). *Fed. Proc., Fed. Am. Soc. Exp. Biol.* **41,** 512.
Wiser, M. F., Eaton, J. W., and Sheppard, J. R. (1983a). *J. Cell. Biochem.* **21,** 305–314.
Wiser, M. F., Wood, P. A., Eaton, J. W., and Sheppard, J. R. (1983b). *J. Cell Biol.* **97,** 196–201.
Wiser, M. F., Leible, M. B., and Plitt, B. (1988). *Mol. Biochem. Parasitol.* **27,** 11–22.
Witman, G. B., Fay, R., and Plummer, J. (1976). *Cold Spring Harbor Symp. Cell Proliferation* **3,** 969–986.
Witman, G. B., Plummer, J., and Sander, G. (1978). *J. Cell Biol.* **76,** 729–747.
Wolfe, J. (1973). *J. Cell. Physiol.* **82,** 39–48.
Wylie, D., and Vanaman, T. C. (1987). *In* "Methods in Enzymology" (A. R. Means and P. M. Conn, eds.), Vol. 139, pp. 50–68. Academic Press, Orlando, Florida.
Yamada, K., and Asai, H. (1982). *J. Biochem.* (*Tokyo*) **91,** 1187–1195.
Yazawa, M., Yagi, K., Toda, H., Kondo, K., Narita, K., Yamazaki, R., Sobue, K., Kakiuchi, S., Nagao, S., and Nozawa, Y. (1981). *Biochem. Biophys. Res. Commun.* **99,** 1051–1057.
Ye, Z., and VanDyke, K. (1988). *Biochem. Biophys. Res. Commun.* **155,** 476–481.
Zarain-Herzberg, A., and Arroyo-Begovich, A. (1985). *Biochim. Biophys. Acta* **816,** 258–266.
Zernig, G. (1990). *Trends Pharmacol. Sci.* **11,** 38–44.

The Molecular Biology of Intermediate Filament Proteins

Kathryn Albers[1] and Elaine Fuchs
Howard Hughes Medical Institute and
Department of Molecular Genetics and Cell Biology
The University of Chicago, Chicago, Illinois 60637

I. Introduction

The cytoarchitecture of eukaryotic cells is composed of three major filamentous networks: the 6-nm diameter actin microfilament network, the 20-nm diameter microtubule network, and the intermediate-sized intermediate filaments (IF), which range in diameter between 8 and 12 nm (for this review, an average size of 10 nm is used). IFs are composed of proteins that self-assemble into filaments that form complex cytoskeletal networks. In contrast to the evolutionarily conserved actin and tubulin proteins, IF proteins are highly diverse and exhibit cell type specificity of expression. IF gene expression is also tightly coordinated with organ development and tissue differentiation and thus IFs provide valuable tools with which to study cell determination and differentiation. In this article, we will examine recent studies on the IF protein family and the current understanding of the role intermediate filaments play within cells and tissues.

II. Characteristics of the Intermediate Filament Proteins

A. Type I and Type II Keratin Proteins

IFs can be classified into types (Table I) either according to the degree of homology in structure and composition or their tissue-specific expression (Steven, 1990). The largest and most complex group of IF proteins consists

[1] Present address: Department of Pathology, Lucille P. Markey Cancer Center, University of Kentucky, Lexington, Kentucky 40536.

TABLE I

Properties and Distribution of the Major Mammalian Intermediate Filament Proteins

IF protein	Sequence type	Average molecular weight ($\times 10^{-3}$)	Estimated number of polypeptides	Primary tissue distribution
Keratin	I	40–56.5	15	Epithelia
Keratin	II	53–67	15	Epithelia
Vimentin	III	57	1	Mesenchymal cells
Desmin	III	53–54	1	Myogenic cells
Glial fibrillary acidic protein (GFAP)	III	50	1	Glial cells and astrocytes
Peripherin	III	57	1	Peripheral neurons
Neurofilament proteins				Neurons of central and peripheral nerves
NF-L	IV	62	1	
NF-M	IV	102	1	
NF-H	IV	110	1	
Lamin proteins				All cell types
Lamin A	V	70	1	
Lamin B	V	67	1	
Lamin C	V	60	1	
Nestin	VI	240	1	Neuronal stem cells

of the type I and type II epithelial keratins. In mammals, at least 30 keratins that range in size from 40 to 70 kDa (Moll *et al.*, 1982a) are expressed as subsets in epithelial cells at various stages of development and differentiation (Fuchs and Green, 1980; Moll *et al.*, 1982b; Wu *et al.*, 1982; Tseng *et al.*, 1982). There are 10 different keratins expressed in the hair- and nail-forming epithelial cells. Twenty other keratins have been identified in the other epithelial cell types. A previously undetected type I keratin referred to as K20 (46 kDa) has been recently discovered and characterized by Moll *et al.* (1990). Immunocytochemistry has shown K20 to be expressed predominantly in intestinal epithelium and Merkel cells of the epidermis.

Keratin proteins are classified as either type I or type II IFs on the basis of charge, immunological relatedness, and similarity of sequences that encode a central α-helical domain (Fuchs *et al.*, 1981; Eichner *et al.*, 1984). Type I keratins are generally smaller in molecular mass and more acidic (M_r 40–56.5K; pK_i 4.5–5.5) than type II keratins, which are generally larger and neutral–basic (M_r 53–67K; pK_i 5.5–7.5). Within the type I and type II families, the helical domains of different keratins share 50–99%

sequence homology while keratins of opposite type share only 25–35% homology in these regions (Hanukoglu and Fuchs, 1982, 1983; Steinert *et al.*, 1983, 1984; Jorcano *et al.*, 1984a,b).

Keratins are expressed in cells as specific type I and type II pairs (Eichner *et al.*, 1986). This specificity of pairing leads to the formation of filaments with different physical properties such as solubility and tensile strength (Eichner and Kahn, 1990). *In vitro* filament assembly studies using combinations of purified subunits have shown keratins to be obligate heteropolymers composed of type I and type II subunits (Steinert *et al.*, 1976; Lee and Baden, 1976; Hatzfeld and Franke, 1985; Eichner *et al.*, 1986; Hatzfeld and Weber, 1990a; Coulombe and Fuchs, 1990; Steinert, 1990).

B. Type III Intermediate Filament Proteins

Type III IF proteins are those of vimentin (M_r 57K), desmin (M_r 53–54K), glial fibrillary acidic protein (GFAP: M_r 50K), and peripherin, or the 57K protein (Portier *et al.*, 1984; Parysek and Goldman, 1988; Parysek *et al.*, 1988; Brody *et al.*, 1989). Vimentin has the most extensive distribution of all cytoplasmic IFs and is expressed in cells of mesenchymal origin and in many cells grown in culture (Bennett *et al.*, 1978; Franke *et al.*, 1979; Virtanen *et al.*, 1981; Lazarides, 1982). Desmin is expressed in adult striated, smooth, and cardiac muscle cells and is localized in myofibrils at the Z disk and in the interfibrillar space (Cooke, 1976; Bennett *et al.*, 1979; Gard and Lazarides, 1980; Tokuyasu *et al.*, 1984). GFAP assembles into filaments as a homopolymer in various glial cell types located throughout the nervous system, whereas peripherin has neuronal localization. Peripherin is found in the peripheral nervous system in neurons of the dorsal root ganglion, sympathetic ganglia, many cranial nerves, and ventral motor neurons (Portier *et al.*, 1984; Parysek and Goldman, 1988; Brody *et al.*, 1989).

While type III IFs share only 25–30% homology with keratins, they share considerable homology among individual members of their own class. For example, hamster vimentin has 60% overall homology with chicken desmin, which is even higher in the central α-helical domain (Quax *et al.*, 1983, 1985). In contrast to the keratin heteropolymer, vimentin, desmin, and glial proteins form homopolymeric filaments composed of a single polypeptide type (Steinert *et al.*, 1981), although they are also capable of coassembling to form heteropolymers when mixed *in vitro* (Steinert *et al.*, 1981; Quinlan and Franke, 1982).

C. Type IV Neurofilament Proteins

The type IV neurofilament (NF) triplet proteins are expressed in neuronal axons, dendrites, and perikarya. The human NF proteins NF-L (light), NF-M (medium), and NF-H (heavy) have molecular weights predicted from their gene sequences of 62, 102, and 110K, respectively (Lewis and Cowan, 1985; Julien *et al.,* 1987b; Levy *et al.,* 1987; Myers *et al.,* 1987; Lees *et al.,* 1988). Mass variation between NFs arises primarily from differences in the length of the carboxy termini. The organization of the NF proteins in the filament structure is unclear. One current model proposes that NF-L comprises the filament backbone and NF-H associates with this backbone to form interfilament crossbridges via its extended carboxy domain (Hirokawa *et al.,* 1984). NF-M is thought to associate with the core of the filament. *In vitro* filament assembly studies support this model to the extent that the NF-L and the NF-M proteins can self-assemble into 10 nm filaments whereas NF-H proteins cannot (Geisler and Weber, 1981; Gardner *et al.,* 1984; Tokutake *et al.,* 1984). In addition, *in vivo* studies in which genes encoding NF-L and NF-M were separately transfected into mouse L cells have shown both NF-L and NF-M to be capable of assembling into the endogenous vimentin network (Monteiro and Cleveland, 1989). Although NF and vimentin proteins were not previously known to coassemble, coexpression of neurofilaments and vimentin in neurons has been shown (Drager, 1983).

D. Type V Lamin Proteins

The type V IF proteins are the nuclear lamins (McKeon *et al.,* 1986; Fisher *et al.,* 1986; Goldman *et al.,* 1986). Lamins are ubiquitously expressed proteins that organize to form the nuclear lamina, a fibrous meshwork on the nucleoplasmic side of the inner nuclear membrane (see Nigg, 1989; Gerace and Burke, 1988). Mammalian somatic cells express three major lamin proteins; A (70 kDa), B (67 kDa), and C (60 kDa) (Gerace and Blobel, 1980). Lamins A and C are encoded by a single gene whose transcript is differentially spliced to produce an extended carboxy domain in the lamin A protein (McKeon *et al.,* 1986; Fisher *et al.,* 1986). Lamin A and C expression is regulated during mammalian development according to the stage of development and tissue type (Rober *et al.,* 1989). Lamin B is encoded by a separate gene (Hoger *et al.,* 1988; Pollard *et al.,* 1990).

Lamin proteins assemble into a matrix on the inner surface of the nuclear membrane that is thought to provide a structural framework for the nucleus and to facilitate chromatin organization (Gerace *et al.,* 1978; Franke *et al.,* 1981; Benevente and Krohne, 1986). This matrix appears as

10-nm orthogonally arranged filaments that interconnect nuclear pore complexes (Aaronson and Blobel, 1975; Aebi *et al.*, 1986; Parry *et al.*, 1987). *In vitro* filament reconstitution studies of purified rat liver lamins have shown that lamins A and C can form 10-nm filaments and, with lowered ionic strength of the reconstitution buffer, thick paracrystals of protein (Aebi *et al.*, 1986). During mitosis, phosphorylation-mediated disassembly of the lamina occurs and lamins A and C become solubilized while lamin B remains associated with the membrane (Gerace and Blobel, 1980; Ward and Kirschner, 1990).

E. Type VI Nestin Protein

The newest protein to be assigned as an IF is the protein nestin. Nestin has been classified as a sixth type of IF protein because of its low amino acid sequence homology (16–29%) over the central helical core domain to other IF types (Lendahl *et al.*, 1990). Although the nestin gene sequence predicts a 200-kDa mass protein, on sodium dodecyl sulfate (SDS)-polyacrylamide gels nestin has an apparent molecular weight of 240K. Nestin expression occurs in proliferating stem cells of the developing mammalian central nervous system and at lower levels in developing skeletal muscle (Hockfield and McKay, 1985; Lendahl *et al.*, 1990).

III. Structure of the Intermediate Filament Polypeptide

Intermediate filament proteins display amino acid sequence homologies and share a highly conserved secondary structure (Fig. 1). All IF proteins have a common tripartite structure in which a central α-helical domain is flanked on either side by nonhelical sequences. The rod domains of two polypeptide chains intertwine in a coiled-coil fashion, a feature that was originally predicted from X-ray diffraction data (Crick, 1953; Pauling and Corey, 1953; Fraser and MacRae, 1973) and later suggested from sequence data and model building (McLachlan and Stewart, 1975; Conway and Parry, 1988). The coiling results from α-helical chains that contain hydrophobic amino acid residues arranged in a repeat pattern, such that the first and fourth of every seven residues are hydrophobic. This provides a coiling hydrophobic seal between two IF proteins.

The central α-helical core of IFs can be subdivided into four smaller regions, referred to as 1A, 1B, 2A, and 2B. Short linker regions interrupt both the heptad repeat and the helix continuity. In cytoplasmic filament types, the entire central domain spans approximately 310 amino acids,

Amino-terminus

```
hu K14 :   1  TTCSRQFTSSSSMKGSCGIGGGIGAGSSRISSVLAGGSCRAPNTYGGGLSVSSSRFSSG-GAYGLGGGYGGG-FSSSSSSFGSG-FGGGYGGGLGAGLGGGFGGGFAGGDGSSVG
hu K5  :   1                                                                       SRQSSVSSGAGGSPSFSTASAITPSVSRTSFTSVSGPGGGGGGGFGRVSLAGA
              CGVGGYGSRSLYNLGGSKRISISISTSGGSFRNRFGAGAGGGYGFGGGAGSGFGFGFGGGAGGGFGLGGGAGFGGGFGGPGFPVCPPGGIQEVTVNQSLLTPLMLQIDPSIQRVRT
hu Vim :   1  STRSVSSSSYRRMFGGPGTASRPSSSRSYVTTSTRTYSLGNLRPSTSRSLYASSP----GGVYVT---------RSSAVR----LRSSVPGVRLLQDSVDFSLADAINTEFKNTRT
ch Des :   1                      QSYSSSQRVSSYRRTFGGGTSPVFPRASFGSRGSGSSVTSRVYQVSRTSAVPTLSTFRTTRVTPLRTYGSAYQGAGELLDFSLADAMNQEFLQTRT
hu NFL :   1                               SSFSYEPYYSTSYKRRYVETPRVHISVRSGYSTARSAYSSYSAPVSSSLSVRRSYSSSSGSLMPSLENLDLSQVAAISNDLKSIRT
hu Lam :   1                                                                                          ETPSQRRATRSGAQASSTPLSPTRITTRL
as IFB :   1                                    SLKQSQESSEYEIAYRSTIQPRTAVRTQSRQSGAYSTGAVSGGGGRVLKMVTEMGSASIGGISPALSANAAKSFLEATD
```

Helical domain 1A

```
hu K14 : 113  SEKVTMQNLNDRLASYLDKVRALEEANADLEVKIRDW-YQRQRPAEIKDY-SPYFK
hu K5  : 170  EEREQIKTLNNKFASFIDKVRFLEQQNKVLETKWTLLQEQGTKTVRQNLEP-LFEQ
hu Vim : 100  NEKVELQELNDRFANYIDKVRFLEQQNKILLAELEQLKGQ----GKSRLGD-LYEE
ch Des :  97  NEKVELQELNDRFANYIEKVRFLEQQNALMVAEVNRLRG-KQP---TRVAE-MYEE
hu NFL :  87  QEKAQLQDLNDRFASFIERVHELEQQNKVLEAELLVL-RQKHSEPSRFRALYEQEI
hu Lam :  30  QEKEDLQELNDRLAVYIDRVRSLETENAGLRLRITESEEVVSREV-SGI-KAAYEA
as IFB :  80  KEKKEMQGLNDRLGNYIDRVKKLEEQNRKLVADLDELRG-RWGKDTSEI-KIQYSD
```

Helical domain 1B

```
hu K14 : 167  TIEDLRNKILTATVDNANVLLQIDNARLAADDFRTKYETELNLRMSVEADINGLRRVLDELTLARADLEMQIESLKEELAYLKKNHEEEMNAL
hu K5  : 217  YINNLRRQLDSIVGERGRLDSELRNMQDLVEDFKNKYEDQINKRTTAKNHFVMLKKDVDAAYMNKVELEAKVDALMDEINFMKMFFDAELSQM
hu Vim : 151  EMRELRRQVDQLTNDKARVEVERDNLAEDIMRLREKLQEEMLQREEAENTLQSFRQDVDNASLARLDLERKVESLQEEIAFLKKLHEEEIQEL
ch Des : 149  ELRELRRQVDALTGQRARVEVERDNLLDNLQKLKQKLQEEIQLKQEAENNLAAFRADVDAATLARIDLERRIESLQEEIAFLKKVHEEEIREL
hu NFL : 142  RDLRLAAEDATTNEKQALRGEREEGLEETLRNLQARYEEEVLSREDAEGRLMERRKGADEAALARAELEKRIDSLMDEISFLKKVHEEEIAEL
hu Lam :  84  ELGDARKTLDSVAKERARLQLELSKVREEFKELKA LHDLRGQVAKLEAALGEAKKQLQDEMLRRVDAENRLQTMKEELDFQKNIYSEELRET
                                   RNTKKEGDLIAAQARLKDLEALLNSKEAALSTALSEKRTLEGE
                                   RYEDVQHRRESDREKINQWQHAIEDAQSELEMLRARWRQLTEE
as IFB : 134  SLRDARKEIDDGARRKAEIDVKVARLRDDLAELRN EKRLNGDNARIWEELQKARNDLDEETLGRIDFQNQVQTLMEELEFLRRVHEQEVKEL
```

```
                 Helical domain 2A
hu K14  :  260  RGQVGGD-VNVEMDAAPGVDLSRILNEMRDQYEKMAEKNRKDAEEWFFTKTEELNREV--ATNSELVQSGKS
hu K5   :  318  QTHVSDTSVVLSMDNNRNLDLDSIIAEVKAQYEEIANRSRTEAESQYQTKYEELQQTA--GRHGDDLRNTKH
hu Vim  :  244  QAQIQEQHVQIDVDV-SKPDLTAALRDVRQQYESVAAKNLQEAEEWYKSKFADLSEAA--NRNNDALRQAKQ
ch Des  :  243  QAQLQEQHIQVEMDI-SKPDLTAALRDIRAQYESIAAKNIAEAEEWYKSKVSDLTQAA--NKNNDALRQAKQ
hu NFL  :  235  QAQIQYAQISVEMDV-TKPDLSAALKDIRAQYEKLAAKNMQNAEEWFKSRFTVLTESA--AKNTDAVRAAKD
hu Lam  :  219  KRRHETRLVEIDNGKQREFESRSADALQELRAQHEDQVEQYKKELEKTYSAKLDNARQSAERNSNLVGAAHE
as IFB  :  269  QALLAQAPADTREFF--KNELALAIRDIKDEYDYIAKQGKQDMESWYKLKVSEVQGSA--NRANMESSYQRE

                 Helical domain 2B
hu K14  :  329  EISELRRTMQNLEIELQSQLSMKASLENSLEDTKGRYCMQLAQIQEMIGSVEEQLAQLRCEMEQQNQEYKILLDVKTRLEQEIATYRRLLEGEDAHL
hu K5   :  384  EISEMNRMIQRLRAEIDNVKKQCANLQNAIADAEQRGELALKDARNKLAELEEALQKAKQDMARLLREYQELMNTKLALDVEIATYRKLLEGEECRL
hu Vim  :  313  ESNEYRRQVQSLTCEVDALKGTNESLERQMREMEENFAVEAANYQDTIGRLQDEIQNMKEEMARHLREYQDLLNVKMALDIEIATYRKLLEGEESRI
ch Des  :  312  EMLEYRHQIQSYTCEIDALKGTNDSLMRQMREMEERFAGEAGGYQDTIARLEEEIRHLKDEMARHLREYQDLLNVKMALDVEIATYRKLLEGEENRI
hu NFL  :  304  EVSESRRLLKAKTLEIEACRGMNEALEKQLQELEDKQNADISAMQDTINKLENELRTTKSENARYLKEYQDLLNVKMALDIEIAAYRKLLEGEETRL
hu Lam  :  291  ELQQSRIRIDSLSAQLSQLQKQLAAKEAKLRDLEDSLARERDTSRRLLAEKEREMAEMRARMQQQLDEYQELLDIKLALDMEIHAYRKLLEGEEERL
as IFB  :  337  EVKRMRDNIGDLRGKLGDLEAKNALLEKEVQNLNYQLNDDQRWYEAALNDRDATLRRMREECQTLVAELQALLDTKQMLDAEIAIYRKMLEGEESRV

                 Carboxy-terminus
hu K14  :  426  SSSQFSSGSQSSRDVTSSSRQIRTKVMDVHDGKVVSTHEQVLRTKN  .  471
hu K5   :  481  SGEGVGPVNISVVTSSVSSGYGSGSGYGGGLGGGLGGGLGGGLAGGSSGSYYSSSSGGVGLGGGLSVGGSGFSASSSRGLGVGFGSGGGSSSSVKFVSTTSSSRKSFKS  .  589
hu Vim  :  410  SLPLPNFSSLNLRETNLDSLPLVDTHSKRTFLIKTVETRDGQVINETSQHHDDLE  .  464
ch Des  :  409  SIPMHQTFASALNFRETSPDQRGSEVHTKKTVMIKTIETRDGEVVSEATQQQHEVL  .  464
hu NFL  :  400  SFTSVGSITSGYSQSSQVFGRSAYGGLQTSSYLMSTRSFPSYYTSHVQEEQTEVEETIEASKAEEAKDEPPSEGEAEEEEKDKEEAEEEEAAEEEEAAKEES
                EEAKEEEEGGEGEEGEETKEAEEEEKKVEGAGEEQAAKKKD  .  543
hu Lam  :  388  RLSPSPTSQRSRGRASSHSSQTQGGGSVTKKRKLESTESRSSFSQHARTSGRVAVEEVDEEGKFVRLRNKSNEDQSMGNWQIKRQNGDDPLLTYRFPPKFTL
                KAGQVVTIWAAGAGATHSPPTDLVWKAQNTWGCGNSLRTALINSTGEEVAMRKLVRSVTVVEDDEDEDGDDLLHHHHVSGSRR  .  572
as IFB  :  434  GLRQMVEQVVKTHSLQQQEDTDSTRNVRGEVSTKTTFQRSAKGNVTISECDPNGKFITLENTHRSKDENLGEHRLKRKLDNRREIVYTIPPNTVLKAGRTMKIYARDQG
                GIHNPPDTLVFDENTWGIGANVVTSLINKDGDERATHTQKTIQTGQ  .  589
```

FIG. 1 Comparison of selected IF sequences aligned for optimal homology. Shown are the amino acid sequences of human epidermal keratin 14 (hu K14) (Marchuk *et al.*, 1984), human epidermal keratin 5 (hu K5) (Lersch *et al.*, 1989), human vimentin (hu Vim) (Ferrari *et al.*, 1986), chicken desmin (ch Des) (Capetanaki *et al.*, 1984), human neurofilament protein L (hu NFL) (Julien *et al.*, 1987a), human lamin C (hu Lam) (McKeon *et al.*, 1986), and the invertebrate IF-B protein isolated from *Ascaris lumbricoides* (as IFB) (Weber *et al.*, 1989). Bars indicate positions of α-helical domains. Circles above the helical domains demarcate amino acids comprising the heptad repeat of hydrophobic residues. These residues are shown in bold for the additional helix 1B sequences of the lamin and invertebrate IF proteins.

whereas for the nuclear lamin proteins A, B, and C, this length is increased by an insertion of 42 amino acids (6 heptads) in helical domain 1B (see Fig. 1). Homology between different IF proteins is particularly high at the start of coil 1A and near the end of coil 2B (Geisler and Weber, 1982; Hanukoglu and Fuchs, 1983; Crewther *et al.*, 1983; Quax *et al.*, 1983). For instance, at the end of coil 2B is a consensus sequence YRR/KL/MLEGE, found in nearly all IF proteins. In this sequence, the third and fourth residues are the least conserved, although variations at these sites are conservative.

In addition to the heptad repeat pattern of the central domain, a conserved periodic distribution of acidic (aspartic acid, glutamic acid) and basic (arginine, histidine, and lysine) amino acids are found in IF proteins (Conway and Parry, 1988). A series of zones of alternating positive and negative charged residues repeat at approximately 9.5-residue intervals along the molecule, or 3 times every 28 residues (Parry *et al.*, 1977; McLachlan and Stewart, 1982; Crewther *et al.*, 1983). This periodicity in positively and negatively charged residues indicates that electrostatic interactions also play an important role in stabilizing the coiled-coil structure (McLachlan and Stewart, 1975). The degree of regularity in the distribution of ionic residues varies between filament type and thus may play a role in the segregation of different IF proteins into distinct networks of filaments (Conway and Parry, 1988).

The central α-helical domain is flanked by nonhelical amino and carboxy termini that are thought to protrude from the filament surface (Geisler and Weber, 1982; Hanukoglu and Fuchs, 1983; Steinert *et al.*, 1983). Termini are variable in length and amino acid composition and may serve to facilitate specific interactions between IFs and other proteins and cellular structures and, in so doing, may act to specify filament function. Length variations are most extreme in the carboxy termini, where sizes range from the complete absence of nonhelical sequence in the epidermal keratin K19 (Bader *et al.*, 1986), to the approximate length of 600 residues predicted for the NF-H protein (Lees *et al.*, 1988). The amino acid composition of ends in the IF family also varies considerably. Epidermal keratins have termini that contain inexact repeats of Gly-Gly-Gly-X (X = Phe, Tyr, or Leu) interspersed with stretches of serine residues (Hanukoglu and Fuchs, 1982, 1983; Steinert *et al.*, 1983,1984) whereas the ends of simple epithelial keratins, such as K7 (Glass *et al.*, 1985), lack these repeat motifs. Lamins A and C have a short amino-terminal region that is strongly basic whereas the larger carboxy-terminal domain is rich in glycine and serine. Desmin, GFAP and vimentin have arginine-rich head domains. All neurofilament tailpieces are highly charged due to an abundance of lysine and glutamic acid residues (Geisler *et al.*, 1985b; Lewis and Cowan, 1986; Lees *et al.*, 1988).

A. Phosphorylation of Terminal Domains

A postsynthetic modification common to all IFs is phosphorylation. Phosphorylation occurs predominantly on serine residues located in end domains and appears to be the primary means of controlling the assembly and disassembly of IF proteins in cells, most notably for the nuclear lamins and the type III IF proteins. During mitosis, phosphorylation of lamin proteins by the cdc 2 kinase initiates the disassembly of the nuclear lamina structure, a necessary prerequisite for cell mitosis (Ward and Kirschner, 1990; Peter *et al.*, 1990). Of particular importance for the lamin A molecule is the phosphorylation of serine residues 22 and 392, which lie just outside the central helical domain of the protein. Transfection of lamin A cDNAs containing double-substitution mutations at these sites causes inhibition of lamina breakdown in mitotic cells expressing the abnormal protein (Heald and McKeon, 1990).

The stability of the cytoplasmic IFs desmin and vimentin has also been shown to be markedly influenced by phosphorylation. Under *in vitro* conditions, these IFs act as substrates for a variety of protein kinases, which include protein kinase A and C and cAMP-dependent kinase (Inagaki *et al.*, 1987; Evans, 1988, 1989; Geisler and Weber, 1988; Kitamura *et al.*, 1989; Ando *et al.*, 1989; Geisler *et al.*, 1989). Subunit phosphorylation blocks polymerization of vimentin (Inagaki *et al.*, 1987) and desmin (Geisler and Weber, 1988), and causes depolymerization of isolated preformed filaments. Serine residues located within a 12-kDa amino-terminal fragment of vimentin are known targets of these kinases (Inagaki *et al.*, 1987; Evans, 1988, 1989). For desmin, amino-terminal residues 29, 35, and 50 have been identified as specific target sites (Geisler and Weber, 1988).

Similar to lamins, vimentin filament reorganization is typically associated with the onset of mitosis. The degree of vimentin reorganization during mitosis is cell-type specific, and varies in phenotype from the formation of a cage of filaments that surrounds the nucleus to total disintegration of the network into cytoplasmic aggregates. Vimentin reorganization may be a prerequisite for mitosis, either in a mechanical sense or perhaps in a regulatory mode. Since vimentin filaments associate at both the nuclear and plasma membranes, regulation could be envisioned to occur via alteration in filament attachments at these membranes. It is also possible that reorganization serves no critical function in cell division, because many cell types divide without any detectable change in the IF network.

Mitotically induced alterations of the vimentin network coincide with the breakdown of the nuclear lamina and have recently been linked to changes in serine residue phosphorylation mediated by the cdc 2 kinase

(Chou *et al.,* 1990). The cdc 2 kinase phosphorylates vimentin on as yet unidentified residues located in the amino terminus that, from preliminary studies, appear to be distinct from residues phosphorylated by protein A and C kinases (Chou *et al.,* 1990). It appears, therefore, that the reorganization of the vimentin network may be regulated in a complex manner by multiple kinases acting at various stages in the cell cycle. Further studies are required to understand the significance and role of each kinase in regulating IF organization and assembly within the cell.

NF-M and NF-H filaments are heavily phosphorylated, containing up to 50 mol phosphate/mol protein (Julien and Mushynski, 1983). Phosphorylation occurs predominantly on carboxy-terminal residues at sites consisting of Lys-Ser-Pro-(Val) repeats (Julien and Mushynski, 1982, 1983; Geisler *et al.,* 1987; Lee *et al.,* 1988). NF phosphorylation does not appear to play a role in regulating filament assembly and disassembly and instead may function to modulate the surface properties of the filaments, which in turn could affect functional interactions. Thus, the functional significance of NF phosphorylation remains to be elucidated.

IV. Assembly of Intermediate Filaments

Although diverse in composition and size, all types of IF proteins assemble to form 10-nm filament structures. Assembly of IF proteins into filaments *in vitro* occurs in a hierarchical manner in the apparent absence of auxiliary proteins or catalytic factors. Instead, assembly relies on the formation of a dimer unit via coiled-coil interactions between the helical domains of two polypeptide chains (Geisler and Weber, 1982; Quinlan *et al.,* 1984; Magin *et al.,* 1987; Hatzfeld *et al.,*1987; Hatzfeld and Weber, 1990a; Coulombe and Fuchs, 1990; Steinert, 1990). Studies of the lamin proteins using rotary shadowing and electron microscopy have enabled visualization of the dimer and show it to be an approximate 50-nm rod with its subunits arranged in register and in parallel, with two amino termini at one end and the two carboxy termini at the other (Aebi *et al.,* 1986; Parry *et al.,* 1987). Cross-linking studies on purified desmin, vimentin, and neurofilament dimers have verified the dimer arranged in parallel and in register as the basic IF structure (Quinlan *et al.,* 1986).

For keratins forming obligate heteropolymers composed of a 1 : 1 ratio of type I and type II subunits, the composition of the dimer as either a homo- or heterostructure has been difficult to determine. Recently, it was shown that coiled-coil dimerization between type I and type II proteins occurs even in the presence of 9 *M* urea plus reducing agent (Coulombe and Fuchs, 1990), explaining why prior isolation of heterodimers had not been achieved. These subunit assembly studies of human K5 and K14 epidermal keratins purified from cloned bacterial expression lines have allowed analysis of the heterodimer unit by gel filtration, chemical cross-linking, and electron microscopy (Coulombe and Fuchs, 1990). Under the conditions used for analysis, K5 and K14 homodimers were not detected. In contrast, studies in which the rod domains of the simple epithelial K8 and K18 dimer were cross-linked at a genetically engineered cysteine residue revealed both hetero- and homodimers of these keratins (Hatzfeld and Weber, 1990a). Subsequent filament assembly studies demonstrated that only the heterodimer was competent for filament formation (Hatzfeld and Weber, 1990a). Collectively, these and one other study (Steinert, 1990) suggest that the major structural subunit of keratin filaments is the very stable keratin heterodimer.

Keratin subunits in the coiled-coil dimer are predicted by model building and biochemical analyses of proteolytic fragments to be arranged with polypeptides aligned in register and in parallel, as predicted for other IFs (McLachlan, 1978; Crewther *et al.*, 1983; Woods and Inglis, 1984; Parry *et al.*, 1985, 1987). This arrangement is supported by EM studies of isolated fractions of K5 and K14 monomers and dimers (Coulombe and Fuchs, 1990). Negative staining of isolated K5 and K14 monomers showed no discernible structure whereas heterodimers in 9 *M* urea were lollipop structures with prominent globular domains at one end and rod domains of approximately 45–50 nm, similar to structures observed with lamin dimers (Aebi *et al.*, 1986; Parry *et al.*, 1987).

Two dimers associate to form highly stable tetramer units that appear to be arranged in an antiparallel manner, thus forming an apolar structure (Geisler *et al.*, 1985a; Quinlan *et al.*, 1984; Ip *et al.*, 1985; Parry *et al.*, 1985; Stewart *et al.*, 1989; Coulombe and Fuchs, 1990). Whether dimer units in the tetramer are arranged in register or are staggered has been controversial. Desmin tetramers decorated with Fab molecules appear as 50-nm dumbbell structures, suggesting that dimers interact in an antiparallel fashion and in register (Geisler *et al.*, 1985a). However, tetramers with partially staggered, antiparallel dimers have also been observed (Ip *et al.*, 1985; Kaufman *et al.*, 1985; Potschka, 1986; Fraser *et al.*, 1987; Stewart *et al.*, 1989), suggesting that both staggered and unstaggered conformations may exist during the assembly process (Aebi *et al.*, 1988; see also Coulombe and Fuchs, 1990). A model depicting the formation of keratin heterotetramers is shown in Fig. 2 and is based on electron microscopic analysis of the dimers and tetramers obtained by *in vitro* association of human K14 and K5 (Coulombe and Fuchs, 1990). Similar structures have been proposed for other IF proteins, with the exception that the subunits can be homopolymeric.

Assembly of the 10-nm IF structure requires the association of tetramers

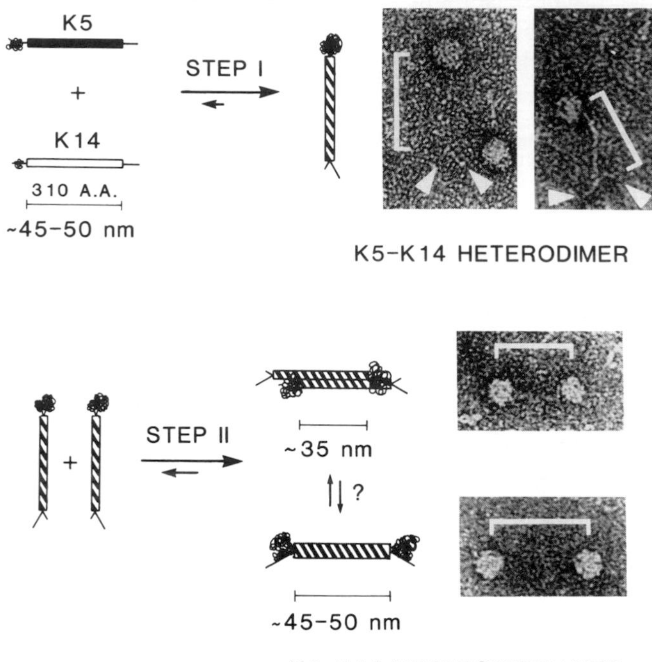

FIG. 2 Formation of keratin heteromeric subunits. The first step in the pathway leading to assembly of keratin 10-nm filaments is the formation of a heterodimer (Coulombe and Fuchs, 1990; Hatzfeld and Weber, 1990a; Steinert, 1990). As judged by electron microscopic analysis of human type II (K5) and type I (K14) heterodimers (Coulombe and Fuchs, 1990), keratin monomers associate in register and in parallel to form a lollipop-shaped structure resembling lamin dimers (Aebi *et al.*, 1986) and minimyosin molecules (McLachlan and Stewart, 1975). The length of the coiled-coil (rod domain), 45–50 nm, is consistent with that predicted for an α-helical domain of 310 amino acids (A.A.) in length, whereas the polar structure of the dimer is consistent with a parallel arrangement of the monomers as predicted from model building (Conway and Parry, 1988). The prominent globular domain at one end of the rod most likely represents the nonhelical amino-terminal ends of both keratins in a folded conformation, because these ends are generally twice the size of their carboxy-terminal counterparts. Its unexpectedly large size is probably due to excessive exclusion of the negative stain. The carboxy-terminal ends apparently do not form any discernable secondary structure, and can be seen extending from the other end of the rod (arrowheads). Electron microscopic analysis of the keratin tetramers indicates that they form as two heterodimers associated in an antiparallel fashion to form an apolar, dumbbell-shaped structure (Coulombe and Fuchs, 1990). Two types of tetramers were observed: one with a 45- to 50-nm rod, and one with a 35-nm rod. The tetramer with the longer rod is consistent with an unstaggered configuration of dimers, while the one with the shorter rod is consistent with overlapping amino ends of rods of the dimers and nonoverlapping carboxy ends of the rods (as also demonstrated in GFAP paracrystals by Stewart *et al.*,1989). Interestingly, Aebi (1988) has proposed the concept of tetramer switching, in which unstaggered and staggered tetramers would be in

into a hierarchy of distinct assembly intermediates (Aebi *et al.*, 1988; Coulombe and Fuchs, 1990). Tetramers assemble into 2- to 3-nm protofilaments, two of which then associate to form 4.5-nm protofibrils. Each 10-nm filament is composed of four intertwined protofibrils (Aebi *et al.*, 1983,1988; Steven *et al.*, 1983; Ip *et al.*, 1985; Eichner *et al.*, 1986). Scanning electron microscopy reveals that approximately 32 polypeptides, or 8 tetramers, contribute to the overall width of the filament (Steven *et al.*, 1983; Engel *et al.*, 1985).

The Role of Terminal Domains in Assembly

Although the end domains of IF polypeptides are quite variable in size and composition, all assemble to form 10-nm structures that are morphologically very similar. Filaments *in situ* in which end domain sequences have been at least partially removed by proteolysis are morphologically unchanged by negative staining (Steinert *et al.*, 1983; Chin *et al.*, 1983; Kaufman *et al.*, 1985), suggesting that the backbone of the filament resides primarily in the coiled coil structure formed by the residues of the central helical domain. Exposed end domains may therefore facilitate interfilament associations and interactions with cellular structures and their variation in composition and size may provide a means to diversify the flexibility and solubility of the filaments.

To examine the role of terminal domains in assembly, studies using truncated IF subunits derived either by proteolytic cleavage or by construction of deletion mutations in isolated IF genes have been conducted. These studies have elucidated the major domain requirements for assembly into the 10-nm filament and indicate that sequence requirements may vary among IF types.

1. Desmin, Vimentin, and Glial Fibrillary Acidic Protein

The role of end domains in filament assembly was initially examined by assembling proteolytically derived fragments of vimentin (Traub and

---

equilibrium. The *in vivo* relevance of that concept, as well as its potential regulatory implications for assembly, are still unknown. The fact that both the heterodimer and heterotetramer of K5 and K14 can form *in vitro* under conditions that would be strongly denaturing for most proteins underscores the extraordinarily strong and yet noncovalent interaction between keratins sharing only 25% homology. The tetramer is currently considered the best candidate for the building block of 10-nm filament assembly *in vivo,* not only for keratins but for intermediate filament proteins in general. (Courtesy of Dr. Pierre A. Coulombe.)

Vorgias, 1983). This truncated vimentin lacked an unknown number of residues from the nonhelical amino terminus and was unable to assemble into 10-nm filaments *in vitro*. Similarly, Kaufman *et al.* (1985) showed that desmin subunits lacking 67 residues from the nonhelical amino terminus could not assemble into IFs *in vitro*. This amino-truncated protein assembled into short, rodlike particles unless copolymerized with wild-type subunits of desmin or vimentin (Kaufman *et al.*, 1985). Recently, Raats *et al.* (1990) examined the behavior of deletion mutants of the amino terminus of hamster desmin. Mutant DNAs were expressed as a consequence of gene transfection into MCF 7 cells, which do not have vimentin or desmin networks. In agreement with *in vitro* assembly studies, desmin polypeptides lacking portions of the amino terminus were incapable of *de novo* filament assembly in these transfected MCF 7 cells. However, amino-truncated desmin subunits did form filaments when expressed in vimentin-expressing HeLa cells, presumably because heteropolymer formation was possible.

In contrast to the amino-truncated type III IF proteins studied thus far, one carboxy-truncated type III IF polypeptide could assemble into seemingly *bona fide* 10-nm filaments. Thus, desmin subunits lacking nearly half (27 residues) of the nonhelical carboxy terminus formed homopolymers efficiently *in vitro* (Kaufman *et al.*, 1985). In contrast, bacterially synthesized GFAP λcII fusion proteins lacking the carboxy domain of GFAP failed to assemble into 10-nm filaments *in vitro* (Quinlan *et al.*, 1989). While more extensive analyses are clearly prerequisite to our understanding of the roles of type III amino and carboxy termini in 10-nm filament assembly, it is clear that in most cases these domains appear to be important.

2. Neurofilament Studies

Deletion mutagenesis and gene transfection studies of the NF-M and NF-L proteins have shown that sequences in both the amino and carboxy nonhelical termini are required for assembly into NFs (Gill *et al.*, 1990; Wong and Cleveland, 1990). Deletion mutants of NF-L and NF-M cDNAs were transfected into mouse fibroblasts and coexpressed with either wild-type vimentin or NF-L proteins. Immunolabeling of the NF proteins demonstrated their ability to incorporate into the existing vimentin network. For the NF-L protein, deletion of amino acids in either the amino- or carboxy-helical domains resulted in subunits that were network incompetent and on integration caused collapse of vimentin filaments (Gill *et al.*, 1990). Coassembly of transiently expressed NF-L and endogenous vimentin into normal IF networks required that NF-L subunits consist of an intact rod domain plus at least 75% of the amino domain and 50% of the carboxy domain. Futhermore, NF-L subunits missing the first 22 residues were capable of integrating into the vimentin network, whereas deletion of an internal segment (between residues 18 and 31) caused disruption of the wild-type vimentin network. Thus, internal sequences of the head domain appear to have a critical role in NF assembly.

Deletion analysis of NF-M has shown that subunits missing up to 90% of the carboxy tail are capable of assembly into a preexisting network without disruption, as long as the relative amount of mutant to wild-type subunits is at a low ratio (Wong and Cleveland, 1990). Subunits with deletions extending into the carboxy terminal α-helical domain caused collapse of the endogenous network. NF-M subunits missing 70% of the amino-terminal domain could assemble into a preexisting network without perturbation, similar to the carboxy-terminal deletion mutants. However, in contrast to the carboxy mutants, subunits containing deletions of amino-terminal helical residues could not incorporate into a preexisting network and caused only slight perturbation in network structure. Thus, residues in the NF-M amino helical domain may be required for integration of subunits into a network.

3. Keratins

Keratin proteins appear to behave quite differently from type III and type IV IFs. *In vivo* studies using deletion mutagenesis and gene transfection of a K14 keratin cDNA have identified sequences required for type I subunit integration into the keratin network of cultured epithelial cells (Albers and Fuchs, 1987, 1989). In contrast to other IFs, K14 proteins missing either the carboxy terminus or nearly all of the amino terminus incorporated into a preexisting network, as did subunits missing both amino and carboxy residues. Thus, subunits assembled with the rod segment of K14 still retained a structure capable of higher ordered interaction within the IF network. Further deletion of residues at either end of the central helical domain resulted in proteins that integrated into the network, but in so doing caused it to collapse. These results demonstrate that (1) the amino and carboxy termini are not required for subunit incorporation into a preexisting network, and (2) the residues of the central helix are critical not only in forming the coiled-coil structure between two polypeptides but also in stabilizing higher ordered subunit interactions within the filament.

To further study assembly of mutated keratins, Coulombe *et al.* (1990) utilized a bacterial expression system to obtain purified K14 mutant proteins. This has enabled *in vitro* assembly studies using K14 mutant proteins and K5, the wild-type partner of K14 (Coulombe *et al.*, 1990). As in the transfection studies, mutant keratins lacking the nonhelical carboxy- or amino-terminal domains were assembly competent, as were subunits missing both the amino and carboxy domains. Deletions that extended into either end of the predicted α-helical rod domain resulted in a slight decrease in filament-forming efficiency for short deletions in the rod, whereas for larger deletions, elongation of filaments was severely restricted.

Although extensive studies on the truncated type I keratin K14 have been reported, analogous studies with K5 remain to be conducted. In the absence of such studies, it is not known whether some type I K14 mutants are tolerated merely because they are compensated by the presence of wild-type type II K5. For instance, can a K14 rod segment form filaments with a K5 rod segment or must it have an intact K5 partner? This issue has been addressed by experiments in which truncated simple epithelial keratins were expressed in fibroblasts (Lu and Lane, 1990). Intact and rod-segment keratin cDNAs were expressed as type I (K18 and K19) and type II (K7 and K8) pairs. The competence of various combinations of headless and/or tailess proteins to form *de novo* filaments was evaluated using light level immunohistochemistry. It was found that type I and type II keratins could polymerize effectively into an extended cellular network only if one of the keratin proteins of the pair had both an intact head-and-tail domain. For instance, assembly of truncated K8 keratin (lacking a portion of the carboxy terminus) with K19, a naturally occurring tailless keratin, resulted in short, crooked filaments in the cell. In contrast to these *in vivo* results, *in vitro* assembly of keratins lacking carboxy termini could produce intact, elongated filaments (Hatzfeld and Weber, 1990b). Specifically, normal-appearing filaments were assembled using a nearly tailless K8 keratin and the tailless K19. One explanation to account for this difference is that conditions required for assembly of subunits into an extended filamentous network in the cell are markedly different from *in vitro* assembly requirements. Alternatively, the conditions used for *in vitro* filament assembly may have been more favorable for the formation of stable IF structures. On the other hand, the short, crooked filaments formed *in vivo* may actually be equivalent to the filaments formed *in vitro*. In any case, the question still remains as to whether homotypic head-to-tail interactions are a prerequisite for efficient keratin filament assembly and elongation.

In summary, what is beginning to emerge from the deletion studies is that, despite having an overall conservation in secondary structure, IF types have different sequence requirements for maintaining filament stability. Whereas the integrity of the central helical domain is crucial for all IFs studied so far, tail domains exhibit varying levels of importance. Type III IFs require nearly all of the amino-terminal domain and perhaps at least 50% of the carboxy termini, whereas type IV filament assembly has varying amounts of dependence on the amino- and carboxy-termini portion. In contrast, type I and type II keratins missing nearly all residues of the amino and carboxy termini are still capable of filament assembly with no detectable perturbation in structure. This increased stability may reflect the more inherent stability of association shown to exist in the keratin heteropolymer structure (Coulombe and Fuchs, 1990). Whether all of the differences in requirements for amino and carboxy termini can be attributed to heteropolymer versus homopolymer interactions remains to be elucidated.

V. Intermediate Filament Associations within the Cell

A. Membrane Association

Studies on the light and electron microscopic levels have shown IFs to have close morphological association with the nuclear (Lehto *et al.*, 1978; Carmo-Fonseca and David-Ferreira, 1990) and plasma membranes (Ramaekers *et al.*, 1982; Granger and Lazarides, 1982). At the nucleus, lamin proteins (A, B, and C) assemble into a fibrillar meshwork that associates closely with the inner nuclear membrane (Aebi *et al.*, 1986). The association of lamins with the inner membrane is thought to be mediated by integral membrane proteins (Senior and Gerace, 1988). One such protein isolated from avian erythrocytes has been identified as a receptor for lamin B (Worman *et al.*, 1988). A cDNA sequence of the lamin B receptor predicts it to have a calculated molecular mass of 73 kDa and to contain a charged amino-terminal region predicted to face the nucleoplasm. It contains two consensus sites for protein kinase A phosphorylation which, assuming lamin B associates with this amino region, may enable regulation of lamin B binding through phosphorylation (Applebaum *et al.*, 1990). In addition, the receptor protein contains eight segments of hydrophobic amino acids thought to be transmembrane domains, and three DNA-binding motifs (Ser/Thr-Pro-X-X) located in the amino terminus. The presence of DNA-binding sites suggests that the lamin B receptor is involved in chromatin organization (Worman *et al.*, 1990).

Associations between cytoplasmic IFs and the nucleus have been documented for the type III IFs, peripherin (Djabali *et al.*, 1991), and vimentin and desmin (Georgatos and Marchesi, 1985; Georgatos *et al.*, 1985, 1987; Georgatos and Blobel, 1987) using filament–membrane binding assays and immunologic evidence. Binding of these proteins to the nuclear envelope appears to be mediated by lamin B. At the plasma membrane of avian erythrocytes, binding of vimentin and desmin was to ankyrin, a constituent of the plasma membrane skeleton. Binding to the nuclear envelope was cooperative and nonsaturable and by electron microscopy was shown to support IF polymerization, whereas binding to the plasma membrane fraction was not saturable and could not support filament polymerization. Futhermore, binding assays using fragments of desmin and vimentin missing either the amino end or the carboxy end of the polypeptide revealed a polarity in IF binding: fragments containing only a carboxy tail bound selectively to the nuclear envelope fraction, while fragments containing only an amino head bound selectively to the plasma membrane fraction. These studies have led to the hypothesis that a functional polarization exists during IF assembly and that initiation of filament assembly occurs at nuclear envelope-associated sites and not at sites on the plasma membrane.

The possibility that initiation of filament assembly occurs at the cell nucleus has been suggested by other experimental results. Solubilized keratin proteins added to triton X-100-extracted PtK1 cells were shown to polymerize into a single long filament that grew out from a discrete region on the nuclear surface postulated to be an IF organizing center (Eckert *et al.*, 1982). Additionally, the recovery behavior of epithelial cells following injection of anti-IF antibodies (Klymkowsky, 1982; Klymkowsky *et al.*, 1983) or transfection of deletion mutations of a K14 keratin cDNA (Albers and Fuchs, 1987, 1989) also supported the role of the nuclear membrane in initiation of filament assembly. Following collapse of the endogenous keratin network by incorporation of K14 mutant proteins, the keratin network began to recover in association with the nucleus. Interestingly, when keratin networks were collapsed due to incorporation of amino helical mutants, punctate nuclear staining of keratin was prevalent and network rebuilding in many transfected cells did not occur (Albers and Fuchs, 1989). This finding suggests that keratin subunits lacking amino helical residues may interfere with the initiation step during *de novo* filament formation.

B. Desmosome Association

Several types of intermediate filaments have close associations with specialized domains of the plasma membrane known as desmosomes (Green and Jones, 1990). Vimentin IFs interact with desmosome structures in human meningiomal cells and arachnoidal tissue (Kartenbeck *et al.*, 1984) whereas desmin filaments expressed in the myocardium and the dendritic follicular cells of the lymph node have attachments at desmosomes (Kartenbeck *et al.*, 1983; Franke *et al.*, 1987). Keratin filaments in epithelial cells, and in particular the epidermis, have numerous desmosomal connections. Filaments appear to loop through the matrix of the desmosomal plaque rather than terminate within it (Kelly, 1966). The nature of the molecular contact between IFs and the constituents of the desmosomal plaque are not yet known, although the proteins desmoplakin I (DP I) and desmoplakin II (DP II) may play a role in this association. Sequence analyses of cDNAs encoding DPI and DP II indicate that the carboxy-terminal domain of these proteins have charge periodicities that match those found in the rod domain of IFs (Green *et al.*, 1990). It is intriguing to speculate that IFs may interact with the DP proteins via their common periodicity.

Another factor which may mediate desmosome–IF association is a 140-kDa protein present as a minor component of purified desmosomal fractions isolated from bovine muzzle epithelium (Cartaud *et al.*, 1989, 1990). The 140-kDa protein was initially identified because of its apparent immunological similarity to lamin B, a nuclear IF protein known to bind vimentin, desmin, and peripherin. Lamin B antibodies were found to label purified desmosomal fractions and on immunoblots to label a 140-kDa component of the desmosome. Using *in vitro* binding assays, vimentin was found capable of binding to the 140-kDa band. Thus, proteins with antigenic similarity to lamin B may mediate attachment of IFs to the desmosomal plaques and other specialized regions of the plasma membrane.

VI. Intermediate Filament-Associated Proteins

The function and supramolecular structure of IFs expressed in cells and tissues are likely to be influenced by filament interactions with a partially defined group of proteins known as intermediate filament-associated proteins (IFAPs). IFAPs have been identified based on their ability to copurify and associate with IF proteins. The function of most IFAPs is unknown, although it is likely they mediate IF interactions with other cellular proteins and structures and regulate IF supramolecular organization. Some IFAPs are known to cross-link IFs laterally and into bundles. Keratin IFs in the cornified cells of the epidermis are noncovalently cross-linked into macrofibrillar aggregates by the cationic protein filaggrin, an IFAP specifically expressed in keratinizing epithelium of mammalian species. In the upper keratinized layers of the skin, bundling of keratin filaments by filaggrin may play a major role in manifesting the protective function of the skin. Filaggrin proteins are species specific and have molecular weights ranging from 26K in the mouse to 64K in the guinea pig (Dale *et al.*, 1990). Filaggrin is derived from a large (>400 kDa), polymeric, phosphorylated precursor known as profilaggrin. During epidermal differentiation, profilaggrin is dephosphorylated and proteolytically cleaved to form functional filaggrin molecules (Resing *et al.*, 1989). Profilaggrin genes are intronless and range in size from 13 kb for the human, 17 kb in the mouse, and 27 kb in rat. DNA sequences of profilaggrin have been obtained for mouse (Rothnagel *et al.*, 1987; Rothnagel and Steinert, 1990), rat (Haydock and Dale, 1986, 1990), and human (McKinley-Grant *et al.*, 1989). Repetitive units in the sequence have confirmed previous peptide mapping studies that showed profilaggrin to consist of repeating filaggrin units interspersed by short linker peptides (Lonsdale-Eccles *et al.*, 1984). These linker peptides are removed during the proteolytic conversion of profilaggrin to filaggrin.

Keratin IFAPs are also expressed in the hair appendage of the skin. During hair follicle growth and differentiation, 50–100 different proteins are expressed (MacKinnon *et al.*, 1990). These proteins have been classified into four groups based on their chemical and physical properties: (1) low sulfur proteins that form 8- to 10-nm IFs, (2) high-glycine/tyrosine proteins, (3) high-sulfur proteins, and (4) ultrahigh sulfur proteins (UHS) (Fraser *et al.*, 1972). Filaments are formed within the differentiating cortical cells of the hair shaft and become embedded in a matrix of the globular-like glycine- and sulfur-rich proteins. Recently, gene sequences have been obtained for UHS proteins of mouse (McNab *et al.*, 1989) and sheep and human (MacKinnon *et al.*, 1990). UHS genes appear to lack introns and their sequence predicts the UHS proteins to range between 31 and 37% in cysteine content and contain amino acid repeats enriched in either glycine, serine, or cysteine. *In situ* hybridization has localized the sheep and human UHS mRNAs to the hair cuticle late in follicle differentiation (MacKinnon *et al.*, 1990). How the UHS proteins associate with keratin IFs in the hair structure remains to be determined.

Another hair follicle-associated protein that may act as an IFAP is trichohyalin (Rothnagel and Rogers, 1986). Trichohyalin is an early marker of differentiation and is expressed in the hair medulla and the inner root sheath cell layers up until these cells undergo tissue hardening, after which they disappear. Sequence data from a partial (40% of the full length sequence) cDNA clone of sheep wool trichohyalin show that the carboxy terminus of trichohyalin has tandem repeats of a 23-amino acid sequence (Fietz *et al.*, 1990). Nineteen of the 23 amino acids of the repeat are charged or polar residues, although the overall net charge is only −1. The repeat also contains residues that may act as substrates for the follicle enzymes peptidylarginine deiminase and transglutaminase. The glutamine residue at position 17 in the repeat is in a charged region, EEQLRR, which is similar to predicted sites of transglutaminase activity in fibrin (Chen and Doolittle, 1971) and involucrin (Eckert and Green, 1986). Secondary structure prediction from the partial trichohyalin sequence shows it to be capable of α-helical formation and has led to the suggestion that trichohyalin may act as either an IF or IFAP in the inner root sheath of the hair follicle (Fietz *et al.*, 1990).

Plectin (300 kDa) is a well-characterized IFAP initially identified as a prominent component of vimentin IF preparations (Pytela and Wiche, 1980). Immunological detection techniques have shown plectin to be an abundant cytomatrix protein with a widespread tissue and cell type distribution (Wiche, 1989). Plectin binds to a variety of cytoskeletal proteins that include vimentin, certain epidermal keratins, GFAP, all three NF proteins, and lamin B. This binding occurs via the conserved central α-helical domains of the IF polypeptide (Foisner *et al.*, 1988) and appears to be regulated by protein kinase A and C-mediated phosphorylation (Foisner *et al.*, 1991a,b). The widespread distribution and diverse protein interactions of plectin have led to the hypothesis that plectin serves as a versatile cytoplasmic cross-linking element that interconnects cytoplasmic IFs with other cytoskeletal filament systems and with the nuclear and plasma membranes (Foisner and Wiche, 1991). Indeed, recent structural and DNA sequence data have revealed a molecule quite suited for these types of interactions. Electron microscopic analysis of rotary shadowed plectin molecules have shown a dumbbell-shaped structure with a filamentous 190-nm rod section flanked by two globular ends. Sequence data obtained from a rat plectin cDNA have confirmed this three-domain structural model (Wiche *et al.*, 1991). The DNA sequence predicts plectin to be 4140 amino acid residues in length and to contain a central α-helical domain in which five 200-residue subregions are present. Each subregion has a repeat of charged amino acids at 10.4 residues, which is identical to the charge periodicity of desmoplakin (Green *et al.*, 1990). In addition, the globular carboxy domain has tandem repeat structures that have significant homology to desmoplakin and bullous pemphigoid antigen.

Other IFAPs that remain to be more fully characterized are the proteins synemin (Granger and Lazarides, 1980), paramenin (Breckler and Lazarides, 1982), epinemin (Lawson, 1983), p95 (Lin, 1981), and NAPA-73 (Ciment *et al.*, 1986). Synemin (230 kDa) protein copurifies with desmin isolated from chicken smooth muscle (gizzard). Synemin binds to vimentin and desmin filaments, as demonstrated by immunofluorescence microscopy, and is postulated to act as a filament cross-linker (Granger and Lazarides, 1982). Paramenin (280 kDa) was originally coisolated with vimentin and desmin from cultured myogenic cells isolated from chick embryos. Paramenin colocalizes by immunolabeling with vimentin and desmin filaments in early myotube structures and at the Z line in late myotubes (Breckler and Lazarides, 1982; Price and Lazarides, 1983). Paramenin has also been detected immunologically in adult avian smooth muscle cells of elastic blood vessels. The function of paramenin remains unknown. Epinemin (44.5 kDa) was identified using a monoclonal antibody made against proteins extracted from chicken gizzard (Lawson, 1983). The epinemin antibody shows immunoreactivity in various cultured cells and skeletal muscle, which by double immunolabeling overlaps with vimentin filament staining. Another IFAP identified using a monoclonal antibody made against proteins isolated from rat skeletal myofibrils is p95 (95 kDa). The p95 monoclonal labels vimentin IFs in nonmuscle cells and reacts with the M line in myofibrils (Lin, 1981). NAPA-73 (73 kDa) is expressed in progenitor cells of the nervous system and heart, and appears to associate with bundles of intermediate filaments in these tissues (Ciment, 1990).

For most IFAPs, a specific role in mediating IF function remains to be determined. IF networks extend throughout the cell and interact with many different cytoplasmic and membrane components. They are also dynamic structures which in some cases undergo major rearrangement during mitosis and cell differentiation. IFAPs are sure to have a crucial role in maintaining functional contacts and perhaps in establishing and maintaining the organization of the network. In addition, because the expression of most IFAPs is cell-type specific, association of filaments with IFAPs may provide a means to further specify the properties and interactions of IFs, and in so doing may define and tailor the IF network to suit the needs of the cell.

VII. Evolutionary Origin of Intermediate Filaments

A. Neuronal versus Nonneuronal Intermediate Filament Types

The striking homology in amino acid sequence and structure among IF proteins predicts that their genes are derived from a common ancestor. For most IF types, common positioning of their introns suggests that this is true (Steinert and Roop, 1990). One exception, however, is the NF gene family (Lewis and Cowan, 1985, 1986; Julien *et al.*, 1985, 1986, 1987a; Levy *et al.*, 1987; Napolitano *et al.*, 1987; Myers *et al.*, 1987; Lees *et al.*, 1988). Thus, while nonneuronal IF genes have seven or eight nearly identically positioned introns, NF genes have only two or three introns, and these are positioned differently from other IF genes. Within the NF gene family, however, intron position is conserved, and it is likely that a common progenitor NF gene gave rise to all NF genes by gene duplication (Lees *et al.*, 1988). Why is there such a difference in gene structure when NF and other IF sequences are so highly conserved? One possibility is that during evolution, IF genes arose from a common primordial gene that may have been closely related in structure to the lamins (see below). With evolutionary pressure being exerted from increasing system complexity and cell specialization, the primordial gene is believed to have undergone gene duplication to form two progenitor lineages, an NF gene lineage and a nonneuronal lineage. Subsequently, the neuronal lineage further diverged to produce the present-day NF genes while the nonneuronal lineage diverged to produce all other IF types.

B. Invertebrate Intermediate Filaments: A Clue to the Primordial Intermediate Filament

Clues to the evolutionary origin of the IF family have been provided by studies of lower eukaryotic IF proteins by Weber and colleagues (Weber *et al.*, 1988, 1989). Isolation and characterization of nonneuronal invertebrate IF proteins have shown these organisms to possess a lower overall IF complexity. Full or partial sequences have been obtained for proteins labeled A (66 kDa) and B (52 kDa), isolated from the esophageal epithelium of the mollusk *Helix pomatia,* and for proteins IF-A (71 kDa) and IF-B (63 kDa), isolated from the body muscle of the nematode *Ascaris lumbricoides* (Weber *et al.*, 1988, 1989). Although the invertebrate IFs lack overt sequence homology to vertebrate IFs in the highly conserved central helical domain, they maintain the basic sequence and structural principles that apply to vertebrate IFs, i.e., the presence of hydrophobic heptad repeats and spacer domains (Weber *et al.*, 1988, 1989). Overt homologies are retained, however, at the conserved consensus sequences located at the ends of the helical domain. The retention of sequence homology in these regions suggests they are required for *in vivo* filament assembly and/or filament organization.

Although the invertebrate epithelial IF proteins are coexpressed, they are not keratin-like obligate heteropolymers because they are capable of assembling into homopolymeric filaments *in vitro* (Bartnik *et al.*, 1985; Weber et al., 1988). Invertebrates lack complex, stratified epithelial structures altogether, which has led Weber *et al.* (1988) to propose that the heteropolymeric structure of keratins evolved only in the vertebrate lineage concurrently with the development of stratified epithelial tissue structure.

The mollusk and nematode IFs are cytoplasmic, and yet they display a number of similarities with the nuclear lamins (Weber *et al.*, 1988, 1989). For both the mollusk and nematode protein pair, sequence comparisons show differences only at the carboxy termini, where the larger protein of each pair contains additional sequence. This arrangement is similar to the mammalian lamins A and C (McKeon *et al.*, 1986; Fisher *et al.*, 1986). In addition, the invertebrate IF proteins also contain (1) a lamin-like insertion in the coil 1b rod segment of six heptads or 42 amino acids and (2) high homology to lamins at the carboxy termini if alignment of residues is optimized by deleting a 15 to 20- residue stretch in the lamin termini. The region of this deletion coincides with the position of the nuclear localization signal sequence of the lamins (Loewinger and McKeon, 1988). These data support the notion that nuclear lamins and cytoplasmic IF proteins are evolutionarily derived from a common lamin-like predecessor, i.e., genes encoding higher eukaryotic cytoplasmic IFs evolved from genes encoding lower eukaryotic nuclear lamins (Osborn and Weber, 1986; Bartnik *et al.*, 1987; Franke, 1987; Myers *et al.*, 1987; Steinert and Roop, 1988).

VIII. What Is the Function of Intermediate Filaments?

A. Do Cells Require Intermediate Filaments?

Even as the diversity of the family of IF proteins continues to grow and more is understood about their structure and expression, the role of these proteins in cell growth and function remains an enigma. The ubiquitously expressed nuclear lamins certainly must contribute to nuclear membrane integrity and structure, although their role in chromatin organization and signal transduction, if any, remains unclear. Cytoplasmic IFs are more diversified and in most cells comprise a major component of the cytoskeletal architecture. Although a global function for cytoplasmic IF networks has not emerged, functions based primarily on structural interactions have been proposed. These include the establishment and maintenance of specific arrangements of organelles and other structural elements in the cell and/or as signal transducers between the cytoplasmic membrane and nucleus. In this regard, IF function may extend across cellular boundaries.

The presence of an IF network does not seem to be essential to cell viability. Cell culture studies in which IF networks are specifically disrupted either by microinjection of anti-IF antibodies (Klymkowsky, 1981; Klymkowsky *et al.*, 1983) or by the forced expression of mutated IF proteins (Albers and Fuchs, 1987, 1989; Wong and Cleveland, 1990; Gill *et al.*, 1990; Raats *et al.*, 1990), have shown that cells retain their shape and are capable of undergoing mitosis and movement. Whether the microinjected and transfected cells could survive for a prolonged period without an extended network is unknown, although cultured cell lines that lack IFs entirely have been isolated (Venetianer *et al.*, 1983; Giese and Traub, 1986; Herdberg and Chen, 1986). Therefore, at least in cultured cells, IFs do not appear to be essential for cell viability and may instead have a more subtle role in the cell.

IF genes and proteins are expressed in a developmental and differentiation specific manner. This fact, coupled with the diversity of the IF family, suggests that the role of the IF network is of a specialized nature and is linked to the differentiative state and function of the tissue. Perhaps the best indication of this is provided by the family of keratin IFs. Expression of different keratin pairs can be correlated with distinctive phenotypic features of epithelial differentiation. While keratin expression does not seem to be involved in epithelial tissue morphogenesis and development (Heid *et al.*, 1988; Kopan and Fuchs, 1989), it may nevertheless be important in imparting to an epithelial cell its integrity and mechanical strength. In this regard, it is known that keratins of simple epithelial cell types differ from keratins of epidermis with regard to filament solubility, tensile strength, and flexibility. These differences may affect filament network structure and interactions with other proteins and cytoskeletal elements. In addition, IFAPs may contribute to IF network differences. Thus, for instance, in epidermis, macrofilament bundles of cross-linked keratin filaments form in the late stages of keratinocyte differentiation and this may enforce the protective role of epidermis by making keratin filaments stronger and more resistant to the massive destructive phases that take place during the later stages of terminal differentiation.

The fact that IF networks exhibit a highly organized structure and establish numerous cellular interactions has led to the suggestion that IF networks function to spatially organize and integrate cellular space (Lazarides, 1980). From such an arrangement, it is also quite conceivable that IFs impart a means of communication not only within the cell but, given their known membrane associations, between the cell surface and nucleus. IFs may also function to integrate mechanical linkage within cells and tissues to strengthen and protect tissues from stress and deformation. This role is easily envisioned for the keratin networks of stratified and lining epithelium and for desmin IFs expressed in smooth, skeletal, and cardiac muscle contractile systems. In striated muscle, desmin is localized by immunofluorescent labeling at Z lines and appears to interconnect them (Lazarides *et al.*, 1982). In smooth muscle, desmin filaments link cytoplasmic dense body structures, which are considered to be structurally analogous to striated muscle Z lines, to specialized regions of the plasma membrane known as dense plaques. This association has recently been examined using confocal microscopy and image analysis of the contractile and cytoskeletal elements. From such analysis, a coupling between desmin filaments and the contractile system via the dense body structure has been proposed (Draeger *et al.*, 1990). The functional significance of this relationship during muscle activity remains to be determined.

For the neurofilament IFs, a specific functional role remains to be identified. NFs are found primarily in axons although not all axons contain them. They are preferentially located in the axon as compared to the cell body and are thought to provide mechanical strength to axons and act as axoplasmic organizers (Lasek *et al.*, 1985). In addition, the density of NFs is postulated to contribute to the axonal architecture by controlling axonal caliber and, in effect, neuronal development and function because the velocity of the action potential traveling down an axon increases as a function of diameter (Lasek *et al.*, 1983; Hoffman *et al.*, 1987). The role of NFs and other IFs in complex systems has recently begun to be analyzed by expressing these proteins in transgenic animals. These studies have provided new insights into the role of IFs in the development and function of the multitude of cells, tissues, and organs of higher eukaryotes.

B. Expression of Intermediate Filament Proteins in Transgenic Mice and Embryonic Stem Cells

Transgenic and embryonal stem cell-derived mice provide powerful systems to study the function of IFs and their possible roles in tissue morphogenesis and development. Transgenic mice expressing IFs have now been reported for vimentin (Krimpenfort *et al.*, 1988; Capetanaki *et al.*, 1989a,b), desmin (Pieper *et al.*, 1989; Dunia *et al.*, 1990), neurofilaments (Julien *et al.*, 1987b; Monteiro *et al.*, 1990), and keratins (Vassar *et al.*, 1989, 1991; Powell and Rogers 1990; Abe and Oshima, 1990; Coulombe *et al.*, 1991b; Baribault and Oshima, 1991). Studies of vimentin expression in transgenic mice were initially carried out using a vimentin–desmin hybrid gene construct in which the 5′ portion of the hamster vimentin gene, encoding the head and rod domain, was linked to the 3′ region of the hamster desmin gene, encoding the tail domain (Krimpenfort *et al.*, 1988). The expression of the hybrid gene was driven by 3.1 kb of the vimentin 5′ flanking region. Multiple copies of the transgene (not specified) were integrated and on Northern analysis gave RNA expression levels comparable to endogenous vimentin levels. The transgenic mice exhibited no abnormalities, and in vimentin-expressing cells normal filaments were formed consisting of endogenous vimentin and the hybrid vimentin-desmin protein. In contrast, overexpression of chicken vimentin in vimentin-expressing cells of transgenic mice (Capetanaki *et al.*, 1989a) led to cataract formation in the lens, apparently as a consequence of impaired fiber cell denucleation and elongation (Capetanaki *et al.*, 1989b). Cataract formation was confined to animals expressing high levels of the transgene RNA (300–1100% relative to endogenous vimentin RNA levels) and protein. Because normal vimentin IFs were detected even in chicken vimentin-expressing mice, which had up to 10 times the level of endogenous vimentin (Capetanaki *et al.,* 1989b), it seems likely that the differences observed in these 2 studies arose from differences in levels of vimentin expression.

In other studies of type III IF expression, the muscle-specific desmin protein was expressed in vimentin-expressing cells of mice using a construct pVDes, in which the coding region of the hamster desmin gene was fused to the 5′ flanking sequences of the hamster vimentin gene (Pieper *et al.,* 1989). Of the strains examined by Southern hybridization, 2 expressed desmin at levels comparable to the endogenous vimentin level whereas another expressed the transgene at an approximately 10-fold lower level. Analysis on the RNA and protein levels revealed tissue-specific expression, with transgenic desmin coassembling with endogenous vimentin filaments. Coexpression of desmin and vimentin did not have any observable effect on normal development, morphology, or physiology. However, prompted by the findings of Capetanaki *et al.* (1989b), which showed cataract formation resulting from vimentin overexpression, further analysis was carried out on lens morphology in the pVDes transgenics (Dunia *et al.,* 1990). Although cataract formation was not evident on the macroscopic level in these mice, microscopic examination revealed perturbation of the enucleation process in the lens fibers, changes in cell shape, fiber fusion, and extensive internalization of the plasma membrane. These alterations were found in all mice examined, with the degree of perturbation corresponding to the level of transgene expression. Interestingly, a number of similarities between the transgenic lens and the early stages of congenital cataract formation could be drawn, i.e., changes in the plasma membrane integrity, alteration in fiber shape and orientation, and the restriction of these changes to individual cells and defined regions of the lens (Dunia *et al.,* 1990). It may be, therefore, that alterations in the control of IF expression in the lens may be a factor in cataract formation.

To study regulation of neurofilament function, transgenic mice expressing the human NF-L gene (Julien *et al.,* 1987b) and the mouse NF-L gene (Monteiro *et al.,* 1990) have been derived. The human NF-L gene contained 14 kb of 5′ flanking sequence and 3.2 kb of 3′ sequence and was expressed in several lines of mice in a neuron-specific manner. Developmental expression of the NF-L transgene did not strictly parallel expression of the mouse NF-L gene and was expressed at high levels only in the postnatal brain. The level of transgene NF-L protein corresponded to the mRNA level and its overexpression did not influence the levels of the endogenous neurofilament proteins. As in many of the tissues of transgenic mice overexpressing type III IF genes, no apparent phenotype was detected in neuronal tissues of the type IV IF-expressing transgenics whose expression was driven by an NF promoter.

The mouse NF-L gene has also been expressed in transgenic mice under the control of the murine sarcoma virus (MSV) promoter (Monteiro *et al.*, 1990). Under the MSV promoter, NF-L was expressed in tissues such as kidney, skeletal muscle, and lens as well as neuronal tissues. Overexpression of NF-L occurred in peripheral axons and led to to a 2.2-fold increase in the density of neurofilaments in axons (Monteiro *et al.*, 1990). Even though filament density increased, no obvious change in axonal caliber was measured. This was somewhat surprising in light of existing hypotheses suggesting that neuronal caliber may be determined by NF concentration within the axon (Hoffman *et al.*, 1987). Further studies are required to determine whether axonal caliber is altered in animals expressing increased levels of NF-M and NF-H proteins. In particular, NF-H proteins are thought to form cross-bridge structures within the filament that may be a prerequisite to increased axonal caliber.

Interestingly, no overt phenotype was apparent in mice expressing NF-L except for lens cataract formation, similar to what had also been observed for transgenic mice overexpressing vimentin IF proteins (Capetanaki *et al.*, 1989a). Because type IV and type III IF proteins can copolymerize (Monteiro and Cleveland, 1989), these observations suggest that the lens is highly susceptible to an increase in IF network density.

Recently, Powell and Rogers (1990) reported that transgenic mice that grossly overexpress a wild-type hair IF keratin exhibit altered patterns of hair growth and fragile hair fibers. In this case, overexpression of the wild-type keratin caused a decreased level of the hair intermediate filament-associated proteins, and a disruption in the array of these proteins in the hair cortex. Whether the disorganization was due to the imbalance in IFs versus IFAPs or, alternatively, to an accumulation of the type II wild-type transgene product in the cortex remains to be shown. Irrespective of mechanism, however, the adverse effects of overexpression of wild-type IF protein in hair differed from those in the lens in that the aberrations appeared to be generated from an alteration in the IF network.

To probe further into the question of IF function, transgenic mice have recently been made that express mutant keratin proteins targetted to the basal cells of stratified squamous epithelia of these mice (Vassar *et al.*, 1991). The mutants chosen for expression were those that had previously been shown to disrupt keratin filament networks in transfected cultured cells (Albers and Fuchs, 1987). Interestingly, these transgenic mice exhibited a high incidence of neonatal mortality, with gross skin abnormalities including epidermal blistering elicited by basal cell cytolysis (Vassar *et al.*, 1991) (Fig. 3). At the ultrastructural level, keratin filaments were disorganized and formed large aggregates throughout the cytoplasm. Suprabasal layers appeared much more normal, with a near-normal wild-type keratin filament network. This could be explained by the fact that the promoter used for gene targeting switches off as cells commit to terminally differentiate, as a new set of keratin genes are switched on (Fuchs and Green, 1980; Tyner and Fuchs, 1986; Vassar *et al.,* 1989). Other stratified squamous epithelia were also affected, but to varying degrees. Collectively, the phenotype of these mice was strikingly similar to that of humans with the genetic skin disease known as epidermolysis bullosa simplex (EBS). Indeed, when different patients with EBS were analyzed, these patients had genetic alterations in their epidermal keratin genes (Coulombe *et al.,* 1991a; Bonifas *et al.,* 1991). This is the first evidence to date demonstrating that mutations in IF genes might lead to human disease.

While additional examples of disrupted IF networks in other tissues will be necessary to ascertain the extent to which disruptions might interfere generally with tissue function, a picture is beginning to emerge whereby disruptions in IF networks may be more deleterious than the density of IFs within a cell. Does this necessarily imply a critical function for IFs? *A priori,* it is tempting to attribute the EBS-like mutant keratin phenotype in transgenic mice to a loss of IF network in the basal cells, leading to a loss of architectural integrity of these cells and subsequent increase in fragility to mechanical stress. However, basal cell cytolysis might also be attributed to the generation of large cytoplasmic protein aggregates, leading to disruption of normal cellular functions. In this scenario, it would not be the compromise of the IF network that caused the aberrations, but rather the appearance of insoluble protein aggregates. To resolve this dilemma, additional keratin mutants have recently been expressed in transgenic mice (Coulombe *et al.,* 1991b). In contrast to mutants that severely disrupted keratin filament assembly and gave rise to mice with severe blistering and the phenotype of Dowling Meara EBS (Vassar *et al.,* 1991), mutants that only mildly perturbed keratin filament assembly gave rise to mice with moderate blistering and the phenotype of milder EBS forms, such as Weber-Cockayne and Koebner (Coulombe *et al.,* 1991b). In the milder cases, basal cells still cytolyzed in response to incidental trauma, but in these cases tonofilaments showed no evidence of clumping. The filaments were distinguishable from wild-type primarily in that they were shorter and disorganized in the basal cytoplasm. Finally, a zone of fragility was detected which was located just beneath the nucleus and above the hemidesmosomes in the columnar basal cells. Collectively, these studies demonstrated a genetic link between multiple subtypes of EBS and the severity with which the keratin filament network is disrupted in a basal cell. Moreover, the data suggest that the function of basal epidermal keratin filaments is to provide mechanical integrity to the basal cell, without which the cell becomes fragile and lyses in response to physical trauma.

Is the function of all IFs to impart mechanical integrity to cells? With regard to nuclear lamins, this is likely to be the case, since nuclear envelopes formed in the absence of lamins are fragile and lyse upon physical trauma (Newport *et al.*, 1990). Given the similarities in IF structure, it is tempting to speculate that such a structural role might be common among all IFs. However, given the apparent diversity in cell and tissue types, certain cells might be more sensitive and more frequently subjected to mechanical stress than others, and hence such a function might be more easily recognized in some cell types and shapes than in others. Finally, because of the diversity in IF sequence, it seems likely that specialized functions will also emerge for given IFs. In this regard, it will be important to analyze each IF type in transgenic animals and in other systems to determine the role(s) of IFs in specific cell types and tissues. For the moment, however, the functions of IFs remain elusive and one of the last frontiers for major structural proteins in eukaryotes.

IX. Conclusions and Perspectives

IF proteins are a complex and still-expanding family of proteins that share a remarkable conservation in structure. Although for most IFs a functional role is still to be defined, the developmental, differentiation, and tissue-specific expression of these proteins provide strong evidence that their 10-nm networks play a crucial role in the maintenance and function of cells and tissues. The isolation and cloning of the genes for these proteins has led to significant progress in the understanding of IF structure, regulation of gene expression, and the nature of IF interactions within the cell. The ability to manipulate the IF genes and express them in cultured cells and in transgenic mice has provided extremely powerful experimental systems in which to dissect IF structure and function. For many IF proteins, gene transfection studies have shown that IFs form an extremely dynamic network that is susceptible to disruption upon incorporation of abnormal subunits. From such studies it has been possible to map specific sequences in the IF molecule that are critical for filament assembly and network formation. In the future, point mutation studies will provide a finer mapping of the residues involved in IF assembly whereas studies using proteins with mixed IF domains will be valuable in identifying regions of the molecule involved in network recognition and formation.

Analysis of IF structure and expression in cultured cells has been complemented by expression of IFs in transgenic animals. Although further studies remain, expression of IFs in transgenics has already provided unique insight into the role of the IF network in tissue structure and organization, particularly in the case of the epidermal keratins. Future studies that examine the consequences of expression of mutated IFs on tissue development, structure, and function, and also the consequences of

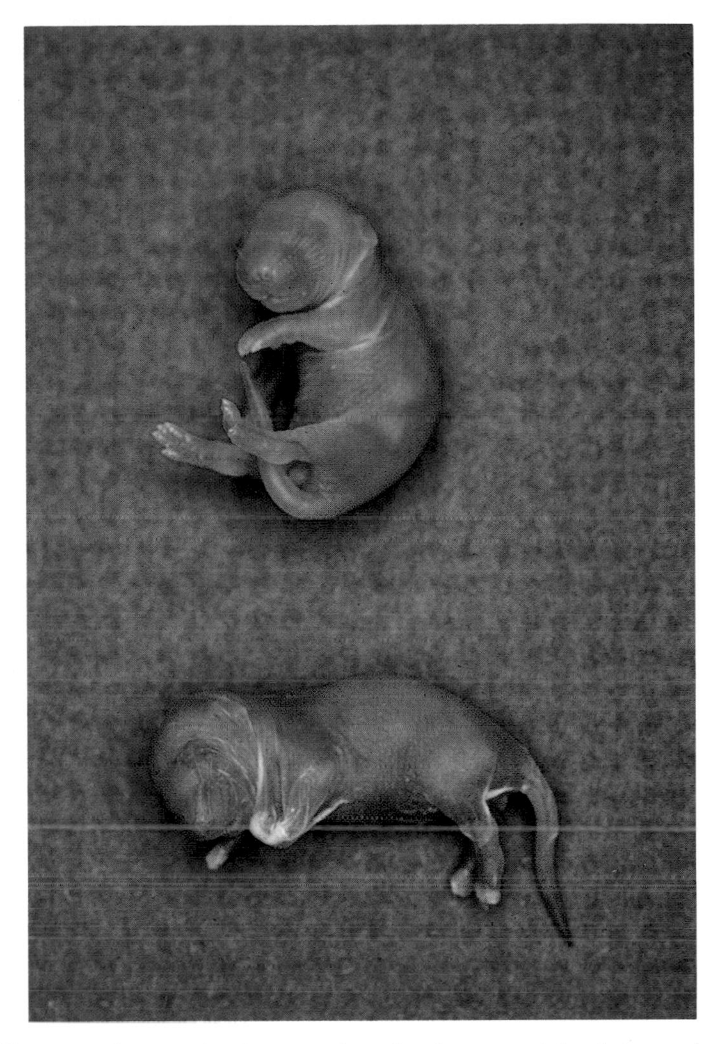

FIG. 3 Phenotype of transgenic mice expressing a keratin gene mutation that severely disrupts 10-nm filament assembly. (Top) Control newborn mouse. (Bottom) F_1 generation transgenic offspring from a mouse expressing a mutant keratin transgene referred to as pgCΔ135K14P(-6000). This gene contains 6000 bp of 5′ upstream sequence from the human K14 keratin gene, linked to a K14 cDNA that is missing sequences encoding approximately 30% of the carboxy end of the α-helical rod domain and the entire nonhelical tail domain. Note the presence of epidermal blistering around the limbs, particularly at the upper left forearm. (From Vassar *et al.*, 1991.)

IF overexpression, are at this point the most promising of directions. These types of studies offer an exciting opportunity to identify and study the role of IFs in human disease.

IF networks exhibit close morphological associations with membranes and other cell structures. These interactions are likely to form the basis of the supramolecular structure and function of the IF network. Proteins that copurify with IFs, or IFAPs, are prime candidates for mediating these interactions and their isolation and characterization will provide essential pieces in the puzzle of IF function. Although the genes for some IFAPs have been isolated and sequenced, most known IFAPs are not well characterized and many more putative IFAPs remain to be identified. Studies of known IFAPs coupled with the isolation and characterization of new IFAPs will be an important step toward determining the role of these proteins in mediating IF interactions, and toward elucidating the significance of the IF network in cellular function.

References

Aaronson, R. P., and Blobel, G. (1975). *Proc. Natl. Acad. Sci. U.S.A.* **72,** 1007–1011.

Abe, M., and Oshima, R. G. (1990). *J. Cell Biol.* **111,** 197–1206.

Aebi, U., Fowler, W. E., Rew, P., and Sun, T.-T. (1983). *J. Cell Biol.* **97,** 1131–1143.

Aebi, U., Cohn, J., Buhle, L., and Gerace, L. (1986). *Nature (London)* **323,** 560–564.

Aebi, U., Haner, M., Troncoso, J., Eichner, R., and Engel, A.(1988). *Protoplasma* **145,** 73–81.

Albers, K., and Fuchs, E. (1987). *J. Cell Biol.* **105,** 791–806.

Albers, K., and Fuchs, E. (1989). *J. Cell Biol.* **108,** 1477–1493.

Ando, S., Tanabe, K., Gonda, Y., Sato, C., and Inagaki, M. (1989). *Biochemistry* **28,** 2974–2979.

Applebaum, J., Blobel, G., and Georgatos, S. D. (1990). *J. Biol. Chem.* **265,** 4181–4185.

Bader, B. L., Magin, T. M., Hatzfeld, M., and Franke, W. W. (1986). *EMBO J.* **5,** 1865–1875.

Baribault, H., and Oshima, R. G. (1991). *J. Cell Biol.* **115,** 1675–1684.

Bartnik, E., Osborn, M., and Weber, K. (1985). *J. Cell Biol.* **101,** 427–440.

Bartnik, E., Stephan-Kossmagk, K., and Weber, K. (1987). *Eur. J. Cell Biol.* **44,** 219–228.

Benevente, R., and Krohne, G. (1986). *J. Cell Biol.* **100,** 1847–1854.

Bennett, G. S., Fellini, S. A., and Holtzer, H. (1978). *Differentiation (Berlin)* **12,** 71–82.

Bennett, G. S., Fellini, S. A., Toyama, Y., and Holtzer, H. (1979). *J. Cell Biol.* **82,** 577–584.

Bonifas, J. M., Rothman, A. L., and Epstein, E. H., Jr. (1991). *J. Invest. Dermatol.* **254,** 1202–1205.

Breckler, J., and Lazarides, E. (1982). *J Cell Biol.* **92,** 795–806.

Brody, B. A., Ley, C. A., and Parysek, L. M. (1989). *J. Neurosci.* **9,** 2391–2401.

Capetanaki, Y., Ngai, J., and Lazarides, E. (1984). *Proc. Natl. Acad. Sci. U.S.A.* **81,** 6009–6013.

Capetanaki, Y., Starnes, S., and Smith, S. (1989a). *Proc. Natl. Acad. Sci. U.S.A.* **86,** 4882–4886.

Capetanaki, Y., Smith, S., and Heath, J. P. (1989b). *J. Cell Biol.* **109,** 1653–1664.

Carmo-Fonseca, M., and David-Ferreira, J. F. (1990). *Electron Microsc. Rev.* **3,** 115–141.

Cartaud, A., Courvalin, J. C., Ludosky, M. A., and Cartaud, J. (1989). *J. Cell Biol.* **109,** 1745–1752.

Cartaud, A., Ludosky, M. A., Courvalin, J. C., and Cartaud, J. (1990). *J. Cell Biol.* **111,** 581–588.

Chen, R., and Doolittle, R. F. (1971). *Biochemistry* **10,** 4486–4491.

Chin, T. K., Eagles, P. A. M., and Maggs, A. (1983). *Biochem. J.* **215,** 239–252.

Chou, Y.-H., Bischoff, J. R., Beach, D., and Goldman, R. D. (1990). *Cell* **62,** 1063–1071.

Ciment, G. (1990). *Ann. N.Y. Acad. Sci.* **588,** 225–235.

Ciment, G., Ressler, A., Letourneau, P. C., and Weston, J. A. (1986). *J. Cell Biol.* **102,** 246–251.

Conway, J. F. and Parry, D. A. D. (1988). *Int. J. Biol. Macromol.* **10,** 79–98.

Cooke, P. (1976). *J. Cell Biol.* **68,** 1539–1556.

Coulombe, P. A., and Fuchs, E. (1990). *J. Cell Biol.* **111,** 153–169.

Coulombe, P. A., Chan, Y.-M., Albers, K., and Fuchs, E. (1990). *J. Cell Biol.* **111,** 3049–3064.

Coulombe, P. A., Hutton, M. E., Letai, A., Hebert, A., Paller, A. S., and Fuchs, E. (1991a). *Cell* **66,** 1301–1311.

Coulombe, P. A., Hutton, M. E., Vassar, R., and Fuchs, E. (1991b). *J. Cell Biol.* **115,** 1661–1674.

Crewther, W. G., Dowling, L. M., Steinert, P. M. and Parry, D. A. D. (1983). *Int. J. Biol. Macromol.* **5,** 267–274.

Crick, F. H. C.(1953). *Acta Crystallogr.* **6,** 689.

Dale, B. A., Resing, K. A., and Haydock, P. V. (1990). *In* "Cellular and Molecular Biology of Intermediate Filaments" (R. D. Goldman and P. M. Steinert, eds.), pp. 393–412. Plenum, New York.

Djabali, K., Portier, M. M., Gros, F., Blobel, G., and Georgatos, S. D. (1991). *Cell* **64,** 109–121.

Draeger, A., Amos, W. B., Ikebe, M., and Small, J. V. (1990). *J. Cell Biol.* **111,** 2463–2473.

Drager, U. C. (1983). *Nature (London)* **303,** 169–172.

Dunia, I., Pieper, F., Manenti, S., van de Kemp, A., Devilliers, G., Benedetti, E. L., Bloemendal, H. (1990). *Eur. J. Cell Biol.* **53,** 59–74.

Eckert, B. S., Daley, R. A., and Parysek, L. M. (1982). *J. Cell Biol.* **92,** 575–578.

Eckert, R. L., and Green, H. (1986). *Cell* **46,** 583–589.

Eichner, R., and Kahn, M. (1990). *J. Cell Biol.* **110,** 1149–1158.

Eichner, R., Bonitz, P., and Sun, T.-T. (1984). *J. Cell Biol.* **98,** 1388–1396.

Eichner, R., Sun, T.-T., and Aebi, U. (1986). *J. Cell Biol.* **102,** 1767–1777.

Engel, A., Eichner, R., and Aebi, U. (1985). *J. Ultrastruct. Res.* **90,** 323–335.

Evans, R. M. (1988). *FEBS Lett.* **234,** 73–78.

Evans, R. M. (1989). *J. Cell Biol.* **108,** 67–78.

Ferrari, S., Battini, R., Kaczmarek, L., Rittling, S., Calabretta, B., Kim De Riel, J., Philiponis, V., Wei, J.-F., and Baserga, R. (1986). *Mol. Cell. Biol.* **6,** 3614–3620.

Fietz, M. J., Presland, R. B., and Rogers, G. E. (1990). *J. Cell Biol.* **110,** 427–436.

Fischer, D. Z., Chaudhary, N., and Blobel, G. (1986). *Proc. Natl. Acad. Sci. U.S.A.* **83,** 6450–6454.

Foisner, R., and Wiche, G. (1991). *Curr. Opin. Cell Biol.* **3,** 75–81.

Foisner, R., Leichtfried, F. E., Herrmann, H., Small, J. V., Lawson, D., and Wiche, G. (1988). *J. Cell Biol.* **106,** 723–733.

Foisner, R., Feldman, B., Sander, L., and Wiche, G. (1991a). *J. Cell Biol.* 112, **397–405.**

Foisner, R., Traub, P., and Wiche, G. (1991b). *Proc. Natl. Acad. Sci. U.S.A.* **88,** 3812–3816.

Franke, W. W. (1987). *Cell* **48,** 3–4.

Franke, W. W., Schmid, E., Winter, S., Osborn, M., and Weber, K. (1979). *Exp. Cell Res.* **123,** 25–46.

Franke, W. W., Scheer, U., Krohne, G., and Jarasch, E. D. (1981). *J. Cell Biol.* **91,** 39s–50s.

Franke, W. W., Cowin, P., Schmelz, M., and Kapprell, H.-P. (1987). *Ciba Found. Symp.* **125,** 26–44.

Fraser, R. D. B., and MacRae, T. P. (1973). *Polymer* **14,** 61–67.

Fraser, R. D. B., MacRae, T. P., and Rogers, G. E. (1972). "Keratins: Their Composition, Structure, and Biosynthesis." Thomas, Springfield, Illinois.

Fraser, R. D. B., Steinert, P. M., and Steven, A. C. (1987). *Trends Biochem. Sci.* **12,** 43–45.

Fuchs, E., and Green, H. (1980). *Cell* **19,** 1033–1042.

Fuchs, E., Coppock, S., Green, H., and Cleveland, D. (1981). *Cell* **27,** 75–84.

Gard, D. L., and Lazarides, E. (1980). *Cell* **19,** 263–275.

Gardner, E. E., Dahl, D., and Bignami, A. (1984). *J. Neurosci. Res.* **11,** 145–155.

Giesler, N., and Weber, K. (1981). *J. Mol. Biol.* **151,** 565–571.

Geisler, N., and Weber, K. (1982). *EMBO J.* **1,** 1649–1656.

Geisler, N., and Weber, K. (1988). *EMBO J.* **7,** 15 20.

Geisler, N., Kaufman, E., and Weber, K. (1982). *Cell* **30,** 277–286.

Geisler, N., Kaufman, E., and Weber, K. (1985a). *J. Mol. Biol.* **182,** 173–177.

Geisler, N. E., Fischer, S., Vandekerckhove, J., Van Damme, J., Plessmann, U., and Weber, K. (1985b). *EMBO J.* **4,** 57–63.

Geisler, N., Vanderkerckhove, J., and Weber, K. (1987). *FEBS Lett.* **221,** 403–407.

Geisler, N., Hatzfeld, M., and Weber, K. (1989). *Eur. J. Biochem.* **183,** 441–447.

Georgatos, S. D., and Blobel, G. (1987). *J. Cell Biol.* **105,** 105–115.

Georgatos, S. D., and Marchesi, V. T. (1985). *J. Cell Biol.* **100,** 1955–1961.

Georgatos, S. D., Weaver, D. C., and Marchesi, V. T. (1985). *J. Cell Biol.* **100,** 1962–1967.

Georgatos, S. D., Weber, K., Geisler, N., and Blobel, G. (1987). *Proc. Natl. Acad. Sci. U.S.A.* **84,** 6780–6784.

Gerace, L., and Blobel, G. (1980). *Cell* **19,** 277–287.

Gerace, L., and Burke, B. (1988). *Annu. Rev. Cell Biol.* **4,** 335–374.

Gerace, L., Blum, A., and Blobel, G. (1978). *J. Cell Biol.* **79,** 546–566.

Giese, G., and Traub, P. (1986). *Eur. J. Cell Biol.* **40,** 266–274.

Gill, S. R., Wong, P. C., Monteiro, M.J., and Cleveland, D. W. (1990). *J. Cell Biol.* **111,** 2005–2019.

Glass, C., Kim, K. H., and Fuchs, E. (1985). *J. Cell Biol.* **101,** 2366–2373.

Goldman, A. E., Maul, G., Steinert, P. M., Yang, H. Y., and Goldman, R. D.(1986). *Proc. Natl. Acad. Sci. U.S.A.* **83,** 3839–3843.

Granger, B. L., and Lazarides, E. (1980). *Cell* **22,** 727–738.

Granger, B. L., and Lazarides, E. (1982). *Cell* **30,** 263.

Green, K. J., and Jones, J. C. R. (1990). *In* "Cellular and Molecular Biology of Intermediate Filaments" (R. D. Goldman, and P. M. Steinert, eds.), pp. 147–171. Plenum, New York.

Green, K. J., Parry, D. A. D., Steinert, P. M., Virata, M. L. A., Wagner, R. M., Angst, B. D., and Nilles, L. A. (1990). *J. Biol. Chem.* **265,** 2603–2612.

Haydock, P. V., and Dale, B. A. (1986). *J. Biol. Chem.* **261,** 12520–12525.

Haydock, P. V., and Dale, B. A. (1990). *DNA Cell Biol.* **9,** 251–261.

Hanukoglu, I., and Fuchs, E. (1982). *Cell* **31,** 243–252.

Hanukoglu, I., and Fuchs, E. (1983). *Cell* **33,** 915–924.

Hatzfeld, M., and Franke, W. W. (1985). *J. Cell Biol.* **101,** 1826–1841.

Hatzfeld, M., and Weber, K. (1990a). *J. Cell Biol.* **110,** 1199–1210.

Hatzfeld, M., and Weber, K. (1990b). *J. Cell Sci.* **97,** 317–324.

Hatzfeld, M., Maier, G., and Franke, W. (1987). *J. Mol. Biol.* **197,** 237–255.

Heald, R., and McKeon, F. (1990). *Cell* **61,** 579–589.
Heid, H. W., Moll, I., and Franke, W. W. (1988). *Differentiation* (*Berlin*) **37,** 137–157.
Herdberg, K. K., and Chen, L.-B. (1986). *Exp. Cell Res.* **163,** 509–517.
Hirokawa, N., Glickman, M. A., and Willard, M. B. (1984). *J. Cell Biol.* **98,** 1523–1536.
Hockfield, S., and McKay, R. (1985). *J. Neurosci.* **5,** 3310–3328.
Hoffman, P. N., Cleveland, D. W., Griffin, J. W., Landes, P. W., Cowan, N. J., and Price, D. L. (1987). *Proc. Natl. Acad. Sci. U.S.A.*. **84,** 3472–3476.
Hoger, T. H., Krohne, G., and Franke, W. W. (1988). *Eur. J. Cell Biol.* **47,** 283–290.
Inagaki, M., Nishi, Y., Nishizawa, K., Matsuyama, M., and Sato, C. (1987). *Nature* (*London*) **328,** 649–652.
Ip, W., Hartzer, M. K., Pang, Y.-Y. S., and Robson, R. M. (1985). *J. Mol. Biol.* **183,** 365–375.
Jorcano, J. L., Franz, J. K., and Franke, W. W. (1984a). *Differentiation* (*Berlin*) **28,** 155–163.
Jorcano, J. L., Rieger, M., Franz, J. K., Schiller, D. L., Moll, R., and Franke, W. W. (1984b). *J. Mol Biol.* **179,** 257–281.
Julien, J.-P., and Mushynski, W. E. (1982). *J. Biol. Chem.* **257,** 10467–10469.
Julien, J.-P., and Mushynski, W. E. (1983). *J. Biol. Chem.* **258,** 4019–4025.
Julien, J.-P., Ramachandran, K., and Grosveld, F. (1985). *Biochim. Biophys. Acta* **825,** 398–404.
Julien, J.-P., Meyer, D., Hurst, J., and Grosveld, F. (1986). Mol. Brain Res. **1,** 243–250.
Julien, J.-P., Grosveld, F., Yazdanbaksh, K., Flavell, D., Meijer, D., and Mushynski, W. (1987a). *Biochim. Biophys. Acta* **909,** 10–20.
Julien, J.-P., Tretjakoff, I., Beaudet, L., and Peterson, A. (1987b). *Genes Dev.* **1,** 1085–1095.
Kaufman, E., Weber, K., and Geisler, N. (1985). *J. Mol. Biol.* **185,** 733–742.
Kartenbeck, J., Franke, W. W., Moser, J. G., and Stoffels, U. (1983). *EMBO J.* **2,** 735–742.
Kartenbeck, J., Schrechheimer, K., Moll, R., and Franke, W. W. (1984). *J. Cell Biol.* **98,** 1072–1081.
Kelly, D. E. (1966). *J. Cell Biol.* **28,** 51–72.
Kitamura, S., Ando, S., Shibata, M., Tanabe, K., Sato, C., and Inagaki, M. (1989). *J. Biol. Chem.* **264,** 5674–5678.
Klymkowsky, M. W. (1981). *Nature* (*London*) **291,** 249–251.
Klymkowsky, M. W. (1982). *EMBO J.* **1,** 161–165.
Klymkowsky, M. W., Miller, R. H., and Lane, E. B. (1983). *J. Cell Biol.* **96,** 494–509.
Kopan, R., and Fuchs, E. (1989). *Genes Dev.* **3,** 1–15.
Krimpenfort, P. J., Schaart, G., Pieper, F. R., Ramaekers, F. C., Cuypers, H. T., van den Heuvel, R. M., Vree Egberts, W. T., van Eys, G. J., Berns, A., and Bloemendal, H. (1988). *EMBO J.* **7,** 941–947.
Lasek, R. J., Oblinger, M. M., and Drake, P. F. (1983). *Cold Spring Harbor Symp. Quant. Biol.* **48,** 731–744.
Lasek, R. J., Phillips, L., Katz, M. J., and Autilio-Gambetti, L. (1985). *Ann. N.Y. Acad. Sci.* **455,** 462–478.
Lawson, D. (1983). *J. Cell Biol.* **97,**1891–1905.
Lazarides, E. (1980). *Nature* (*London*) **238,** 249–256.
Lazarides, E. (1982). *Annu. Rev. Biochem.* **51,** 219–250.
Lazarides, E., Granger, B. L., Gard, D. L., O'Connor, C. M., Breckler, J., Price, M., and Danto, S. I. (1982). *Cold Spring Harbor Symp. Quant. Biol.* **46,** 351–378.
Lee, L. D., and Baden, H. P. (1976). *Nature* (*London*) **264,** 377–379.
Lee, V. M.-Y., Otvos, L., Carden, M. J., Hollosi, M., Dietzschold, B., and Lazzarini, R. A. (1988). *Proc. Natl. Acad. Sci. U.S.A.* **85,** 1998–2002.
Lees, J. F., Shneidman, P. S., Skuntz, S. F., Carden, M. J., and Lazzarini, R. A. (1988). *EMBO J.* **7,** 1947–1955.
Lehto, V. P., Virtanen, I., and Kurki, P. (1978). *Nature* (*London*) **272,** 175–177.

Lendahl, U., Zimmerman, L. B., and McKay, R. D. G. (1990). *Cell* **60,** 585–595.
Lersch, R., Stellmach, V., Stocks, C., Giudice, G., and Fuchs, E. (1989). *Mol. Cell. Biol.* **9,** 3685–3697.
Levy, E., Liem, R. H. K., D'Eustachio, P., and Cowan, N. J. (1987). *Eur. J. Biochem.* **166,** 71–77.
Lewis, A. S., and Cowan, N. J. (1985). *J. Cell Biol.* **100,** 843–850.
Lewis, A. S., and Cowan, N. J. (1986). *Mol. Cell. Biol.* **6,** 1529–1534.
Lin, J. J. C. (1981). *Proc. Natl. Acad. Sci. U.S.A.* **78,** 2335.
Loewinger, L., and McKeon, F. (1988). *EMBO J.* **7,** 2301–2309.
Lonsdale-Eccles, J. D., Resing, K. A., Meek, R. L., and Dale, B. A. (1984). *Biochemistry* **23,** 1239–1245.
Lu, X. and Lane, E. B. (1990). *Cell* **62,** 681–696.
MacKinnon, P. J., Powell, B. C., and Rogers, G. E. (1990). *J. Cell Biol.* **111,** 2587–2600.
Magin, T. M., Hatzfeld, M., and Franke, W. W. (1987). *EMBO J.* **6,** 2607–2615.
Marchuk, D., McCrohon, S., and Fuchs, E. (1984). *Cell* **39,** 491–498.
McKeon, F., Kirschner, M. W., and Caput, D. (1986). *Nature (London)* **319,** 463–468.
McKinley-Grant, L. J., Idler, W. W., Bernstein, I. A., Parry, D. A., Cannizzaro, L., Croce, C. M., Huebner, K., Lessin, S. R., and Steinert, P. M. (1989). *Proc. Natl. Acad. Sci. U.S.A.* **86,** 4848–4852.
McLachlan, A. D. (1978). *J. Mol. Biol.* **124,** 297–304.
McLachlan, A. D., and Stewart, M. (1975). *J. Mol. Biol.* **98,** 293–304.
McLachlan, A. D., and Stewart, M. (1982). *J. Mol. Biol.* **162,** 693–698.
McNab, A. R., Wood, L., Theriault, N., Gierman, T., and Vogeli, G. (1989). *J. Invest. Dermatol.* **92,** 263–266.
Moll, R., Franke, W. W., Schiller, D. D., Geiger, B., and Krepler, R. (1982a). *Cell* **31,** 11–24.
Moll, R., Moll, I., and Wiest, W. (1982b). *Differentiation (Berlin)* **23,** 170–178.
Moll, R., Schiller, D. L., and Franke, W. W. (1990). *J. Cell Biol.* **111,** 567–580.
Monteiro, M. J., and Cleveland, D. W. (1989). *J. Cell Biol.* **108,** 579–593.
Monteiro, M. J., Hoffman, P. N., Gearhart, J. D., and Cleveland, D. W. (1990). *J. Cell Biol.* **111,** 1543–1557.
Myers, M. W., Lazzarini, P. A., Lee, V. M. Y., Schlaepfer, W. W., and Nelson, D. L. (1987). *EMBO J.* **6,** 1617–1626.
Napolitano, E. W., Chin, S. S. M., Colman, D. R., and Liem, R. H. K. (1987). *J. Neurosci.* **7,** 2590–2599.
Newport, J. W., Wilson, K. L., and Dunphy, W. G. (1990) *J. Cell Biol.* **111,** 2247–2259.
Nigg, E. A. (1989). *Curr. Opin. Cell Biol.* **1,** 435–440.
Osborn, M., and Weber, K. (1986). *Trends Biochem. Sci.* **11,** 469–472.
Parry, D. A. D., Crewther, W. G., Fraser, R. D., and MacRae, T, P. (1977). *J. Mol. Biol.* **113,** 449—454.
Parry, D. A. D., Steven, A. C., and Steinert, P. M.(1985). *Biochem. Biophys. Res. Commun.* **127,** 1012–1018.
Parry, D. A. D., Goldman, A. E., Goldman, R. D., and Steinert, P. M. (1987). *Int. J. Biol. Macromol.* **9,** 137–145.
Parysek, L. M., and Goldman, R. D. (1988). *J. Neurosci.* **8,** 555–563.
Parysek, L. M., Chisholm, R. L., Ley, C. A., and Goldman, R. D. (1988). *Neuron* **1,** 395–401.
Pauling, L., and Corey, R. B.(1953). *Nature (London)* **171,** 59.
Peter, M., Nakagawa, J., Doree, M., Labbe, J. C., and Nigg, E. A. (1990). *Cell* **61,** 591–602.
Pieper, F. R., Schaart, G., Krimpenfort, P. J., Henderik, J. B., Moshage, H. J., van de Kemp, A., Ramaekers, F. C., Berns, A., and Bloemendal, H. (1989). *J. Cell Biol.* **108,** 1009–1024.

Pollard, K. M., Chan, E. K. L., Grant, B. J., Sullivan, K. F., Tan, E. M., and Glass, C. A. (1990). *Mol. Cell. Biol.* **10,** 2164–2175.
Portier, M.-M., deNechaud, B., and Gros, F. (1984). *Dev. Neurosci.* **6,** 335–344.
Potschka, M. (1986). *Biophys. J.* **49,** 129–130.
Powell, B. C., and Rogers, G. E. (1990). *EMBO J.* **9,** 1485–1493.
Price, M. G., and Lazarides, E. (1983). *J. Cell Biol.* **97,** 1860–1874.
Pytela, R., and Wiche, G. (1980). *Proc. Natl. Acad. Sci. U.S.A.* **77,** 4808–4812.
Quax, W., Vree Egberts, W., Hendriks, W., Wuax-Jeuken, Y., and Bloemendal, H. (1983). *Cell* **35,** 215–223.
Quax, W., van den Broek, L., Vree Egberts, W., Ramaekers, F., and Bloemendal, H. (1985). *Cell* **43,** 327–338.
Quinlan, R. A., and Franke, W. W. (1982). *Proc. Natl. Acad. Sci. U.S.A.* **79,** 3452–3456.
Quinlan, R. A., Cohlberg, J. A., Schiller, D. L., Hatzfeld, M., and Franke, W. W. (1984). *J. Mol. Biol.* **178,** 365–388.
Quinlan, R. A., Hatzfeld, M., Franke, W. W., Lustig, A., Schulthess, T., and Engel, J. (1986). *J. Mol. Biol.* **192,** 337–349.
Quinlan, R. A., Moir, R. D., and Stewart, M. (1989). *J. Cell Sci.* **93,** 71–83.
Raats, J. M. H., Pieper, F. R., Vree Egberts, W. T. M., Verrijp, K. N., Ramaekers, F. C. S., and Bloemendal, H. (1990). *J. Cell Biol.* **111,** 1971–1986.
Ramaekers, F. C. S., Dunia, I., Dodemont, H. J., Beneditti, E. L., and Bloemendal, H. (1982). *Proc. Natl. Acad. Sci. U.S.A.* **79,** 3208–3212.
Resing, K. A., Walsh, K. A., Haugen-Scofield, J., and Dale, A. D. (1989) *J. Biol. Chem.* **264,** 1837–1845.
Rober, R.-A., Weber, K., and Osborn, M. (1989). *Development* **105,** 365–378.
Rothnagel, J. A., and Rogers, G. E. (1986). *J. Cell Biol.* **102,** 1419–1429.
Rothnagel, J. A., and Steinert, P. M. (1990). *J. Biol. Chem.* **265,** 1862–1865.
Rothnagel, J. A., Mehrel, T., Idler, W. W., Roop, D. R, and Steinert, P. M. (1987). *J. Biol. Chem.* **262,** 15643–15648.
Senior, ■., and Gerace, L. (1988). *J. Cell Biol.* **107,** 2029–2036.
Steinert, P. M. (1990). *J. Biol. Chem.* **265,** 8766–8774.
Steinert, P. M., and Roop, D. R. (1988). *Annu. Rev. Biochem.* **57,** 593–625.
Steinert, P. M., and Roop, D. R. (1990). *In* "Cellular and Molecular Biology of Intermediate Filaments" (R. D. Goldman and P. M. Steinert, eds.), pp. 353–370. Plenum, New York.
Steinert, P. M., Idler, W. W., and Zimmerman, S. B. (1976). *J. Mol. Biol.* **108,** 547–567.
Steinert, P. M., Idler, W. W., Cabral, F., Gottesman, M. M., and Goldman, R. D. (1981). *Proc. Natl. Acad. Sci. U.S.A.* **78,** 3692–3696.
Steinert, P. M., Rice, R. H., Roop, D. R., Trus, B. L., and Steven, A. C. (1983). *Nature (London)* **302,** 794–800.
Steinert, P. M., Parry, D. A. D., Racoosin, E., Idler, W., Steven, A., Trus, B., and Roop, D. (1984). *Proc. Natl. Acad. Sci. U.S.A.* **81,** 5709–5713.
Steven, A. C. (1990). *In* "Cellular and Molecular Biology of Intermediate Filaments" (R. D. Goldman and P. M. Steinert, eds.), pp. 233–263. Plenum, New York.
Steven, A., Hainfeld, J., Trus, B., Wall, J., and Steinert, P. (1983). *J. Cell Biol.* **97,** 1939–1944.
Stewart, M., Quinlan, R. A., and Moir, R. D. (1989). *J. Cell Biol.* **109,** 225–234.
Tokutake, S., Liem, R. K. H., and Shelanski, M. L. (1984). *Biomed. Res.* **5,** 235–238.
Tokuyasu, K. T., Maher, P. A., and Singer, S. J. (1984). *J. Cell Biol.* **98,** 1961–1972.
Traub, P., and Vorgias, C. E. (1983). *J. Cell Sci.* **63,** 43–67.
Tseng, S. C. G., Jarvinen, M. J., Nelson, W. G., Huang, J.-W., Woodcock-Mitchell, J., and Sun, T.-T.(1982). *Cell* **30,** 361–372.
Tyner, A. L., and Fuchs, E. (1986). *J. Cell Biol.* **103,** 1945–1955.

Vassar, R., Rosenberg, M., Ross, S., Tyner, A., and Fuchs, E. (1989). *Proc. Natl. Acad. Sci. U.S.A.* **86,** 1563–1567.
Vassar, R., Coulombe, P. A., Degenstein, L., Albers, K., and Fuchs, E. (1991). *Cell* **64,** 365–380.
Venetianer, A., Schiller, D. L., Magin, T., and Franke, W. W. (1983). *Nature* (*London*) **305,** 730–733.
Virtanen. I., Lehto, V.-P., Lehtonen, E., Vartio, T., Stenman, S., Kurki, P., Wager, O., Small, J. V., and Bradley, R. A. (1981). *J. Cell Sci.* **50,** 45–63.
Ward, G. E., and Kirschner, M. W. (1990). *Cell* **61,** 561–577.
Weber, K., Plessmann, U., Dodemont, H., and Kossmagk-Stephan, K. (1988). *EMBO J.* **7,** 2995–3001.
Weber, K., Plessmann, U., and Ulrich, W. (1989). *EMBO J.* **8,** 3221 3227.
Wiche, G. (1989). *CRC Crit. Rev. Biochem.* **24,** 41–67.
Wiche, G., Becker, B., Luber, K., Weitzer, G., Castanon, M. J., Hauptmann, R., Stratowa, C., and Stewart, M. (1991). *J. Cell Biol.* **114,** 83–89.
Wong, P. C., and Cleveland, D. W. (1990). *J. Cell Biol.* **111,** 1987–2003.
Woods, E. F., and Inglis, A. S. (1984). *Int. J. Macromol.* **6,** 277–283.
Worman, H. J., Yuan, J., Blobel, G., and Georgatos, S. D. (1988). *Proc. Natl. Acad. Sci. U.S.A.* **85,**8531–8534.
Worman, H. J., Evans, C. D., and Blobel, G. (1990). *J. Cell Biol.* **111,** 1535–1542.
Wu, Y.-J., Parker, L. M., Binder, N. E., Beckett, M. A., Sinard, J. H., Griffiths, C. T., and Rheinwald, J. G. (1982). *Cell* **31,** 693–703.

Addendum to
Dynamics of Nucleotides in Plants Studied on a Cellular Basis

K. G. Wagner and A. I. Backer

NOTE ADDED IN PROOF: The authors urgently want to mention an excellent review on the biogenesis and metabolism of the ureides as well as on enzymology and enzyme localization (Schubert and Boland, 1990). In another review, Ries (1991) stressed the occurrence of the L(+) enantiomer of adenosine, whereas all known nucleosides and nucleotides have the D configuration of the ribose. In rice roots about 12% of the total Ado content (124 μg/g DW) were claimed to have the L configuration. It was further shown that L(+)Ado was elicited by an ubiquitous growth substance, triacontanol, a long chain fatty alcohol; the mode of elicitation, synthesis of the Ado enantiomer, and its mechanism of action, however, remain to be elucidated. Studies on the cellular localization of hemicellulose synthesis (glucuronoxylan), performed with etiolated pea seedlings, revealed high activities of GlcUA-Tase (with UDPCGlcUA as substrate) on the cis-face of the Golgi apparatus, although lower activities were also found in the plasma membrane fraction. The enzyme activity was stimulated by UDPXyl (Hobbs *et al.*, 1991).

Recent reports on nucleotide transport into amyloplasts isolated from suspension-cultured sycamore cells (Pozueta-Romero *et al.*, 1991a) described an adenylate-specific carrier active in the uptake of ATP, ADP, AMP, and, interestingly, also ADPGlc with K_m values in the range from 25 to 47 μM. Transfer of Glc from ADPGlc to starch was shown with intact and broken amyloplasts; thus the authors suggested that in the studied amyloplasts starch synthesis is coupled to the transport of ADPGlc. They further argued that ADPGlc may be synthesized in the cytoplasm by sucrose synthase, an enzyme which was believed previously to use UDP (cf. Table VII). With enzyme extracts from cultured sycamore cells and spinach leaves, Pozueta-Romero and co-workers further found low K_m values for ADP (5.3 and 16.8 μM) and high V_{max} values for the synthesis of ADPGlc, provided the standard sucrose–Sase assay was modified by omitting Tris buffer and ATP which inhibited the enzyme noncompetitively (Pozueta-Romero *et al.*, 1991b).

Kaiser and Spill (1991) described a reversible modification, most probably phosphorylation, of nitrate reductase (inactivated when phosphorylated) which they suggested to be regulated by the ATP/AMP ratio of the cytoplasm. With crude extracts from spinach leaves (but not with the purified enzyme) an inactivation of nitrate reductase by ATP and an activation by AMP was reported. Determination of metabolite levels in the leaf showed that under photosynthetic conditions (high CO_2) the ATP/AMP ratio was lower than at conditions of low CO_2 (stomatal closure), although the rather low AMP levels did not change very much.

Heineke *et al.* (1991) investigated the thermodynamics of redox transfer across the inner chloroplast envelope membrane by measuring the metabolite concentrations in the different compartments of spinach leaf cells. To maintain the existing high redox gradient between the NADPH/NADP in the stroma and the NADH/NAD in the cytoplasm, the responsible metabolite shuttles were claimed to be strictly controlled. Calculating free energies of the different reactions, the control steps were assigned to the chloroplastic PGA reduction and the cytoplasmic trioseP oxidation (trioseP/PGA shuttle, see Fig. 14) and to the chloroplastic NADP-malate dehydrogenase and the translocating step of the malate–oxalacetate shuttle.

References

Heineke, D., Riens, B., Grosse, H., Hoferichter, P., Peter, U., Flügge, U.-I., and Heldt, H. W. (1991). *Plant Physiol.* **95,** 1131–1137.

Hobbs, M. C., Delarge, M. H. P., Baydoun, E. A.-H., and Brett, C. T. (1991). *Biochem. J.* **277,** 653–658.

Kaiser, W. M., and Spill, D. (1991). *Plant Physiol.* **96,** 368–375.

Pozueta-Romero, J., Frehner, M., Viale, A. M., and Akazawa, T. (1991a). *Proc. Natl. Acad. Sci. U.S.A.* **88,** 5769–5773.

Pozueta-Romero, J., Yamaguchi, J., and Akazawa, T. (1991b). *FEBS Lett.* **291,** 233–237.

Ries, S. (1991). *Plant Physiol.* **95,** 986–989.

Schubert, K. L., and Boland, M. J. (1990). *Biochem. Plants* **16,** 197–282.

INDEX

B

C

D

F

G

N

O

P